PLANT BREEDING

PLANT BREEDING

Ricky Bates
Stuart Walters

Kruger Brentt
Publishers

2024

Kruger Brentt Publishers UK. LTD.
Company Number 9728962

Regd. Office: 68 St Margarets Road, Edgware, Middlesex HA8 9UU

ISBN: 978-1-78715-270-0

For information on all our publications visit our website at http://krugerbrentt.com/

PREFACE

Plant breeding is a dynamic and essential field in agricultural science dedicated to the enhancement of plant traits for the benefit of humanity. It involves the controlled manipulation of plant species to develop new and improved varieties with desirable characteristics such as increased yield, resistance to pests and diseases, tolerance to environmental stresses, and enhanced nutritional content. Through a combination of traditional and modern breeding techniques, plant breeders work to achieve genetic improvements that address the ever-evolving challenges faced by agriculture. This multidisciplinary science draws upon genetics, genomics, physiology, and biochemistry to unlock the potential of plants, contributing to global food security, sustainable agriculture, and the development of crops tailored to diverse climates and agricultural practices. As a pivotal force in shaping the future of agriculture, plant breeding plays a crucial role in meeting the growing demands of a rapidly expanding global population while ensuring the resilience and adaptability of crops in the face of changing environmental conditions.

It presents a dynamic exploration into the forefront of agricultural innovation, offering a comprehensive panorama of the evolving landscape of plant breeding. As the global population burgeons and climate change challenges traditional agricultural paradigms, the imperative to enhance crop productivity, quality, and resilience becomes increasingly urgent. This compendium comprises four intricately woven sections, each delving into pivotal aspects of plant breeding. From the cultivation intricacies of high-density planting systems in the cotton fields of India to the molecular intricacies of marker-assisted selection and genome editing techniques, the book traverses the spectrum of breeding strategies. It scrutinizes stress conditions, investigates the interplay between plants and microbiomes, and explores the promise of crop biotechnology. By marrying the wisdom of conventional breeding with the cutting-edge tools of genomics and biotechnology, this book not only encapsulates the current state of plant breeding but also casts a discerning gaze into the future, where innovative strategies and technologies hold the key to sustainable and resilient agriculture. With regional insights, practical strategies, and a global perspective, this compilation serves as an indispensable guide for researchers, plant breeders, and agricultural professionals navigating the challenges and opportunities of contemporary and future plant breeding endeavors.

We are grateful to all those persons as well as various books, manuals, periodicals, magazines, journals etc. that helped in the preparation of this book. In spite of the best efforts, it is possible that some errors may have occurred into the compilation and editing of the book. Further queries, constructive suggestions and criticisms for the improvement of the book are always welcome and shall be thankfully acknowledged.

Ricky Bates

Stuart Walters

CONTENTS

SECTION-1

DEVELOPMENT OF CROP VARIETIES

CHAPTER-1

THE CULTIVATION METHOD OF COTTON WITH A HIGH-DENSITY PLANTING SYSTEM IN INDIA: STATUS AND BREEDING STRATEGIES

1. INTRODUCTION: AN OVERVIEW

The agricultural landscape in India has witnessed transformative approaches to cotton cultivation, with a notable shift towards the implementation of high-density planting systems. This innovative method involves strategically optimizing the spatial arrangement of cotton plants to achieve a higher plant population per unit area. As cotton holds significant economic importance in India, adopting advanced cultivation techniques like high-density planting is aimed at enhancing crop yield, resource efficiency, and overall agricultural productivity.

Cotton (*Gossypium hirsutum* L.) is the major fiber and cash crop, not only in India but for the entirety. Cotton is the only crop which travels with the human being in each and every part of his/her life. It is cultivated in tropical as well as sub-tropical regions of more than seventy countries of the World. Cotton is a crop of global significance playing a significant role in the agricultural and industrial economy. Around 60% of fiber to Indian textiles is from cotton. The recent statistics released from USDA - Foreign Agricultural Service (September 2020) indicates that India (13.40 million ha) has more than one third of the World's area (32.94 million ha) under cotton with a productivity of 487 kg/ha, which is far below than the World's productivity of 775 kg/ha. Many of the Countries like Brazil, China, Turkey, Australia have the productivity of more than 1500 kg/ha. China with 3.25 million ha of cultivation (less than one fourth of the area of India) could result in a production of 27.25 million 480 lb. bales (as projected by USDA) compared to the

Indian production projected at 30.00 million 480 lb. bales from 13.40 million ha. This clearly signifies the productivity gap prevailing at India for cotton.

If Indian production is juxtaposed against China during 2019/20 marketing year, China accounted for just over 22 percent of total world cotton production, of which 86 percent of that was produced in Xinjiang province (just under 20 percent of the world total). China exports only a small amount of cotton lint, half-of-one-percent of production. Other than some minor exports to North Korea, China is the world's largest importer of cotton. This provides China with a supply of cotton normally greater than one-third of world use and nearly 40 percent larger than India [1]. This scenario also clearly explains the gap which is existing in India compared to other Countries which could make the country to progress further in the yield front of cotton provided newer technologies and cropping systems are adopted *in toto*.

2. INDIAN COTTON SCENARIO

About 59 per cent of the raw material requirement of the Indian textile industry is met by cotton. It plays a major role in sustaining the livelihood of An estimated 5.8 million cotton farmers' livelihood is sustained by cultivating cotton. Besides, this crop engages 40–50 million people in one or the other related activities. As seen, the area under cotton in India is also tremendous which is around 13.40 million hectares. Among the 53 species of *Gossypium* available, Indians cultivate all the four species of cotton namely *Gossypium arboreum* and *herbaceum* (Asiatic cotton), *G. Barbadense* (Egyptian cotton) and *G. hirsutum* (American Upland cotton) with *G. hirsutum* being cultivated over the entire Country. It is about 88% of the hybrid cotton being cultivated in India belongs to *hirsutum* type and almost all the *Bt* cotton hybrids belong to G.*hirsutum* type.

Cotton is grown in all the three different agro-ecological zones of India *viz.,* Northern, Central and Southern zones. Nearly 70 percent of the crop is cultivated under rainfed condition in the Central and Southern regions of the country. Among the cotton producing states, Maharashtra is the largest producer with an area of 38.06 lakh ha followed by Gujarat (24 lakh ha) and Telangana (17.78 lakh ha). In India, the production of cotton is recorded in bales which are of 170 kg. The production is highest in Gujarat with 95 lakh bales followed by Maharashtra (89 lakh bales) and Telangana (59.50 lakh bales). Karnataka stands first in productivity with 769 kg ha^{-1} followed by Andhra Pradesh (719 kg ha^{-1}) and Rajasthan (692 kg ha^{-1}) [2].

Majority of the cotton produced in India is derived from nine major cotton growing states and these States fall under three diverse agro-ecological zones.

- Northern Zone-Punjab - Haryana and Rajasthan.

- Central Zone-Gujarat - Maharashtra and Madhya Pradesh.
- Southern Zone-Telangana - Andhra Pradesh and Karnataka.

In addition, cotton is also grown in the States of Tamil Nadu and Odisha. Recently, cotton is also being cultivate in small scale in non-traditional States such as Uttar Pradesh, West Bengal, Tripura, etc. Nevertheless, India is the largest producer cum leading consumer of cotton in the World. It's very clear now that albeit having higher area under cotton, the productivity of cotton is very low compared to many of the Countries which warrants attention mainly on developing newer genotypes that would yield better on higher management condition. Strategies that could maximize the per unit area yield in cotton would include.

- Developing ideotypes in cotton that would suit mechanized cultivation starting from sowing to lint collection.
- Standardized agro-management systems for exploiting more unit area productivity.
- Robust management procedures to ward off pests, diseases and other nutritional disorders.
- Assured price for quality produce.

Primarily, the productivity enhancement in any crop depends on the development of suitable genotype and cotton is not an exception. Many of the wild species available in the crops are exploited for transferring the segments (QTLs) that harbor pests and diseases resistance *vis-a-vis* high yield. Though about 53 species of *Gossypium* are available including the four cultivated species, only very few diploid and tetraploid wild species of *Gossypium* are crossable with the cultivated species. Among the species of *Gossypium*, seven species are with AD genome measuring 2400 Mb genome size, three species with A genome (1700 Mb), four species with B genome (1350 Mb), three species in C genome (1980 Mb), 13 species with D genome (885 Mb), seven species in E genome (1560 Mb), one species belonging to F genome (1300 Mb), three species under G genome (1785 Mb) and 12 species under K genome (2570 Mb) [3]. Since cotton is being available in the field for more than 5–6 months before harvest, per day productivity of the crop also receives much attention.

Another statistical prediction provided by Dr. M. V. Venugopalan of CICR, Nagpur [4] is that the cotton productivity during 2018–2019 would be the lowest despite the fact that almost 90% of the farmers have adopted the state of art BG II hybrids. This had been exclusively due to the increase in cost of cotton cultivation from Rs. 2233/q of seed cotton in 2002–2003 to Rs.4803/q in 2015–2016, mainly due to increase in labour wages and increased use of inputs like fertilizers and pesticides. Considering these facts, primary aim of the plant breeders has to be in designing a genotype that would fit for the given situation.

Moreover, the present day hybrids put forth biomass enormously and are of speed and spread growing in nature. Thus, the ratio of bolls to biomass if worked out would be much lesser. For having a match between the growth, water requirement, duration, yield, per unit and day productivity etc. a system was arrived at by the Central Institute of Cotton Research, Nagpur which is High Density Planting System (HDPS) with early maturing, semi compact genotypes for realizing higher yields with low production costs under rainfed condition primarily. The main tenets of this proposition covers tailoring a genotype suiting to high density planting (more than one lakh plants per hectare), its uniformity in boll development, maturation and bursting, its adoptability to the given condition and efficiency in the nutrient utilization etc.

In the forth coming discussions, let us see about the genetic, agronomic and plant protective interventions that would help in developing a suitable genotype fitting to HDPS.

2.1 Genetic interventions

2.1.1 Genetic enhancement

The term "enhancement" was first used by [5] for defining the transfer of useful genes from exotic or wild types into agronomically acceptable background, preferably a cultivar of choice. This term of enhancement was later [6] rechristened as pre-breeding or developmental breeding to describe the same activity. Thus, having varied terminologies basically refer to the transfer or introgression of genes or gene combinations from unadapted sources, mostly the wild sources into the breeding materials, preferably an adapted background [7]. Normally, genetic enhancement is complementary to that of traditional breeding. However, these activities, as name suggests, form the base of any plant breeding programme where the gene transfer from wild species /related species is targeted. Thus, enhanced germplasm can be more readily used in breeding programmes for cultivar development. Thus, pre-breeding qualifies as prior step of sustainable plant breeding which normally starts with identifying a useful character in unadapted or wild genotypes, capturing its genetic diversity and extracting the genes/QTLs that govern these variations for exploitation. Thus, pre-bred materials may also be an intermediary with a value addition which could be further exploited.

Normally, wild species are exploited for transferring traits related to the improvement of yield, quality, pests and diseases resistance. In cotton, for altering the plant types, the wild species are not that much useful for the reasons that most of them are perennial with spreading habit. Hence, for breeding an ideal genotype with less/shy branching, zero monopodia, minimalistic bolls with uniform weight, luster, shape and bursting etc. which would also suit mechanized cultivation, exploring the available variability among the germplasm must be the pre-requisite.

Research on HDPS on cotton gained momentum under the leadership of ICAR – Central Institute for Cotton Research, Nagpur in 2010. Shortly thereafter, in 2012, the All India Coordinated Cotton Improvement Project (AICCIP, now AICRP on Cotton) started a separate trial on the evaluation of compact genotypes for HDPS under rainfed and irrigated situation to facilitate the release of compact genotypes suitable for HDPS. Variety CSH 3075 was the first cotton variety released for HDPS in India. [4]. Subsequently, research got momentum and Tamil Nadu Agricultural University (TNAU) has also released two varieties *i.e.*, TCH 1705 as CO 15 and TCH 1819 as CO 17 for HDPS. Cotton CO 17 has been performing well under rice fallow conditions of Cauvery Delta Zone of Tamil Nadu (Figure 1).

Figure 1.HDPS cotton (CO 17) being cultivated at farmer's holding.

2.1.2 Varietal evaluation

A high yielding *G.hirsutum* variety CO 15 developed from Department of Cotton, Centre for Plant Breeding and Genetics, Tamil Nadu Agricultural University, Coimbatore was evaluated in All India Coordinated Cotton Improvement Project trials on cotton during the year 2012–2013. This culture registered a significant seed cotton yield (2346 kg/ha) which is 111.2% increase over the local check Suraj (normal spacing) (1111 kg/ha). In south zonal trials, during 2013–2014, CO 15 (3582 kg/ha) out yielded the zonal check Suraj (3158 kg/ha) by 13.4% and stood at third place and during 2014–2015 registering 5.3% increased yield (2378 kg/ha) than the zonal check Suraj (2259 kg/ha). The seed cotton yield was maximum to the tune of 3226 kg/ha as against 2443 kg/ha of LH 2298 under.

60 x 10 cm. This culture is found to be fertilizer responsive in all the location. Increasing the spacing results yield loss in almost all the centres indicating its suitability for high density planting. It registered moderate resistance to Bacterial blight and Gray mildew. The overall performance (2012–2015) revealed its superiority in mean seed cotton yield (2807 kg/ha) as against the local check Suraj (normal spacing) (2146 kg/ha). The increase in kapas yield was 30.8% over local check. Besides high seed cotton yield, it possessed higher ginning out turn of 36.6% than zonal check Suraj (34.1%). This culture comes under the medium long staple category with 2.5% span length of 27.1 mm, fiber strength of 21.5 g/tex and micronaire value of 4.3. It can spin up to 30–40's counts.

Cotton variety CO 17 is a short duration compact plant type with synchronized boll maturity suitable for high density planting system (HDPS) released by the University during 2020. This culture was developed at Department of Cotton, TNAU, Coimbatore from the parental hybridization involving Khandwa 2 and LH 2220 followed by pedigree breeding. It matures in 125–135 days and possesses zero monopodia with short sympodial length and is highly suited for high density planting system. It recorded an average seed cotton yield of 2361 kg/ha which is 18.9% increase over the check variety Suraj (National check entry identified for HDPS). Culture TCH 1819 recorded seed cotton yield of 3427 kg/ha which was 21.7% increase over Suraj and 29.0% increase over MCU 7 under rice fallow condition. It recorded a seed cotton yield of 2051 kg/ha which was 13.8% increase over Suraj under summer irrigated conditions and also recorded 1604 kg/ha of seed cotton yield under winter rainfed which was 20.1% increase over the check Suraj.

It was also evaluated in All India Coordinated Cotton Improvement Project trials for two years during 2016–2017 and 2017–2018 across ten locations. It registered seed cotton yield of 1850 kg/ha which was 37.9% increase over Suraj. Adaptive Research Trials (ARTs) were conducted under three different cotton growing seasons *viz.*, rice fallow, winter rainfed and summer irrigated conditions during 2016, 2017 and 2018. It recorded the highest mean seed cotton yield of 4530 kg/ha which was 17.2% increase over the check Suraj under rice fallow condition. It recorded Upper Half Mean Length (UHML) of 27 mm with bundle strength of 26.9 g/tex. It can spin upto 40's counts.

Considering the descriptors of these two varieties, few features can be noticed in common and they are short intermodal length with lesser distance of boll from main stem, bolls of 4–5 g in weight and lesser plant surface coverage of not exceeding 0.25 m^2 ground area and synchronized maturity. Research work undertaken at the Department of Cotton, TNAU during one decade has resulted in identifying these genotypes which could exclusively fit in HDPS. Moreover, research being undertaken in the entire country had resulted in the release of varieties which are meant for HDPS which are detailed (Table 1).

Table 1. Details of cotton varieties released for HDPS (courtesy: Central Institute of Cotton research, Regional Station, Coimbatore).

Name	Year	Center/State variety Release	Institution
F 2383	2016	State	PAU, Faridkot
CSH 3075	2017	Central	CICR, Sirsa
Cotton CO 15 (TCH 1705)	2018	Central	TNAU, Coimbatore
F 2381	2016	Central	PAU, Faridkot
ARBC 19	2016	Central	UAS, Dharwad
CO 17	2020	State	TNAU, Coimbatore
RS 2818	2020	Central	SKRAU, Sriganganagar
ARBC 1601	2020	Central	UAS, Dharwad
ARBC 1651	2020	Central	UAS, Dharwad
ARBC 1651	2020	Central	UAS, Dharwad
DSC 1651	2020	Central	UAS, Dharwad

2.1.3 Genetics of the traits governing HDPS

Majority of the traits that define a genotype fitting for HDPS include shy branching, 10–15 bolls per plant, boll setting nearer to the main stem, boll of 4–5 g in weight, uniform in size and shape, completion of bursting of all bolls in 3–4 days time, medium to superior medium fiber length with appreciable strength, 120 to 125 days of crop duration for fitting into various cropping programmes. These traits have been extensively studied using the available germplasm and prominent crosses have been affected to identify the genotypes that would fit in HDPS. Studies taken at the Department of Cotton by effecting crosses with genotypes fitting with HDPS and heavy yielders have indicated that crosses CO 17 x CO 14 and TCH 1926 x RB 602 showed high *per se* performance and positive significant *sca* effect for single plant yield. The hybrids C -10-8 x GISV 310 and CO 17 x GISV 310 which involves compact lines showed high *per se* performance, positive significant *sca* affect and positive standard heterosis for single plant yield. The hybrid CO 17 x TCH 1926 had high *per se* performance and positive significant *sca* effect for number of sympodial branches per plant and single plant yield and *per se* performance for number of bolls per plant.

Another study was taken up at the Department of Cotton during 2018–2019 to assess the spectrum of variability realized from differently yielding compact hybrids. Among the 900 observed plants in F2 population, surface covers of 689 plants were recorded as lower than the check (CO 15). The crosses viz., 343–1-1 x CO 14, 343–1-1 x RB 602, TCH 1926 x RAH 1070 and CO 17 x RB 602 were identified as elite combinations as they had a greater number of individuals whose plant surface was considerably lower than the check (CO 15).

Effect of Okra Leaf Shape in HDPS:

- Considering the bigger leaf lamina which is available with CO 17, more pronounced leaf hoppers problem had been observed and breeding research to develop plants with HDPS traits along with okra leaf type had been the tailored programme which was started to function from 2015 to 2016. The okra leaf shape character, on an average in varieties over locations, caused a significant reduction in the incidence of boll rot in comparison with normal leaf cotton [8]. Altering, rather reducing the leaf lamina was significantly associated with an increase in yield, earliness, lint percentage and micronaire value, and a substantial increase in fruiting rate. However, it was also observed that the okra leaf shape had no effect on boll weight and fiber related attributes *viz.,* fiber length, fiber length uniformity, or fiber strength, but with a reduction in fiber elongation and total leaf area.
- At present in the Department of Cotton, work initiated on the development of compact plant type with okra leaf shape suitable for HDPS resulted in two F2 populations viz., TCH 1819 x PBH 115 and TCH 1819 x F 2382. Okra-leaf types of the upland cotton have the potential to be competitive to the normal-leaf types in yield and fiber quality, in addition to its potential resistance to insect pests and drought. In cotton, okra leaf type plants confer resistance/non-preference against insect pests. Form these two F2 populations, a total of 85 single plants were selected for compactness with okra leaf. Reciprocal crosses of above two cross combinations were also made. Forwarding these progenies would help in identifying compact plant types with the okra leaf type.
- Prominent feature being considered for HDPS is that the genotype must have occupy an area of <0.25 m^2 on the ground and invariably possesses the traits like optimum plant height plant (around 100 cm) with shorter sympodia, shorter intermodal length with lesser distance of boll from main stem. Albeit the fact that HDPS was worked out for rainfed eco system, it fits well in the irrigated scenario also.

2.2 Agronomic management and interventions

In India, though the area under cultivation of cotton is higher, the seed cotton yield per unit area is very low compared to many other cotton growing countries in the world. The primary factors that attribute for this low realized yield besides the non-availability of choices of genotypes is the low plant population density. Various techniques like maintaining suitable plant density, use of optimum dose of fertilizers, growth regulators etc., are being suggested to overcome these constraints in cotton production. The optimum level of cotton productivity would, however, depends on the plant type being grown. The present-day cotton genotypes are of long duration (180–200 days), late maturing, tall growing with spreading nature leading to bushy appearance besides with lesser number of bolls compared to the crop canopy. They also require wide spacing for the expression of the crop resulting in the production of netted canopy resulting in various problems in taking up plant protection measures, machine picking, and inefficiency in trapping of solar energy, physiological efficiency and harvest index.

Because of longer duration, these varieties require a greater number of pickings, as a result, cost of cotton cultivation upsurge especially owing to manual picking which warrants increased labour and fluctuating prices resulting in lesser realistic income. The availability of labour for clean picking is also a serious constraint. At present, in India, entire cotton is picked manually which is labour intensive and is becoming expensive day by day. About 30 per cent of world cotton being produced in Australia, Israel and USA and other developed countries are picked using machines which were sown as per the requirements of the machine. Such picking through machines would result in "Machine maximum, Man minimum" in cultivating cotton which will not only minimize cost of cultivation, but also reduce the dependency on labour. Machine picking esseentially depends on cultivation of cotton genotypes having short stature, earliness, compactness, sympodial growth habit and synchronous boll opening. Under these circumstances, compact cotton genotypes are ideally suited. They offer great scope for reducing not only row width, but also spacing between the plants in a row.

2.2.1 Ultra narrow row (UNR) planting of cotton

Over the last few decades, many cotton growing countries like China, USA, Australia, Brazil, Uzbekistan and Greece were able to enhance cotton yields by increasing the planting density. The planting geometry, in general adopted varies from 8 to 10 cm distance between plants in a row with the row spacing ranging from 18 to 106 cm, ideal being worked out to 100 cm at our conditions. This planting system is referred as narrow row (NR) if the row-to-row spacing is less than 75 cm and ultra-narrow-row (UNR) if the spacing is less than 45 cm. Currently in India, depending on the local conditions, hybrid cotton is planted at row spacing ranging from 90 to 120 cm and plant spacing ranging from 30 to 90 cm resulting in 15,000 to 25,000 plants/ha. In HDPS, short duration, semi-compact cotton varieties are planted at populations ranging from 1.1 lakh to 2.45 lakh plants/ha by planting at a distance of 45–90 cm between rows depending upon the soil type and growing conditions and 10 cm between plants in a row. It aims to establish around 7–8 plants/m row length. The objective is to limit the boll number to around 10/plant, maximize the number of bolls/unit area and realize high yield in the shortest possible time. If the number of bolls/plants is few, the fruiting window (or flowering period) is short (4–5 weeks) and the plant matures early, producing fibers with good quality.

Ultra narrow row (UNR) cotton production is considered as a potential strategy for reducing production costs by shortening the growing season [9]. By cultivating the genotypes that would fit for UNR, it provides the scope for increasing per unit area of plants *vis-a-vis* the productivity. Being shorter and earlier to mature, these genotypes under UNR provides scope for double cropping and mechanical harvesting. Since the number of bolls are less with uniformity in maturation and bursting, these compact types require few pickings only. This results in savings of labour cost as well as seed cost as it provides farmers an opportunity to reuse the varietal seeds for few sowing seasons.

Adoption of HDPS amicable compact and early maturing cotton varieties offer an alternate to sustainable production at decreased production cost under Indian condition. However, availability of more determinate cultivars, more efficient options of weed control and insect pest management (including transgenics), growth regulations to modify morpho frame, planting and harvesting equipments etc., has made high density cotton planting system popular in several countries. The concept on high density cotton planting, more popularly called Ultra Narrow Row (UNR) cotton was initiated by [10] and this concept has been one of the most researched topics during the last 15 years. Availability of early maturing, compact sympodial plant types with more fruiting bodies closer to the main stem is a pre-requisite for successful HDPS.

2.2.2 Planting density

Theoretically, higher planting density ensures earlier crop canopy cover, higher sunlight interception leading to higher and earlier yields at reduced cost. The obvious advantage of this system is earliness [11] since UNR needs less bolls/plant to achieve the same yield as conventional cotton and the crop does not have to maintain the late formed bolls till maturity. In general, it was observed that lower plant densities produced higher values of growth and yield attributes per plant, but yield per u--nit area was also higher with higher plant densities [12, 13, 14]. Fertilizer and pest management are important consideration for increased yields under high density planting system. Changes in plant density modifies the microclimate and this may alter the incidence of pests and diseases as well [15]. Studies taken up using the genotypes AKH 081,Suraj and NH 615 under HDPS revealed that these entries could yield better at 60 x10 cm spacing under medium depth soils with a planting density of 1.66 lakh plants per hectare on broad bed furrow (BBF) with 125 per cent of recommended fertilizers (75:37.5:37.5 NPK + 2.5 Zn kg/ha) along with a foliar spray of 1% urea and 1% magnesium sulphate at boll development stage [16].

2.2.3 Growth manipulation

The very purpose of evolving genotypes suitable for HDPS would be achieved only by manipulation of row spacing, plant density and the spatial arrangements of cotton plants for obtaining higher yields. The most commonly tested plant densities range from 5 to 15 plants per sq. m [17] resulting in a population of 50,000 to 150,000 plants per ha. The UNR cotton plants produce a smaller number of bolls per plant than conventional cotton but retain a higher percentage of the total number of good opened bolls per unit area in the first sympodial position and a lower percentage in the second position [18]. The other advantages include better light interception, efficient leaf area development and early canopy closure which shades out the weeds and reduce their competitiveness [19]. The early maturity in soils that do not support excessive vegetative growth [20] can make this system ideal for shallow to medium soils under rainfed condition where conventional late maturity hybrids experience terminal drought.

Cotton growth must be regulated and eventually terminated by chemical means, due to the plants' intrinsic indeterminate growth habit. Plant growth retardants are natural or synthetic organic compounds that control or modify one or more physiological events in plants. These synthetic compounds are widely used in cotton for reducing plant height. The plant growth retardants affect many physiological functions in plants. The crop growth regulator Mepiquat Chloride (MC) is commonly used in cotton production in China and elsewhere to maximize cotton yield and fiber quality [21, 22]. The application of MC increases leaf thickness, reduces leaf area [23], shortens internodes [24] and decreases plant height [25], and thus results in a more compact plant architecture [26] which had been witnessed in the CO 17 culture as well (Figure 2).

Modification of cotton structure increases the light interception in the middle part of the canopy [27]. In addition, Light Use Efficiency (LUE) of cotton is increased by MC application [28]. Furthermore, cotton canopy structure is affected by population density [29] and practices such as wheat–cotton intercropping [30] which influence the crop light interception and fruit formation, thereby biomass growth and yield.

Figure 2. Application of Mepiquat chloride in cotton variety CO 17.

High population densities increase leaf area index (LAI) but reduces the individual leaf area [29]. Like most species, cotton plant height increases with population density [31]. Field experiments raised with CO 17 wherein cotton plant structure was obviously affected by MC and plant density. As MC could bring in changes in the architecture of the applied plant resulting structural changes, it leads to a challenge in maintaining the cotton plant's architecture suiting to the mechanized cultivation. Application of MC is essentially responsible for controlling cell elongation and shoot and stem growth [32]. When plant growth retardants are applied to plants, internodes become shorter and leaves become thicker and greener which leads to altered plant morphology and altered assimilate partitioning resulting in reduced plant growth. The response of plants to PGR applications can differ with plant growth stage, rates of application, and environmental conditions during the applications [33]. Cotton producers and researchers have, therefore, used plant growth retardants as a means to manage the balance between vegetative and reproductive growth for efficient cotton production. But research on application of growth retardants in conjunction high density planting will pave way for synchronized maturity of the crop with uniform plant height that may help in harvesting of seed cotton mechanically at large scale level. This research is at its nascent level in India.

Mepiquat Chloride (1,1-dimethyl-piperidinium Chloride), a plant growth regulator is widely used to manage cotton structure, regulate plant development and hasten maturity under high plating densities [34]. Although plant growth regulators have been thoroughly widely tested in cotton in India, specific recommendations regarding their dose and timing for modifying the architecture at high planting densities are yet to be arrived for adoption on a holistic perspective. Reduction in plant height, decrease in height/node ratio, an increase in boll weight and a delay in maturity with the application of growth regulators were observed with non-significant effect on yield. Application of Mepiquat Chloride reduced the leaf area and increased the number of bolls per unit area at high plant density. It also helped in retention of bolls on lower sympodia and increased the synchrony of boll maturation [29].

However, the effect of Mepiquat Chloride on cotton is affected by environmental conditions, particularly temperature [35]. Studies taken up at the Department of Cotton indicated the differential response of cultures and its performance across centres. At Coimbatore, there was a variety dependent response to Mepiquat Chloride application @ 75 g ai/ha in three splits on 45, 60 and 75 DAS in winter irrigated cotton planted at 45 x 15 cm spacing and the effect was more pronounced in CO 17 (Figure 3).

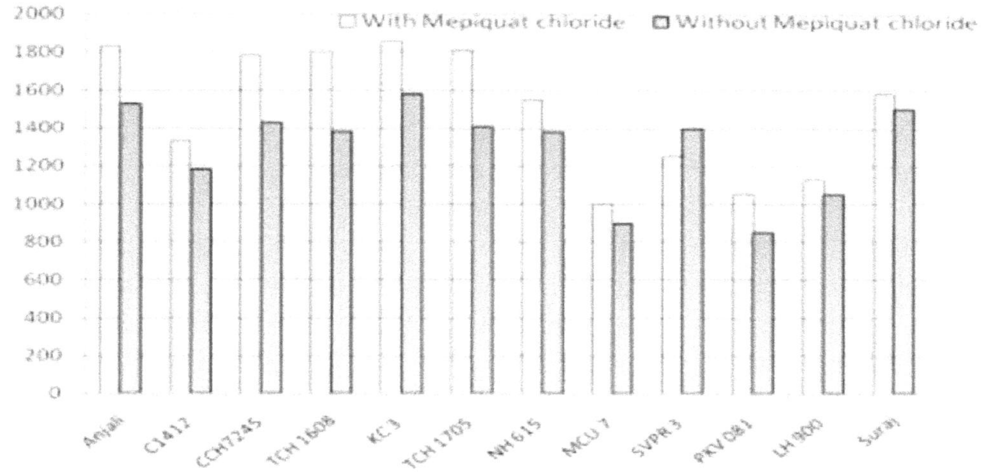

Figure 3. Effect of Mepiquat chloride on cotton varieties.

There was a reduction in plant height, sympodial length and LAI and an enhanced the number of burst boll/sq. m leading to an increase in yield at Coimbatore. Across the cultivars, application of Mepiquat Chloride increased seed cotton yield from 1330 kg/ha to 1530 kg/ha. Interaction effect of cultivars and application of Mepiquat Chloride was significant. Taller cultivars namely TCH 1608 and TCH 1705 (CO 15) benefitted more from the application of Mepiquat Chloride compared to the other cultivars having a compact growth habit. Cultivars with a more indeterminate growth habit responded more positively to Mepiquat Chloride application [36]. There is a need for detailed investigations on this aspect before any recommendations are given.

2.2.4 Effect of Mepiquat chloride application on seed cotton yield in different genotypes

2.2.4.1 Defoliation

Arresting the growth of the cotton plant followed by the defoliation of leaves at the physiological maturity stage of the plant would facilitate in harvesting the fully opened bolls through mechanical instruments. As we know that killing and fixation of specimen are required for having cytological observations by employing a mix of acid and alcohol, the two events namely growth arrest and defoliation have to occur for effecting the harvest of bolls in cotton. Thus, cotton defoliation requires the application of certain chemicals to help prepare a cotton crop for harvest. Benefits of proper cotton defoliation include: reduction of the main sources of stain and trash (leaves), increased harvest efficiency, quicker drying of dew, potential for increased boll opening and a reduction of boll rots. As a cotton plant matures, a physiological process takes place which separates the living tissue near the leaf petiole called the abscission zone. A regulated enzyme activity under the influence of plant hormones making the leaf to fall through the creation of abscission zone is the need.

Two types of defoliants, by and large are available. A herbicidal defoliant can be used to cause injury to the leaf, upset the hormone balance resulting in the abscission process and finally the leaf drop. The other one is the application of a hormonal defoliant which increases ethylene synthesis in a plant causing the leaves to fall off. Correct application rates are important, especially with herbicidal defoliants, as over application can cause the leaf to die before the abscission process, resulting in "stuck" leaves. Conversely, when too little defoliant is applied, the abscission process may not begin, resulting in no leaf defoliation.

2.2.4.2 Factors affecting defoliation

When applying a defoliant or desiccant or boll opener, many factors must be taken into consideration for successful application. Best results from an application occur when the factors like the type of cotton being grown, its duration, weather/climatic conditions, soil conditions, availability of inorganic fertilizers to the plant, bursting of bolls, water availability to the plants, spacing between the rows and plants, etc. are taken into consideration.

Studies taken up at TNAU utilizing the application of MC and Defoliant indicated that the combination of application of MC 50 g ai/ha at 45 and 65 DAS and Dropp ultra (Thidiazuron and Diuron) @ 200 ml/ha at maturity resulted in 90 per cent defoliation in CO 17 and prepared the crop for harvest by 135 days. This is in the preliminary stage of testing and would be tested in the years to come to arrive at a comprehensive package (Figure 4).

Figure 4.Application of Mepiquat chloride and defoliant in CO 17.

2.3 Pests management

2.3.1 International level

Because of various types of insects that attack the cotton crop, especially from the young seedlings (leaf miner, gall formers etc.) till the crop attains maturity (various borers, weevils etc.), cotton crop receives excessive rounds of pesticides spray. This results in excessive consumption of plant protection products. Many research are being taken up towards developing holistic packages including chemical, physical, biological and IPM techniques to contain these pests. Cotton is under cultivation in 69 countries and the production had exceeded 20 million tones of lint in the recent years where the cultivable area spread on 30–35 million hectares. In spite of improvement attained through chemical control strategies, harvest losses remain very high which dwindles around 30% [37, 38]. Occurrence of varied insects in the cotton system during varied crop growth stages makes it as an experimental model crop for devising plant protection strategies to be practiced under various agronomic conditions.

Albeit very many newer molecules are synthesized and tested to contain the pests, harvest losses remained high. All the pest management strategies aims to keep the pest population below the Economic Threshold Level (ETL) which is normally attained by having a judicious mix of appropriate methodologies. Total pest management is achievable only when the pest prefers a single crop, say cotton and there are no significant alternate hosts available in the vicinity of the crop system. However, the application of IPM principles greatly depends on the concept of an intervention threshold and the limitations of many of the specific non-chemical techniques proposed but the application of IPM modules/principles have the advantage of taking into consideration the full pest complex in a cropping system [37].

Biological control by introducing beneficial arthropods has not been notably successful in all the crop-based systems, which is true for cotton also. This is because of the difficulty in identifying and acclimatizing the predators/parasites, developing a bunch of beneficial organisms capable of responding effectively, the nature of the crop grown and the disrupting effects of chemical control measures directed against the remaining pests [39]. More benefit is to be obtained from the active conservation of the indigenous fauna of beneficial organisms. In spite of an increased general environmental awareness, the practice of using insecticides could not be resisted as pests evolve resistance to pesticides and a combinatory approach to contain the pest is the need of the hour.

Suggested strategies were adopted throughout the growing season in Australia. Primary target of reducing the pesticides consumption in cotton ecosystem was brought out by the introduction of Bt gene engineered cotton hybrids which allows the co-existence of natural enemies. However, the least affected species by the Bt toxins, the sucking pests took a prominent place in cotton based production system displacing the vegetative and fruit feeding caterpillars as key pests of Bt cotton [40]. The spatio-temporal dimension of natural population regulatory factors has led to changes in agricultural practices and production systems. In cotton, for example, production systems maintaining a permanent ground cover, are having increasing success. Many a times, farmers leave the crop in the field after harvest of bolls alone, especially for getting a second flush with the onset of rains resulting in enhanced outbreak. Intercropping and trap cropping have been favorable to the maintenance of beneficial arthropod complexes and unfavorable to the growth of pest populations. Thus for having an effective control over the pests, a changed strategy towards a total systems approach, characterized by a movement from a paradigm of pest control field-by-field, through farm-by-farm and agroecosystem-by-agroecosystem, to a landscape by landscape approach is required as reported by [38].

The rich and diverse insect fauna found in cotton harbors more than one thousand species. However, very few are designated as significant potential pests. These pests damage the flowers and fruiting bodies or consumes the leaves or mine the leaves and sucks the juice of the leaves of young plants. Some of them are monophagous species, restricted to the genus Gossypium (Anthonomus, Diaparopsis) while oligophagous feeding on plants in the family of Malvaceae and closely related families (*Pectinophora, Dysdercus, Earias*) or polyphagous feeding behavior (*Helicoverpa, Heliothis, Cryptophlebia, Spodoptera, Helopeltis*) were also reported. The heliothine lepidopteran species complex (*Heliothis virescens, Helicoverpaarmigera, Helicoverpazea*) is considered as the most dangerous, found attacking numerous other cultivated plants which are often associated with cotton in a range of cropping systems [41]. However, as indicated earlier, the leaf hoppers and white fly are becoming a menace nowadays.

2.3.2 National level

Cultivating cotton by adopting closer spacing with the available cultivars and with those having short branches was conceived and implemented in India since sixties of the 20th century. Visits, exposures, dialogs and discussions could improvise the learning and the team's visit to Brazil had sown the idea of researching on High Density Planting System (HDPS) and the churned idea was implemented by reorienting research through All India Coordinated Cotton Improvement Project (AICCIP) and through other schemes like Technology Mission of Cotton (TMC) or National Food Security Mission (NFSM) in India [42]. It is reiterated that the success of HDPS at varied locations solely depends on the availability of genotypes, appropriate spacing and nutrition for adopting more plants per hectare to achieve more productivity per unit area and sound pest management criteria. This necessitates the evaluation of available genotypes in varied spacing and nutrition level to the incidence of insect pests. As farmers tend to grow more Bt which are normally of spreading in nature, evaluation of both Bt and not Bt compact genotypes for their suitability to HDPS is the need of the hour. The adoption of HDPS along with better genotype and fertilizer management is a viable approach to break the current stagnation of yield.

More the synthetic fertilizer application, especially nitrogen (N) fertilizer, more the serious insect herbivores occurrence and crop damage from these insects by reducing plant resistance, the concept which has been conceived clearly [43, 44]. Reducing fertilizer applications can reduce the production costs for cotton growers, as well as nitrogen (N) leaching into the soil and contamination of surface and ground water, but altered N fertilization may also affect pests and their natural enemies [45]. Occurrence of insects and their abundance are heavily dependent on the micro climate available in the system which is primarily based on the biomass production by the plants and their nearness (spacing). Hence, it is utmost important to study the pest dynamics under closer spacings as well as increased levels of nitrogen applications.

A study taken up using GSHV-01/1338 and GBHV-164 genotypes among others revealed their ability as promising genotypes of the region suited to high density planting system due to its compact nature [42]. At two levels of closer spacings (60x15 and 45x15 cm) against the normal spacing of 120x45 cm along with two increased level of nitrogen application (i.e. 125 and 150% of RDN), these two genotypes performed better. The studies were carried out in factorial randomized block design *Kharif* 2013–2014. Closer spacings (45x15 and 60x15 cm) attracted more thrips as compared to the recommended spacing (120x45 cm). The mean population of thrips was found significantly high on GBHV-164 than GSHV-01/1338. Higher dose of nitrogen application on crop (125 and 150% RDN) attracted more thrips as compared to 100 per cent recommended dose of nitrogen (RDN).

[46] by quoting a field experiment that was conducted to study the mean incidence of major cotton insect pests during two consecutive seasons i.e. during kharif, 2010–2011 and 2011–2012 at CICR, Nagpur under high density planting system (HDPS) using different genotypes of G. hirsutum with different spacings indicated that pest incidence was not altered by closer spacing. The main objectives of the work was to identify lines of G. hirsutum which have less infestation of major insect pests under HDPS system and to investigate whether the incidence is influenced by plant density. In 2010–2011, the minimum mean population of leafhopper was observed in NISC-50 (1.82 nymphs / 3 leaves /plant) which was grown at a spacing of 45x13.5 cm followed by PKV-0811 (1.91 nymphs /3 leaves/plant) grown at a spacing of 45x13.5 cm and these genotypes are significantly superior over the others. The injury grade was I in both NISC-50 and PKV-081 genotypes. The mean per cent square damage was low in CNH-120 MB (2.76%) followed by PKV-081 (3.82%), both being statistically on par with each other and significantly superior over other genotypes. The mean pink bollworm population was low on PKV-081 (2.53 larvae/25 green bolls). The lowest per cent locule damage due to pink boll worm was noticed on PKV-081 (8.48%). However, the performance of genotypes and geometry against all the insect pests in 2011–2012 was not significantly different leading to a conclusion that pest incidence was not altered by closer spacing.

[47] reported in a study undertaken during 2015–2016 on High density planting demonstrations (50) which were taken up in farmers' fields at varied close spacings (75x10 and 90x10cm) with available compact genotypes (Suraj and G.Cot.16) compared to normal spacing (120x45 cm) under Insecticide Resistance Management (IRM) umbrella in rainfed regions of Bharuch district. Aphids, thrips and leafhopper were found above ETL whereas whitefly and mealybug were found below ETL. The mean larval population of pink bollworm was 4.41 and 3.14 larvae/20 green bolls in Suraj and G.Cot.16 spaced at closed spacings respectively. The pink bollworm population was 2.51 and 2.68 larvae/20 green bolls in Bt-IRM and non IRM plots respectively. Suraj variety spaced at 75x10 and 90x10 cm required 4.21 and 3.33 sprays and G.Cot.16 spaced at 75x10 and 90x10 cm required 4.40 and 3.60 sprays against sucking pests and 2.37 and 2.38 and 3.20 and 2.40 sprays against bollworms respectively as against 5.00 and 5.60 sprays against sucking pests and 2.00 and 3.80 sprays against bollworms in Bt-IRM and Bt-Non IRM cotton respectively. These

results suggest the need for excessive sprays in ultra closer spacing than the normal closer spacing for both the cultivars. The net return was found higher in G.Cot.16 HDPS at both the spacing (Rs. 22,966 and 17,456/acre) than the Suraj HDPS (Rs. 16,461 and 8235/acre). The net return for Bt-IRM farmers was higher (Rs.21527/acre) than non IRM-Bt farmers (Rs. 17,919/acre). Thus, it has been concluded that HDPS offer viable option to increase productivity especially under rainfed region.

The cotton crop is attacked by 1326 species of insect pests throughout the world, of which about 130 different species of insects and mites found to devour cotton at different stages of crop growth in India. Among the bollworms, pink bollworm assumed major pest status in recent past [48]. The pink bollworm, Pectinophoragossypiella (Saunders), a pest which received more attention in almost all the cotton growing states of India (except Tamil Nadu as of now), is identified as the most destructive pest of cotton and causes 2.8 to 61.9 per cent loss in seed cotton yield, 2.1 to 47.1 per cent loss in oil content and 10.7 to 59.2 per cent loss in normal opening of bolls [49]. Locule damage was noted to an extent of 55 per cent and 35–90 per cent reduction in seed cotton yield has been reported by [50, 51] estimated the yield loss to an extent of 6525 MT annually.

Losses caused by pests vary by 10–30% depending on the intrinsic genetic factors and its rigidity in expressing the inherent resistance. Pests are supposed to evolve in a short and strategic cycle to circumvent the problems being arisen and judicious use of insecticides along with physical, biological control methods is the need of the hour. Ignoring pests can lead to complete crop failure. In the overall crop protection program under the National Agricultural Policy, The Government's IPM is a time-tested, eco-friendly approach, socially acceptable and economically viable that is widely accepted across the country. Appropriate control measures should be taken when insect populations cross the ETL [52].

2.4 Diseases management

High density planting which entails closer planting may have every chance of microclimate getting altered due to which the propensity of infectious diseases in cotton may also vary. The high density planting in cotton is a recent phenomena which opened avenues for research in various domains including plant pathology. The plant pathologists have been trying to understand the nexus between the incidence of various cotton diseases and the change in microclimate of the plant coupled with external atmosphere [53, 54].

The life cycle of pathogens is amenable for changes in line with the changes in plant canopy and the microclimate mediated through weather parameters. Space between plants and rows is bound to have a say in the mode of dispersal, the intensity of infection and the production of secondary inoculum of plant pathogens. Cotton, being a commercial crop, is no exception to this phenomena of infection and the high density planting in cotton need to be carefully contemplated taking into account the changed plant geometry and the corresponding incidence of cotton diseases. Despite the research on influence of high density planting in cotton on the incidence of diseases is in the nascent stage, an attempt has been made to take stock of striking developments in the management of important cotton diseases in the succeeding pages.

The major diseases of cotton which are prevalent in most part of the cotton growing countries in the world were reported to inflict a damage ranging from 10 to 30% and it may be more when favorable conditions prevail for the spread of the pathogens which culminates in cotton farmers spending huge cost to keep the biotic stress under control [55].

Fungal diseases are predominant in cotton followed by very few bacterial and viral diseases. The prominent bacterial disease which inflict major damage in cotton crop is bacterial blight, caused by *Xanthomonas citri* pv. *malvacearum* [56]. Abundant literature is available on major fungal diseases of cotton namely *Fusarium* wilt caused by *Fusarium oxysporum* f. sp. *vasinfectum*, *Verticillium* wilt caused by *Verticillium dahliae*, anthracnose caused by *Colletotrichum gossypii*, Ramularia gray mildew caused by *Mycosphaerella areola*, root rots caused by *Rhizoctonia solani* and *R. bataticola*, leaf blight caused by *Alternaria macrospora* and leaf spot caused by *Cercosporagossypina* [57, 58, 59, 60, 61, 62]. The cotton leaf curl disease is the only virus disease documented in cotton which belongs to the genus Begomovirus and transmitted by insect vectors [63].

Plethora of studies were conducted for the management of cotton diseases which reported solitary or combination of management practices to control them. Besides chemical control, significant research work has been carried out on the biological control of cotton diseases [64, 65, 66, 67]. Similarly, there were prominent studies on use of plant extracts [68, 69, 70] and essential oil [71, 72] for the management of cotton diseases. Cultural methods [73] and organic amendments [74] also form part of the strategy to control cotton diseases.

In the recent past, several research studies have documented the efficacy of Endophytic bacteria [75] in suppressing the incidence of cotton diseases. Molecular level studies namely transcriptomic, proteomic and metabolomic studies [76, 77, 78] and studies on gene editing [79] were on the rise for the past two decades. As there are several methods and molecules have been designed for effecting the control of diseases, their role in containing the diseases that occur under HDPS is in infant stage as the genotypes which have been evaluated at Coimbatore were not found to have adequate disease expressions under HDPS.

2.4.1 Soil borne diseases

Soil borne fungal diseases of cotton namely damping off, root rot and wilt have been reported to cause extensive damage in cotton crop. Juxtaposing chemical control with biological method, the latter was found to be effective which is evidenced from the finding of [80, 81, 82] who reported that the combined application of *T. harzianum* and *P. lilacinus* showed the best result by inhibiting the growth of pathogen than alone. A recent study of [83] reported that Trichodel®, based on *Trichoderma* spp., reduced the incidence and severity of wilt caused by *F. oxysporum* f. sp. *vasinfectum*. Besides, a score of agronomic practices namely fine tilth of the soil, adjusting sowing season, crop rotation, soil solarization, amending soil for altering pH of the soil and use of resistant varieties have been reported to reduce the incidence of soil borne diseases. Biological control of *V. dahliae* in cotton with a mixture of lignin and *Trichoderma viride* [84] has been reported. Thus, biological control of soil borne diseases is the viable option which had been arrived by various authors in the normal cotton growing situations. However, the same might hold good under HDPS also.

2.4.2 Foliar diseases

Among the foliar diseases, Alternaria leaf spot, gray mildew, boll rot, rust, anthracnose and bacterial leaf blight were reported in cotton and they were reported to inflict damage significantly. The chemical fungicides mancozeb (0.3%), propiconazole (0.1%), propineb (0.3%) were found more effective against *Alternaria macrospora*, propiconazole (0.1%) and copper oxychloride (0.25%) against *Myrothecium roridum* [85, 86]. Moreover, a decadal analysis of *Alternaria* occurrence among the various genotypes at Coimbatore indicated that Bunny Bt cotton, NCEHBT, Dhannu BGII and 1037 BGII genotypes were found to be more susceptible and the disease incidence ranged from 0.5 to 10.53 PDI compared to the types which are resistant/field tolerant. In addition, the sowing taken up during 29th–30th Standard Meteorological Week (SMW) resulted in lesser incidence of the disease irrespective of the cultures evaluated [87].

Among the biocontrol agents studied, *Pseudomonas fluorescens* strains and *Bacillus subtilis* and the botanicals derived from *Azadirachata indica*, *Lantana camera*, *Calotropis procera*, *Ocimum sanctum*, *Allium cepa* and *Allium sativum* have been reported to significantly reduce the mycelial growth of the pathogenic fungus [88]. The methanol extracts of *Polyalthia longifolia* and *Terminalia chebula* and chloroform extract of *Zingiber officinale*, *Datura alba*, *Moringa olifera*, *Azadirachta indica* and *Syzgyiumcumini* have showed significant biological control of cotton bacterial blight in greenhouses and in fields [89, 90]. This indicates that principles available in various plants are having sizable influence in containing the growth of disease causingmicro organisms.

2.4.3 Viral diseases

Among the viral diseases infecting cotton, cotton leaf curl virus and tobacco streak virus are important. The TSV disease was reported to be spread through mechanical means, infected seeds and through thrips species. Parthenium, a widely distributed and symptom less carrier of TSV, plays a major role in perpetuation and spread of the disease [91, 92, 93, 94, 95].

A study carried out by [96] uncovered the application of *Bacillus* species which possess diverse anti microbial peptide (AMP) genes which are responsible for the biosynthesis of antibiotics like iturin, bacilysin, bacillomycin, fengycin, surfactin, mersacidin, ericin, subtilin, subtilosin, and mycosubtilin in curtailing the infection of TSV. Genetic Engineering studies to evolve transgenic cotton using an antisense RNA approach [97] could be a potential option for managing the disease. Interestingly, transgenic cotton plants that over express miR166 also show potential in reducing *Bemisiatabaci* populations and, more importantly, the spread of whitely transmitted plant viruses [98, 99]. Gene editing technology *i.e.,* CRISPR/Cas9 system has recently been used to confer molecular immunity against several eukaryotic viruses, including cotton DNA geminiviruses [100].

3. CONCISE SUMMARY

Though evolving a suitable ideotype remains the basics of successful adoption of any HDPS, the evolved genotype's suitability to the various management practices results in its adoptability. Nowadays, many of the farmers are going for Bt cotton and that too under rainfed situations. Considering this as a backdrop, the genotypes under development either through conventional method employing crossing between the selected genotypes and progeny selection or mutation breeding or selection of desirable genotypes from the pre-bred materials or by any of the molecular methods must fit for one or more situations as described below.

- Considering the occurrence of weeds in both irrigated and rainfed systems, the HDPS genotype must possess inherent herbicide tolerant/resistant behavior either through inheritance or through infusion using biotechnological tools.
- Transplanting of cotton have been identified as one of the viable option for maintaining the sufficient plant population. Preliminary studies taken up at the Department of Cotton also indicated the survival of 15–20 days old seedlings of CO 17 variety upto 25 per cent. However, the growth observed in the transplanted plants was stunted. Studies taken up by [101] indicated that date of sowing, age of seedling for transplanting bears a relationship on the seed cotton yield. Moreover, Bt cotton sown directly or through tray nursery or through polythene

nursery of varied duration had varied impact on seed cotton yield. Polythene nursery with 3 weeks old seedlings upon transplanting could result in a yield of 4727 kg/ha. Thus, clear cut studies on the evolved genotype for its suitability to transplanting through any one of the methods are required.

- It has been observed that cotton plants raised under dense paired sowing produced the highest seed cotton yield and water use efficiency [102]. Hence, the evolved genotypes need to be tested for its water use efficiency under fertigation studies which is the need of the hour. Preliminary experiments conducted with CO 17 in this regard with varied level of population and varied nutrient levels under drip fertigation revealed that spacing 75 cm × 10 cm spacing coupled with STCR based drip fertigation recorded net return of Rs. 131,302/ha.
- Another interesting observation made by [103] was weed density and weed dry matter production were lesser at closer spacing of 30 x 30 and 45 x 30 cm as compared to widely spaced cotton. Though this is to be taken care of, more closer spacing would attract pest and diseases and hence an ideal spacing has to be arrived for evolved genotypes.
- The evolved genotype (preferably as a variety) must have an yield advantage and increased per day productivity compared to the *Bt* hybrids which are normally bred and released by the private companies.
- It has to be ascertained, that by all means, the evolved genotype should contain the deregulated gene/QTLs conferring resistance to bollworms or genes that are of native origin which are freely available. This is because that the non Bt cotton variety F 2383 released by Punjab Agricultural University during 2015 for cultivation under HDPS had the susceptibility to boll worms besides smaller boll size [104].
- In USA, it took 30 years to achieve 100 per cent mechanization, while it was 45 years in Brazil. Turkey reached 75 per cent mechanization in 15 years and China took 20 years to reach 15 per cent mechanization. However, in India, as seen, harvesting is completely labour oriented and all the activities need to be mechanized [105]. It has been also indicated that the mechanical picking with single row picker under HDPS provided 25–40 per cent yield increase compared to the farmer's practice. Preliminary studies done in this aspect in the Department of Cotton with CO 17 variety indicated that mechanizing the sowing, spraying, weeding operations alone had resulted in a savings of Rs. 51,000(698 US $).

- This gives a clear message that mechanizing cotton cultivation is of essential one and the evolved genotype must fit for mechanized cultivation in India.

AT A GLANCE

Cotton, a crop of choice, occupies the second premier position next to food crops in providing clothing. Though 53 species of Gossypium are available, only four species are cultivable and among the four, the major cultivable area falls under G. hirsutum. Though varieties with medium, superior medium, long and extra long staple cotton were released earlier, with the advent of machineries, ginning facilities, mills were literally requiring cotton fiber of any length. With the advent of Bt technology and the release of hybrids during 2002, cotton productivity had a momentum. However, considering the duration, cost involved in manual harvesting etc., farmers were looking for alternate option and High Density Planting System (HDPS) offered a promise in this direction. Farmers were looking for genotypes that could yield better under higher planting densities with fewer bolls per plant, synchronized maturity with uniform bursting. Efforts have been taken all over the World in this direction and India is not an exception. Handful of varieties fitting to this situation has been released from many of the Universities. This chapter essentially summarizes the genetic, agronomic, plant protection interventions and the futuristic requirements for achieving at least 700 kg of lint per hectare.

REFERENCES

1. Latha,P,Premalatha,N,Sakthivel,N,Senguttuvan,K,Mahalingam,L,Amalabalu,P and M.Kumar. 2017. A decade analysis of Alternaria leaf blight in cotton. ISBN: 978-93-83799-61-9.pp.87
2. Dhoble, M. V., Shaikh, M., Patil, V. D., Giri, D. G., and Pawar, B. R. 1992. Performance of different cotton genotypes as influenced by varying plant row spacings under rainfed conditions. Journal of Cotton Research and Development, 6(1), 128-134
3. Yuan Yuan , Hongjie Feng , Lingfei Wang , Zhifang Li, Yongqiang Shi, LiHong Zhao, Zili Feng, Heqin Zhu. 2017. Potential of Endophytic Fungi Isolated from Cotton Roots for Biological Control against Verticillium Wilt Disease. PLoS One 20;12(1):e0170557
4. Jagtap GP, Jangama AM and Deya U. 2012. Management of bacterial blight of cotton caused by Xanthomonas axonopodispv. malvacearum. Scientific Journal of Microbiology, 1(1) :10
5. Craig, C.C. and O. Gwathmey. 2005. Variety response to mepiquat chloride applications. In Proc. Beltwide Cotton Conf. New Orleans, LA. 4-7. Natl. Cotton Counc. Am., Memphis, TN. p. 1946

6. Mekonnen, Z.A., William Riley, James T. Randerson and R.F. Grant. 2019. Expansion of high-latitude deciduous forests driven by interactions between climate warming and fire. Nature plants 5(9)
7. Reddy, A.R.; Reddy, K.R.; Hodges, H.F. 1996.Mepiquat chloride (PIX)-induced changes in photosynthesis and growth of cotton. Plant Growth Regulation 20: 179-183
8. Jones, Q. 1983. Germplasm needs of Oilseed Crops. Economic Botany, 37: 418-422
9. Rossi, J., Novick, G., Murray, J., Landivar, J., Zhang, S.,Baxevanos, D., Mateos, A., Kerby, T., Hake, K. and Krieg, D. 2004. Ultra Narrow Row Cotton: Global Perspective and Reduce Pesticide use Proceedings of the Technical Seminar of the 3rd Plenary Meeting of the ICAC: How to improve yields. Mumbai, India. Nov 2004, pp. 7-11
10. Briggs, R. E., Patterson, L. L., and Massey, G. D. (1967). Within and between-row spacing of cotton. Arizona Annual Report, P. 6_7. University of Arizona Agriculture Extension Service, Arizona
11. Latha, P, T Anand, V Prakasam, EI Jonathan, M Paramathma. 2011. Combining Pseudomonas, Bacillus and Trichoderma strains with organic amendments and micronutrient to enhance suppression of collar and root rot disease in physic nut. Applied soil ecology 49, 215
12. Oosterhuis, D.M. and Egilla, J.N. 1996. Field evaluation of plant growth regulators for effect on the growth and yield of cotton.p.1213-1215. In P. Dugger and D. Richter (ed.) Proc. Beltwide Cotton Conf. National Cotton Council, Memphis, TN
13. Ram Prasad,B, Sundashanam,A, Malathi,S and Raghu Rami Reddy,P. 2019. Influence of high density planting system (HDPS) on cotton sucking pests and natural enemies in rainfed alfisols. Journal of Entomology and Zoology studies 7(5):1328-1331
14. Zhang, Z., Ge, X., Luo, X., Wang, P., Fan, Q., Hu, G., et al.(2018). Simultaneous editing of two copies of GH14-3-3D confers enhanced transgene-clean plant defense against Verticillium dahliae in allotetraploid upland cotton. Frontiers in Plant Science, 9(June), 1-13
15. Jagtap G P and Utpal Dey. 2013. Natural occurrence of Tobacco streak virus (ilar virus) on Bt cotton hybrids. International Journal of Agricultural Sciences, 9 (1):151
16. Cia, E., Fuzatto, M. G., Kondo, J. I., Carvalho, L. H., Ito, M. F., Dias, F. L. F., et al. (2016). Response of cotton genotypes to the incidence of Alternaria leaf spot. Summa Phytopathologica, 42(4), 357-359

17. ICAC (2006) Cotton World Statistics, ICAC, Washington, 127 p
18. Vinodkumar S. Nakkeeran S., Renukadevi P. Mohankumar S.2018. Diversity and antiviral potential of rhizospheric and endophytic Bacillus species and phyto-antiviral principles against tobacco streak virus in cotton. Agriculture, Ecosystems and Environment 267:42-51
19. Amudha, J., Balasubramani, G., Malathi, V. G., Monga, D., and Kranthi, K. R. 2011. Cotton leaf curl virus resistance transgenics with antisense coat protein gene (AV1). Curr. Sci. 101, 300-307
20. Jean-Philippe Deguine, Pierre Ferron, Derek Russell. 2008. Sustainable pest management for cotton production. A review. Agronomy for Sustainable Development, Springer Verlag/EDP Sciences/INRA, 2008, 28 (1), pp.113-137
21. Shaban, M., Miao, Y., Ullah, A., Khan, A. Q., Menghwar, H., Khan, A. H., et al. 2018. Physiological and molecular mechanism of defense in cotton against Verticillium dahliae. Plant Physiology and Biochemistry, 125, 193-204
22. Yaheia Omer Fatouh, Farid Abd-El-Kareem, Faten, M. Abd-El-Latif and Riad S. El-Mohammedy. 2011. Effects of citrus essential oil compounds on management leaf spot diseases on sugar beet under field conditions. Journal of Agricultural Technology. Vol.7(5): 1389-1396
23. Sathish,S, Raveesha,K.A. and Janardhana,G.R. 1999. Antimicrobial activity of plant extract on Phytopathogenic Xanthomonas campestris pathovars. Letters in Applied Microbiology, 28:145-147
24. Fernandez C J, Cothren J T, Mcinnes K J. 1991. Partitioning of biomass in well-watered and water-stressed cotton plants treated with mepiquat chloride. Crop Science, 31, 1224-1228
25. Pangga,I, Hanan. J. and Chakraborty. S.2011. Pathogen dynamics in a crop canopy and their evolution under changing climate. Plant Pathology 60(1):70-81
26. Pushavathi, D., Shilpa, M., Petkar,T., Siddidha,A. and PrashithKekuda, Y.R. 2017. Evaluation of antifungal activity of some lants against seed borne fungi. Scholars Journal of Agriculture and Veterinary Science, 4 (4): 155-159
27. Gwathmey, C.O. and Clement, J.D. 2010. Alteration of cotton source–sink relations with plant population density and mepiquat chloride. Field Crops Res. 116 : 101-07
28. Gwathmey, C. O., Wassel, O. M., and Michaud, C. 1995. Pix effects on canopy light interception by contrasting cotton varieties. In: Proceedings of Beltwide Cotton Conference. pp. 4-7

29. Gonias E D, Oosterhuis D M, Bibi A C. 2012. Cotton radiation use efficiency response to plant growth regulators. Journal of Agricultural Science, 150, 595-602

30. Paramjit Singh. 2018. Optimum plant type for high density planting system in cotton for North Indian conditions. In: Intl. congress on "Cotton and other fibre crops" (Eds. M.S.Chauhan and R.S. Sangwan), at ICAR Research Complex for NEH Region, Umaim, Mghalaya organized by Cotton Research and Development Association, CCS Haryana Agricultural University, Hisar-125004, India.pp.35-38 (ISBN-978-81-936782-0-6)

31. Huang X, Zhang N, Yong X, Yang X, Shen Q .2012. Biocontrol of Rhizoctonia solani damping off disease in cucumber with Bacillus pumilus SQR-N43. Microbiol Res 167:135-143

32. de Farias, O., de Nascimento, L. C., Lima Cruz, J. M.,Oliveira Silva, H. A., Oliveira, M., Alcântara Bruno, R.,et al. (2019). Biocontrol potential of Trichoderma and Bacillus species on Fusarium oxysporum f. spvasinfectum. Journal of Experimental Agriculture International, 34(1), 1-11

33. Rajendran, T. P., A. Birah, and P. S. Burange. 2018. "Insect Pests of Cotton." In Pests and Their Management, 361-411. Singapore: Springer

34. Prashant B, Sandipan GR, Bhanderi RD, Patel and Solanki BG. 2017. Survey and Status of Different Diseases of Cotton under South Gujarat Region, India. Int. J. Curr. Microbiol. App. Sci. 6(9): 1362-1367

35. Latha.P.2020. Low cost and No cost disease Management strategies in cropping systems perspective for ensuring sustainable production. Ecofriendly disease management for food security (ISBN: 978-81-945631-3-6. First Edition, Chapter-I.p.134-155

36. Vories, E. D. and Glover, R. E. 2006. Comparison of Growth and Yield Components of Conventional and Ultranarrow Row Cotton - The Jour. Cotton Sci. 10 : 235-43

37. Siebert J D, Stewart A M. 2006. Influence of plant density on cotton response to mepiquat chloride application. Agronomy Journal, 98, 1634-1639

38. Rosolem, C.A., Oosterhuis, D. M., Souza, Fabio S. de 2013. Cotton response to mepiquat chloride and temperature - Scientia Agricola 70 : 82-87

39. Chinna Babu Naik, V., P.L. Dabhade, S. Kranthi, M.V. Venugopalan and Subbireddy, K.B. 2018. Intensity of Insect Pests in Gossypium hirsutum Cultivars under High Density Planting System (HDPS) in India. Int.J.Curr. Microbiol.App.Sci. 7(12): 125-134

40. Zohora.U.S., Takashi Ano, Mohammad shahedur Rahman. 2016. Biocontrol of Rhizoctonia solani K1 by IturinA Producer Bacillus subtilis RB14 seed treatment in tomato plants Advances in Microbiology. Vol.06. pages.8

41. Mohan Venkata Siva Prasad B, SL Bhattiprolu, V Prasanna Kumari and Anil Kumar P. 2017. In Vitro Evaluation of Bacterial Biocontrol Agents and Botanicals against Alternaria Leaf Spot Caused by Alternaria macrospora in Cotton. Int.J.Curr.Microbiol.App.Sci, 6(11): 750-758

42. Salaheddin K, Valluvaparidasan V, Ladhalakshmi D and Velazhahan R. 2010. Management of bacterial blight of cotton using a mixture of Pseudomonas fluorescens and Bacillus subtilis. Plant Protection Science. 46(2):41-50

43. Brodrick, R. , Michael P. Bange, Stephen P. Milroy, and Graeme. L. Hammer. 2010 . Yield and Maturity of Ultra-Narrow Row Cotton in High Input Production Systems. Agronomy Journal, Volume 102(3): 843-848

44. Mondino, M H, Peterlín, O. A. and Garay F. 2004. Response of late-planted cotton to the application of a growth regulator (chlorocholine chloride, cycocel 75). Australian Journal of Experimental Agriculture, 40, 381-38

45. Sharma, J. K., Upadhyay, A., Mishra, U. S., Khamparia, S. K. and Mandloi, K. C. 2001. Effect of spacings and fertility levels on growth and yield of hirsutum genotypes. J. Cotton Res. Dev. 15 : 151-53

46. Venugopalan, M. V., Kranthi, K. R., Blaise, D., Lakde, S., and Sankaranarayana, K. 2014. High density planting system in cotton-The Brazil Experience and Indian Initiatives. Cotton Research Journal, (2), 172-185

47. Jost, P.H., and Cothren, J.T. 2001. Phenotypic alterations and crop maturity differences in ultra-narrow row and conventionally spaced cotton. Crop Sci.41:1150-59

48. Reddy, V.R., Baker, D.N., Hodges, H.F. 1990. Temperature and mepiquat chloride on cotton canopy architecture. Agro. Jour. 82 : 190-95

49. Rothrock, C. S., Woodward, J. E., &Kemerait, R. C. 2015. Diseases In: Cotton, 2nd ed., Agron. Monogr. 57. ASA, CSSA, and SSSA, Madison, WI. 465-508

50. United States Department of Agriculture : Foreign Agricultural Service: Cotton: World Markets and Trade, Global Market Analysis, September 2020

51. Cox, K. L., Babilonia, K., Wheeler, T., He, P., & Shan, L. (2019). Return of old foes - Recurrence of bacterial blight and Fusarium wilt of cotton. Current Opinion in Plant Biology, 50, 95-103

52. Mohan, S., D Monga, Rishi Kumar, V Nagrare, Nandini Gokte-Narkhedkar, S Vennila, R K Tanwar, O P Sharma, Someshwar Bhagat, Meenu Agarwal, C Chattopadhyay, Rakesh Kumar, Ajanta Birah, N Amaresan, Amar Singh, S N Sushil, Ram Asre, K S Kapoor, P Jeyakumar and K Satyagopal. 2014. Integrated Pest Management Package for Cotton. p. 84

53. Shete, P. P., Kasal, Y. G., &Perane, R. R. 2018. Screening of the cotton genotypes against Ramularia areola atk. Under field condition. Plant Archives, 18(1), 734-736

54. Manickam, K. and Rajappan, K. 1999. Field efficacy of plant extracts and chemicals against greengram leaf curl disease. Indian Journal of Virology,15:35-37

55. Wamiq G, Khan JA. 2018. Overexpression of ghr-miR166b generates resistance against Bemisiatabaci infestation in Gossypium hirsutum plants. Planta.247:1175-89

56. Stuart, B.L., Isbell, V.R., Wendt, C.W., Abernathy, J.R., 1984. Modification of cotton water relations and growth with mepiquat chloride - Agron. J. 76 : 651-55

57. Paslawar, A.N., T.H. Rathod, V.M. Bhale, G.J. Bhagat, P.W. Nemade and A.S. Deotalu. 2018 Integrated management approaches for improving non Bt and Bt cotton productivity through high density planting system under rainfed condition of Vidarbha. In: Intl. congress on "Cotton and other fibre crops" (Eds. M.S.Chauhan and R.S. Sangwan), at ICAR Research Complex for NEH Region, Umaim, Mghalaya organized by Cotton Research and Development Association, CCS Haryana Agricultural University, Hisar-125004, India.pp.296-301. (ISBN-978-81-936782-0-6)

58. Kerby, T.A., Cassman, K.G. and Keekey, M. 1990. Genotypes and plant densities for narrowrow cotton system. I. Height, nodes, earliness, and location of yield . Crop Sci., 30 : 645

59. Ghosh, S. K. 2001. GM Crops: Rationally irresistible, Curr. Sci., 6: 655-660

60. Latha.P, M.Karthikeyan and E.Rajeswari.2019. Endophytic Bacteria: Prospects and Applications for the Plant Disease Management In book: Plant Health Under Biotic Stress p.1-50

61. Zaidi, S. S., Mansoor, S., & Paterson, A. (2018). The rise of cotton genomics. Trends in Plants Science, 23(11), 953-955

62. Mihajlović,M., Emil Rekanović, Jovana Hrustić, Mila Grahovac and Brankica Tanović. 2017. Methods for management of soil borne plant pathogens. Pestic. Phytomed. (Belgrade), 32(1), 9-24

63. Watkins, G.M. 1981. Compendium of cotton diseases. The Disease compendia series. St Paul MN, USA, Amer Phytopathological Society. www.apsnet.org › APS › APS Store. Accessed on April 24, 2016

64. Malone L. A., Gatehouse A. M. R., Barratt B. I. P. 2008. Beyond Bt: alternative strategies for insect-resistant crops. In Integration of insect-resistant genetically modified crops within integrated pest management programs (eds

Romeis J., Shelton T., Kennedy G.). Springer Series on Progress in Biological Control Berlin, Germany: Springer

65. Vaissayre, M. (1995) Ecological attributes of major cotton pests : implications for management, in: Constable G.A., Forrester N.W. (Eds.), Proc. World Cotton Research Conf.–1, Challenging the Future, Brisbane 1994, CSIRO, Melbourne, pp. 499-510
66. Umamaheswari, T, Beura, S.K and Mary, D.S. 2018. In vitro evaluation of phyto-extracts in the management of Myrothecium leaf blight of cotton incited by Myrothecium roridum. International Journal of Chemical Studies. 6(6): 571-574
67. Vinodkumar, S., Nakkeeran, S., Renukadevi, P., Malathi, V.G., 2017a. Biocontrol potentials of antimicrobial peptide producing Bacillus species: multifaceted antagonists for the management of stem rot of carnation caused by Sclerotinia sclerotiorum. Front. Microbiol. 8, 446. https://doi.org/10.3389/fmicb.2017.00446
68. Desai HR, Bhanderi GR, Patel RD, Sankat KB and Patel RK. 2019. High density planting with insecticide resistance management approach for sustainable and profitable cotton production in rain fed region. Journal of Entomology and Zoology Studies, 7(5): 453-458
69. Waghunde,R., Rahul Mahadev Shelake and Ambalal N. Sabalpara. 2016. Trichoderma: A significant fungus for agriculture and environment. African journal of agricultural research. 11(22):1952-196
70. Ashraf, J., Zuo, D., Wang, K., Malik, W., Zhang, Y., Abid, M. A.,et al. 2018. Recent insights into cotton functional genomics:Progress and future perspectives. Plant Biotechnology Journal,16(3), 699-713
71. Patil, S. B. 2003. Studies on management of cotton pink bollworm Pectionophoragossypiella (Saunders) (Lepidoptera: Gelechiidae). Ph. D. Thesis (unpublished) submitted to University of Agricultural Sciences, Dharwad (India)
72. Ali, Z., Ali, S., Tashkandi, M., Zaidi, S. S. E. A., & Mahfouz, M. M.2016. CRISPR/Cas9-mediated immunity to geminiviruses: Differential interference and evasion. Scientific Reports, 6
73. Chen, Y. and Ruberson, J. R., 2008. Impact of variable nitrogen fertilisation on arthropods in cotton in Georgia, USA. Agric. Ecosyst. Environ., 126 : 281-288
74. VenugopalanM.V.,KrantiK.R.,BlaiseD.,ShubhangiLakdeandSankaranarayana K.2013. High density planting system in cotton-The Brazil experience and initiations. Cotton Res. J. 5 : 172-85

75. Ge, F., Liu, X., Li, H., Men, X. and Su, J., 2003. Effect of nitrogen fertilizer on pest population and cotton production. Chin. J. Appl. Ecol., 14 : 1735-1738

76. Food and Agriculture Organization, 2006

77. Buttar, G.S., Sekhon, K.S., Sidhu, B.S., Thind, H.S. and Kaur, Anureet. 2011. Effect of seedling transplanting systems on seed cotton yield and profit in Bt cotton under semi arid conditions of Punjab. J.Cotton Res. Dev. 25: 207-09

78. Arunvenkatesh, S., Rajendran, K. and Baskaran, R. 2017. Effect of plant geometry on light interception and weed density in cotton under rainfed vertisol. Agric update, 12 (TEACHSEAR_2): 421-25

79. Narayana, E. and N.V.V. Durga Prasad. 2010. High density planting system and mechanical harvesting in India. In: Intl. congress on "Cotton and other fibre crops" (Eds. M.S.Chauhan and R.S. Sangwan), at ICAR Research Complex for NEH Region, Umaim, Mghalaya organized by Cotton Research and Development Association, CCS Haryana Agricultural University, Hisar-125004, India.pp.66-73. (ISBN-978-81-936782-0-6)

80. Zhang, L., W.Vander Werf, L. Bastiaans, S. Zhang, B.Li, J.H.Spiertz. 2008. Light interception and utilization in relay intercrops of wheat and cotton. Field Crop Research. Volume 107 (1). Pp. 29-42

81. Narayanan, E. S. (1962). Biology and method of control of some important insect pests of cotton in India. Indian Central Cotton Committee Publication, Bombay, p. 44

82. Aujla, M.S., Thind, H.S. and Buttar, G.S. 2008. Response of normally sown and paired sown cotton to various quantities of water applied through drip system. Irr.Sci. 26: 357-88

83. Wright, D. L., Marois, J. J., Sprenkel, R. K. and Rich J. R. 2011. Production of Ultra Narrow Row Cotton.University of Florida (UF), IFAS Extension.SSAGR-83

84. Naqvi, R. Z., Zaidi, S. S., Shahid Mukhtar, M., Amin, I.,Mishra, B., Strickler, S., et al. 2019. Transcriptomic analysis of cultivated cotton Gossypium hirsutum provides insights into host responses upon whitefly-mediated trans-mission of cotton leaf curl disease. PloS One, 14(2), 1-2

85. Ren X M, Zhang L Z, Du M W, Evers J B, Werf W V D, Tian X L and Li Z H. 2013. Managing mepiquat chloride and plant density for optimal yield and quality of cotton. Field Crops Research, 149, 1-10

86. Dighule,SB, Perane, RR, Amle,KS and More, PE. 2011. Efficacy of chemical fungicides and bio-agents against major cotton fungal foliar diseases in vitro. International Journal of Plant Sciences, Vol.6, 2, 247-250

87. Bi, J. L., Ballmer, G. R., Hendrix, D. L., Henneberry, T. J. and Toscano, N. C. 2001. Effect of cotton nitrogen fertilization on Bemisiaargentifolii populations and honeydew production. EntomologiaExperimentalis et Applicata, 99 (1) : 25-36
88. Agarwal R. A. and Katiyar, K. N. 1979. An estimate of losses of Kapas seed due to bollworms on cotton in India. Indian J. Ent., 41(4): 143-148
89. Kim SK, Lee SC, Lee BH, Choi HJ, Kim KU, Lee IJ. Bulbil formation and yield responses of Chinese yam to application of gibberellic acid, mepiquat chloride and trinexapac-ethyl.J Agron Crop Sci. 2003; 189:255-260
90. Yan, R., Liang, C., Meng, Z., Malik, W., Zhu, T., Zong, X., et al. 2016. 3.Biotech, 6, 217
91. Spitzer T, Matusinsky P, Klemova Z, Kazda J. Management of sunflower stand height using growth regulators. Plant Soil Environ. 2011; 57:357-363
92. Krishan Chandra, Greep,S. and Srivathsa, R.S.H. 2005. Liquid biofertilizers-Solution for longer shelf-life. Spice India. 29-32
93. Khan, M.A. and Gangodhyay.S. 2008. Efficacy of Pseudomonas fluorescens in controlling root rot of chickpea caused by Macrophominaphaseolina. J.mycol. Pl.Pathol., vol.38.3.p.582-587
94. Rageshwari, S., Renukadevi, P., Malathi, V.G and Nakkeeran, S. 2016. Occurrence, biological and serological assay of tobacco streak virus infecting cotton at Tamil Nadu. Journal of Mycology ad Plant Pathology, 46(2): 159-168
95. Vinodkumar, S., Nakkeran, S., Malathi, V.G., Karthikeyan, G., Amalabalu, P.,Mohankumar, S., Renukadevi, P., 2017b. Tobacco streak virus: an emerging threat to cotton cultivation in India. Phytoparasitica 45, 729
96. Rageshwari, S., Renukadevi, P., Malathi, V.G., Amalabalu, P. and Nakkeeran, S. 2017. DAC-ELISA and RT-PCR based confirmation of systemic and latent infection by tobacco streak virus in cotton and parthenium. Journal of Plant Pathology. 99(2): 469-475
97. Shim, J., P.K. Mangat and R.A.Shim. 2018. Natural variation in wild Gossypium species as a tool to broaden the genetic base of cultivated cotton. J Plant Sci Curr Res, 2(5):1-9
98. Venugopalan M.V. (2019). Avenues to Improve Farm Income from Cotton under Changing Climatic Scenario. PKV Research Journal (Accepted)
99. Amudha, J., Balasubramani, G., Malathi, V. G., Monga, D., and Kranthi, K. R.2010. Cotton transgenics with antisense AC1 gene for resistance against cotton leaf curl virus. Electron. J. Plant Breed. 4, 360-369
100. Cotton Corporation of India.2018

101. Namdeo, K. N., Sharma, J. K., Choudhary, S. K., and Mandloi, C. 1991. Effect of planting dates and geometry on growth and yield of hirsutum cotton under rainfed condition. Journal of Cotton Research and Development, 5(1), 59_62

102. Andries, J.A., J. E. Jones, L. W. Sloane and J. G. Marshall. 1969. Effects of Okra Leaf Shape on Boll Rot, Yield, and Other Important Characters of Upland Cotton, Gossypium hirsutum L. Crop Science. https://doi.org/10.2135/cropsci1969

103. Rick, C.W. 1984. Plant Germplasm Resources. Pp. 9-37.In:D.A. Evens, W.R. Sharp, P.V. Ammirato and Y.Yamada (eds.). Handbook of cell culture, Macmilan, New York

104. Mehta, Y., Marangoni, M. S., R Bocatti, C., Rodrigues, H. P., Cunha, T. S., and Galbieri, R. 2015. Systemic acquired resistance of cotton, soybean and common bean to Rhizoctonia solani and Sclerotium rolfsii induced by shale water seed treatment. American Journal of Plant Sciences, 06(09), 1493-1500

105. Jalloul, A., Sayegh, M., Champion, A., & Nicole, M. 2015. Bacterial blight of cotton. PhytopathologiaMediterranea, 54(2), 241-252

CHAPTER-2

SYSTEM FOR THE RELEASE, NOTIFICATION, AND DENOTIFICATION OF VARIETALS IN INDIA

1. INTRODUCTION: AN OVERVIEW

In the dynamic realm of agricultural development in India, the process of varietal release, notification, and denotification stands as a crucial mechanism shaping the landscape of crop cultivation. Varied crops play a pivotal role in the nation's agrarian economy, and their continuous improvement through the introduction of new varieties or the removal of outdated ones is managed by a structured system.India is a fast-growing economy and agrarian country. Almost 70 percent of the Indian population depends on agriculture and its allied sectors to obtain employment and sustain livelihood. The seed is considered as a basic and key input in agriculture. High-quality seed production was the major concern in the Indian subcontinent till the 1960s. Before that India was mostly dependent on the USA for food grain (PL480) to mitigate its hunger [1] and feed large human population. In order to reduce the dependence of food on foreign countries and to meet the food and nutritional demand of burgeoning population and to become self-reliant in food grain production, Indian Government established All India Coordinated Crop Research Projects (AICCRPs) and other institutes in a systemic manner to produce a large number of varieties with assured seed quality in all major crops. The production of high-quality seeds was one of the pillars to change the position of Indian agriculture into the new world order. The ultimate intention was to introduce the newly evolved high yielding cultivars to the resource-poor farmers for broad-spectrum cultivation in the area of their adoption.

By seeing this scenario, the Government of India acknowledged seed an essential commodity under the Essential Commodities Act, 1955. On October 1964

Varietal Release System (VRS) came into existence with the formation of the Central Variety Release Committee (CVRC) at the national level, and State Variety Release Committees (SVRCs) at each state level. A Central Seed Committee (CSC) was established under the Ministry of Agriculture, Cooperation and Farmers Welfare provided in the Seeds Act, 1966. The functions of the CVRC were taken over by the CSC in 1969 to ensure the quality of seeds on sale and notification of the kinds/varieties. To perform the function at central level to release/notification, provisional notification and de-notification of cultivars, CSC constituted a Central Sub-Committee on Crop Standards, Notification & Release of Varieties for Agricultural Crops and Horticultural Crops, while to perform similar functions at state level, State Seed Sub-Committee (SSSC) was constituted [2].

2. DEVELOPMENT OF PLANT GENETIC MATERIAL (IN HOUSE BREEDING)

Entries (pure lines/open pollinated varieties/composites/synthetics/hybrids etc.) are developed by the concerned plant breeders/agencies through scientific temperament and extensive breeding programs for the benefit (food and nutritional security) of humankind. Different conventional (Introduction, selection, hybridization followed by selection etc.) and advanced (tissue culture-based techniques like somaclonal variation, anther and pollen culture; mutation, marker assisted breeding, transgenic or genome editing techniques) breeding methods are being used by the different agencies (ICAR or non-ICAR national institutes, SAUs, private national or multinational companies etc.) to generate elite material for high yield potential, nutritional quality and other associated traits. Developed elite materials are being tested by the concerned plant breeder/s at their research station for three to four years in replications for stability and selected superior cultivars enter into the All India coordinated crop improvement projects (AICCIPs) trials for further testing in multi-environments across the country.

3. ALL INDIA COORDINATED CROP IMPROVEMENT PROJECTS SYSTEM OF VARIETAL TESTING

First AICCIP was started in way back of 1957 by ICAR on maize crop for systemic testing of entries and for release of high yielding new maize varieties. In general, the three-tier system of multi-location evaluation is used for three years except perennial fodder crops (requires four years-one for crop establishment and three for evaluation) in India. Multilocational trials are conducted by the Project Coordinator (PC)/Project Director (PD) of AICCIPs [3] with the help of concerned principle investigators. The AICCIPs have been developed for all the major crops including forage crops. The AICCIPs come under the umbrella of ICAR, has great role in the development of improved crop varieties and generation of production and protection technologies that directly benefit farmers for their economic amelioration. All AICCIP trials are well organized, systemic and conducted through a uniform testing procedure across the centers as per crop standard. It is a powerful system to screen large number of entries and recommend well-tested, superior, and adapted new cultivars to the end users. The flow chart of varietal release and notification in India is illustrated in Figure 1.

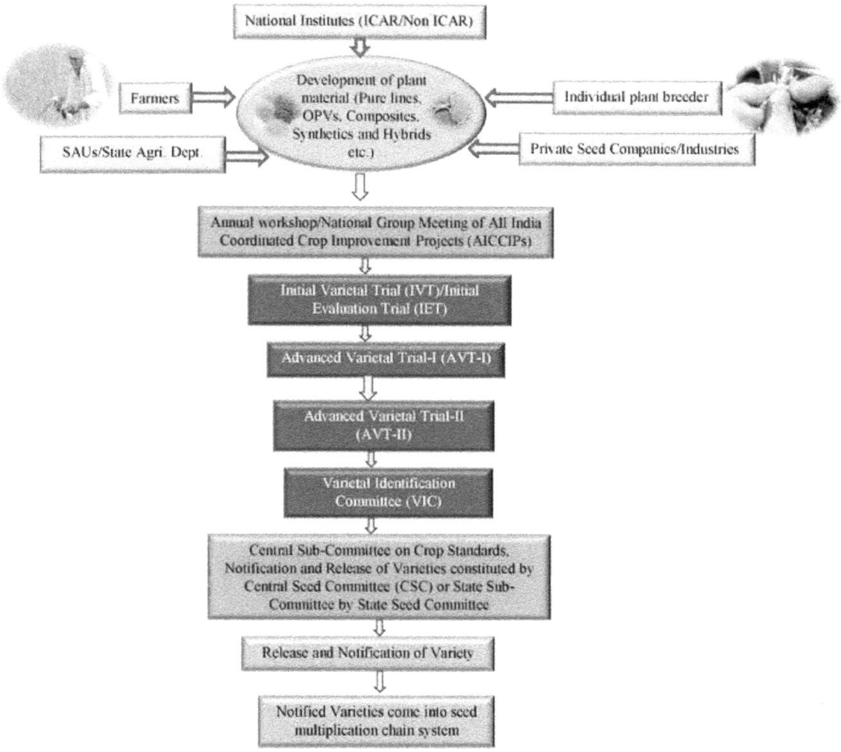

Figure 1. Flow chart of plant varietal release and notification system in India.

Newly entered material/entries into the three-tier system must have the following requirements.

- Station trial or preliminary yield trial-Concerned plant breeder must perform station or regional trial and proposed entry must have undergone censorious evaluation process or screening (insect pests and diseases). Crop based quality parameters and tolerance to key abiotic stresses are also to be screened as per the requirement. Pre-coordinated trial data on yield, trait stability and other related agronomic traits must be available to the PC/PD in support of the relevance of his/her entry [4].
- The entry must have a high degree of genotypic stability, phenotypic uniformity, germination percentage and physical purity (as per the minimum seed certification standards).

- The entry must have few distinct diagnostic traits which make it different to all remaining varieties. These distinct traits help to identification of variety during legal infringement (DUS testing) [5].
- All the information related to the development of entry *i.e.* parentage or pedigree should be available to the PC/PD by the concerned plant breeder/agency. If the performance of entries are same in the coordinated trials, then preference will be given to the variety which has been developed by the using of diverse parents in breeding program.
- Private companies can enter their material into the coordinated trial system as similar to other agencies but have to pay the prescribed fee for their entries as per guideline of the Government of India.
- All the material (product of selection, hybridization followed by selection, synthetics, composites, and hybrids, etc.) shall be subjected to the same system.

4. THREE TIER SYSTEM

The AICCIP centers for various crops are located at ICAR institutes or State Agricultural Universities (SAUs) or other volunteer centers recommended by AICCIP workshop based on covered crop area, adaptability, and agro-climatic condition etc. It involves various steps [6].

4.1 Initial varietal trial (IVT)

The time duration of the initial varietal trial (IVT) is one year. All the entries, which were superior to their respective station trials, would be introduced into the IVT. These entries would be used for multi-location trials along with checks. In general, three checks (national, zonal and local checks) are being used for efficient evaluation of entries across the centers. The national check (a crop variety which had been released previously for whole country) would be used for a long period but zonal (a crop variety which had been released previously for specific zone) and local checks (high yielding local variety) can be replaced based on the requirement. These checks cannot be replaced after the IVT. Maintenance of genetic purity, germination and physical purity of new material are the prime objectives of the concern plant breeder/agency. The IVT trials are conducted in such a manner that minimum difference of yield (5–10%) and other ancillary traits can be measured. Experimental layout (experimental design, number of replications and treatments) is the prime responsibility of the PC/PD through concerned principle investigator. The cultural practices *i.e.* seed rate, date of sowing, row to row and plant to plant spacing; weed, fertilizer and water management etc. shall be strictly followed by the IVT centers as per guideline of PC/PD. The plot size of IVT is smaller than advanced trials. An IVT includes the maximum number of locations across the country to evaluate varietal adaptation and performance. The number of testing locations varies with crop across the zones. A team of scientists (plant breeder, agronomist, pathologist, entomologists etc.) will monitor all the trials as per the recommendation of the PC. Project coordinator constitutes monitoring teams (includes scientists from various disciplines such as plant breeding, agronomy and plant protection) for evaluation of trials. Each member of monitoring team submits their report to the PC based on their observation during trial monitoring. Each crop has different objective and requires different technical requirement for their evaluation. For example, The IVT centers will generate information related to days to flowering, physiological maturity, plant height, lodging, threshability, disease, and pest tolerance, green fodder yield, dry matter yield, and nutritional quality traits for forage crops. Technical program is formulated during workshop or national group meet and PC will specify characters on which data shall be recorded. Entries which are superior over the best check in terms of yield and other related traits will be promoted into the advance varietal trial-I. The superiority is primarily decided based on yield potential and other related important traits such as quality traits. In some agricultural crops, where a large number of entries enters into the IVT system then; IVT is preceded by testing of these entries for one year in the National Screening Nursery.

4.2 Advance varietal trial-I (AVT-I)

Based on superiority (5–10%) over the best performing check, superior entries will enter into the AVT-I from IVT. The number of tested entries in the AVT-I will be less than IVT. The plot size is large in AVT-I as compared to IVT, therefore data generated on yield and other ancillary traits will be more realistic, accurate and minimal chances of error. The number of testing locations should be more as compare to IVT in a given zone for more realistic data on yield and other economically important traits, varietal adaptation, biotic and abiotic tolerance, quality parameters, etc. National, zonal and local checks (which were used in IVT) shall be used for critical analysis along with the entries. During AVT-I, additional data on disease and or insect pest tolerance under artificial epiphytotic condition must be generated by the experts. Same as IVT, monitoring team would be deputed by the PC/PD at different growth stages of crop and observed data shall be submitted to the concerned PC/PD. Based on the performance of entry over the best performing check-in the respective zone, the superior entries would enter into the AVT-II.

4.3 Advance varietal trial-II (AVT-II)

All the requirements shall be fulfilled as similar to AVT-I. However, few additional data will be generated at AVT-II stage *i.e.* response of entries to different dates of sowing, seed rate, spacing between plant to plant and row to row (population density), behavior in different level of fertilizer and irrigation by sponsored agronomists; response of diseases and pests by the plant pathologists, crop quality parameters by the biochemists. The seed technology center will develop descriptors which help in the seed certification process. All the processed and analyzed data on yield and other related traits, across the locations/centers (cooperating and volunteer) shall be submitted to the PC. On the basis of these data, annual reports are being made in each crop. All the data of superior entries are comprehensively discussed in the annual workshop/national group meetings by the PC/project director. After completion of the AVT-II, the concerned breeders are informed to submit varietal proposal based on the performance of their entries during three years of evaluation.

5. PROCEDURE FOR VARIETAL IDENTIFICATION

Based on three year performance, best performing test entries shall be identified in the annual crop workshop or national group meet at the pre-defined institute/university. The Zonal Coordinators and Principal Investigators attend the national group meet to provide wider aspects of information on the varieties. After the approval from Deputy Director General (Crop Science) of Indian Council of Agricultural Research (ICAR), a "Varietal Identification committee (VIC)" constituted in advance of annual workshop or national group meet. All the committee members (Table 1) shall be informed well in advance by the PC or PD [3]. Principal investigators (PIs) of different disciplines can assist in the process of discussion but they do not have the right to vote. Only committee members have the right to cast vote. The VIC provides detailed information on recommended entries to the Central Sub-Committee on Crop Standards, Notification, and Release. This committee has sole right to release and notify the best-performing entry into national wise or zonal wise based on the recommendations of the VIC.

Table 1: Organizational setup of varietal identification committee (VIC). The committee comprises one chairman and nine members.

S.N.	Representative	Organizational position
1	DDG (Crop Science)/ his or her nominee	Chairman
2	Project Coordinator/Project Director of AICCIP	Member Secretary
3	Director of Research of institute/SAUs of that region where the meeting is held	Member
4	Agricultural Commissioner (Department of Agriculture)	Member
5	One nominee of Seed organization (NSC, SSC)	Member
6	One representative of private seed agencies	Member
7	One representative of crop-based industries	Member
8	Project coordinator (seed technology)	Member
9	Two eminent scientists of that institute	Member

Source: [3] Tandon et al., 2015.

6. ELIGIBILITY CRITERIA FOR IDENTIFICATION OF THE VARIETY

- The candidate variety must have a minimum of three years of yield and other ancillary trait data from multi-location coordinated trials.
- At least two-year data on disease and pest reaction at a hot spot or artificial epiphytotic condition.
- The candidate variety must have at least one-year data on agronomic performance like seed rate, dates of sowing, planting density, irrigation, and fertilization. In forage crops, three year rigorous evaluation must be done for annual crops (seed yield data for third year only) and four year for perennial crops (one year for crop establishment and other three years for evaluation).
- The concerned breeder must have at least a minimum requirement of nucleus seed so that breeder seed can be generated easily.
- The concerned plant breeder should have pure seed for planting of 5 ha area. If he or she did not match the requirement, then identification can only be postponed for one year.

All these issues shall be discussed by the project coordinator in the annual workshop itself. The candidate variety must be phenotypically uniform (plant height, maturity, etc.) and stable in performance throughout the years.

7. CENTRAL SUB-COMMITTEE ON CROP STANDARDS, NOTIFICATION, AND RELEASE

Central Sub-Committee on Crop Standards, Notification, and Release of Varieties appointed by Central Seed Committee under Section 3 of the seed act, 1966 during 1994. The committee comprised one chairman and 17 members (Table 2).

Table 2. Organizational setup of central sub-committee on seed standards, release, and notification of variety (agricultural crops). The committee comprised one chairman and 17 members.

S.N.	Representative	Organizational position
1	Deputy Director General (Crop Sciences), ICAR	Chairman
2	Deputy Commissioner (QC) DAC & FW, GOI	Member Secretary
3	Directors of State Seed Certification Agencies, or their representatives	Member
4	Project Directors of Departments of Agriculture of all states, or their representatives	Member
5	Project Coordinators/Directors of AICCIPs	Member
6	Agricultural Commissioner, GOI	Member

S.N.	Representative	Organizational position
7	Representatives of the seed industry, NSC, State Seed Corporations, private seed companies	Member
8	Representatives of ICAR, ICAR institutes, NGOs	Member
9	Progressive farmers	Member

Source: [7] SeedNet India Portal; QC-Quality Control, DAC & FW-Department of Agriculture, Cooperation and Farmers Welfare, GOI-Government of India.

Central Sub-Committee releases varieties as per the benefit of the stakeholders and need of regional, zonal or national importance, and the State Seed Sub-Committee releases varieties beneficial for particular state. Notification of variety is compulsory on regulating the seed quality under the provision of Seed Act, 1966. Notification usually authorizes certified seed production throughout the country, by private or public seed multiplication organizations. Once the Central Sub-Committee accepts the proposal, the varieties/hybrids will be released for the concerned agro-climatic zone/s (may cover one or more number of states or nationally). Simultaneously, it must be notified for seed certification purpose in the country. During the release, the concerned breeder must have a minimum amount of seed which can be sown at least ten-hectare area [3]. Later on, seed multiplication is the responsibility of various seed agencies (NSC, SSC, private seed companies and progressive farmers, etc.). The significant differences between released and notified varieties are illustrated in Table 3.

Table 3. Critical differences between released and notified varieties.

S.N.	Released variety	Notified variety
1	It is not a statutory function under the Seed Act, 1966	Statutory function and variety will be registered under Section 5 of seed act 1966.
2	It cannot be used for seed certification	Only notified varieties to come under seed certification
3	No guarantee on seed quality for farmers	Assured seed quality
4	Seed law enforcement agencies (seed inspector etc.) cannot draw and test seed samples	They have the right to draw and test seed samples
5	These are not assets of Govt. of India	Notified varieties are assets of Govt. of India
6	Its main purpose is to make available the information of cultivar to the public and its area of adoption	The main purpose is seed quality regulation
7	Difficult to trace out the genesis	The notification of the varieties will help to trace out its genesis.

8. CENTRAL SEED COMMITTEE (CSC)

It is a legal body constituted by the Department of Agriculture, Cooperation and Farmers' Welfare (DAC&FW), Ministry of Agriculture and Farmers' Welfare (MoA&FW), Government of India to advise central and state government on matters related to the implementation of seed act, 1966 and other related functions. The core committee includes one chairman and eight members to be nominated by the central government and one person to be nominated by the Governments of each State (Table 4). State Seed Committee (SSC) has a similar role at the state level. The CSC and SSC are empowered to release varieties but only CSC can notify those [8].

Table 4. Organizational setup of Central Seed Committee (CSC). The committee comprises one chairman and nine other members by the Central Government.

S.N.	Representative	Organizational position
1	Secretary, DAC&FW, MoA& FW, GOI	Chairman
2	Additional Secretary (In charge Seeds), MoA& FW, GOI	Member
3	Agricultural Commissioner, MoA& FW, GOI	Member
4	Deputy Director General (Crop Sciences), ICAR	Member
5	Joint Secretary (In charge Seeds), MoA& FW, GOI	Member
6	Progressive farmers/ seed growers (4) nominated by the Central Government	Member
7	One representative from each State Govt.	Member
8	Director of National Seeds Project, MoA & FW, GOI	Member Secretary

Source: [9] SeedNet India Portal; MoA& FW- Ministry of Agriculture and Farmers Welfare.

9. EMPOWERMENT OF CENTRAL SEED COMMITTEE

- The CSC has authority to release varieties (pure lines/hybrids/ composites/synthetics) developed by central research institutes (ICAR/ non-ICAR), AICCIPs, private or corporate sector, and other organization as per the scientific data authenticity for zonal basis (which may include more than one state) or at national level.
- The CSC has authority to approve proposals received from the State Variety Release Committees/State Seed Sub-Committees for varieties developed by the State Research Institutes but is considered suitable for areas outside the state (based on their performance).
- The CSC can delimit the regions or tracts for the cultivation of varieties approved for release.
- The CSC can advise the ICAR regarding the manner in which the National Register of Approved Varieties may be maintained, and to suggest the standard description of crop varieties.

- The CSC can notify kinds/varieties for the purpose of the Seeds Act and the areas of their notification.
- The CSC specifies minimum limits of germination percentage and purity for the notified kinds/varieties of seeds as per minimum seed certification standards.
- The CSC specifies the "mark" or "label" in respect of notified kinds/varieties.

The State Seed Sub-Committees are constituted by Central Seed Committee and are authorized to set up a State Seed Laboratory, State Seed Certification Agency (SSCA) and an Appeals Authority, and to appoint seed inspectors and seed analysts. The differences between Central Sub-Committee and State Sub-Committee are given in Table 5.

Table 5. Critical differences between central sub-committee and state sub-committee.

S.N.	Central sub-committee	State sub-committee
1	Authorized by Central Seed Committee	Authorized by State Seed Committee
2	Releases varieties for regional/zonal/national level	Only for concerned state/regions within the state
3	Statutory body for varietal notification	It cannot notify varieties
4	Notification followed by seed certification	Certified seed cannot be produced without notification
5	Members in the committee are appointed by the Central Govt.	Members in the committee are appointed by the State Govt.

10. EMPOWERMENT OF STATE SEED SUB COMMITTEE

There are some rights which have been provided by the Central Seed Committee for proper functioning of seed chain in respective state in India. These empowerments are-

- The State Seed Sub Committee will advise the state government on all matters related to the execution of the Seeds Act, 1966.
- Reviewing the implementation of the Seeds Act in the state and send periodic reports to the state government and the Central Seed Committee.
- Inspect, analyze and report on the State Seed Testing Laboratory.
- Advise on educational and promotional measures for proper enforcement and understanding of the Seed Act.

- Planning for different crop varieties to be grown in different regions of the state, and to review the assessment of seed requirements.
- Considering the release of new varieties for the state and recommend their notification to the Central Seed Committee.
- Monitoring the performance of newly released varieties in the state.

Being agriculture as a state subject in India, centrally released varieties are not directly accepted by all the states for which they have been released. Each state has its own regulatory system which they have to follow for varietal release in the state. They have to pass through all the steps of the concerned state release procedure before they approve for cultivation in the state *viz.*, state wise multi-location trials for three years and adaptive trials based on the requirement. The notification requires that the variety must have been tested at least for one year in the AICCIP trials and recommended for release in the state by the AICCIP Varietal Identification Committee.

11. NECESSITY OF NOTIFICATION

Since only notified varieties will be under the purview of Seed Law Enforcement, hence it is necessary to bring the seed of a particular crop variety under notification system. The seed inspector can only draw a sample from notified variety for analysis and ensure the seed quality [10, 11]. A released variety cannot come under seed chain without notification by the Gazette of India. Therefore, these issues will make the notification as necessary requirement for other things to act on it. The notification is made by the Central Government on the recommendation of the Central Seed Committee. Thus, notification is prerequisite for production of certified seed which ensures high quality of seeds to the farmers. After notification, variety becomes asset of government of India. The breeder seed can only be produced after the notification of variety and notified varieties enter into seed chain. Notification also helps in the genesis of original variety based on its pedigree and also regulates any kind of infringement in the later stages of varietal promotion.

12. DENOTIFICATION OF VARIETIES

Released varieties can be denotified if they are not performing well in the area of their adoption or have been in cultivation for more than 15 years or are not much in demand. Denotification can be done based on the recommendation of central seed committee by the government of India.

13. CONCISE SUMMARY

There are several ways and means to increase the crop production and productivity, however using genetically pure and high-quality seed is first and prime objective in agriculture. Therefore the variety which will be used by farmers must have undergone several evaluations in order to ensure its stable yield potential, tolerance to biotic and abiotic stresses and these criteria are being fulfilled by a legal varietal release system. The main objective of the varietal release system in India is to introduce newly developed, high yielding varieties to the farmers for broad-spectrum cultivation in the area of their adoption and only those varieties will be notified which are superior to existing one. It provides choice to the farmers to cultivate a specific variety, based on their need for crop diversification. In India, the systemic framework has helped farmers to get high quality of seed from market and production has increased many folds since the inception of AICCPs. Notification is mandatory to release a variety, though the release process itself does not have legal cover.

AT A GLANCE

Agriculture is the backbone of India and improved agricultural practices principally depend on the use of newly evolved improved variety. In the Indian scenario, a statutory varietal release system is working where notification and denotification process are playing a crucial role in quality regulation of seed. Crop research institutes of ICAR (Indian Council of Agricultural Research), SAUs (State Agricultural Universities) and private seed companies are the main pillars to develop improved varieties in India. The thumb rule is, the improved variety must have a higher yield compared to the existing one (national and state check varieties) and this is ensured via several multilocational evaluations at a different level. This article covers the Indian regulatory system of variety release, evaluation process at a different level, and the importance of notification and denotification. This information will help the scientific community in regards to suggesting improved variety for general cultivation by farmers.

REFERENCES

1. Spielman, D. J., and Smale, M., Policy options to accelerate variety change among smallholder farmers in South Asia and Africa South of the Sahara. International Food Policy Research Institute (IFPRI) Discussion Paper Series. 2017.
2. Virk, D. S., The regulatory framework for varietal testing and release in India. Seeds of Choice: Making the Most of New Varieties for Small Farmers. 1998. Pp. 69-84.
3. Tandon, J.P., Sharma, S.P., Sandhu, J.S., Yadava, D.K., Prabhu K.V., and Yadav, O.P., Guidelines for Testing Crop Varieties under the All-India Coordinated Crop Improvement Projects, Indian Council of Agricultural Research, New Delhi, India. 2015.

4. Singh, B.D., A textbook of plant breeding, Kalyani publishers. ISBN 81-272-2408-1. 1999. Pp.391-400.
5. Chopra, V. L., Plant Breeding: Theory and Practice, Oxford & IBH Publishing Company. 1989. Pp. 110.
6. Tandon, J.P., and Sharma, S.P., Guidelines for Crop Variety Testing under All India Coordinated Crop Imrovement Projects, Seed Science Technology Division, IARI, New-Delhi, India. 2002.
7. SeedNet India Portal; QC-Quality Control, DAC & FW-Department of Agriculture, Cooperation and Farmers Welfare, GOI-Government of India.
8. Tonapi, V.A., Bhat, B.V., Kannababu, N., Elangovan, M., Umakanth, A.V., Kulakarni, R., Tonapi, K.V., Rao, K.V.R. and Rao T.G.N., Indian regulatory system of plant variety testing, identification, release, and notification (part-III). ICAR-Indian Institute of Millet Research, Rajendra Nagar Hyderabad, Telangana, India. 2015. Pp. 279-310.
9. SeedNet India Portal; MoA& FW- Ministry of Agriculture and Farmers Welfare.
10. Kaul, J., Kumar, R., Nara, U., Jain, K., Olakh, D., Tiwari, T., Yadav, O.P. and Dass, S., Development of Dtabase of Maize Hybrids and Open Pollinated Varieties Released and Notified for Cultivation in India. Jurnal of Agricultural Science. 2017. 9 (10).
11. Spielman, D.J. and Kennedy, A., Towards better metrics and policymaking for seed system development: Insights from Asia's seed industry. Agricultural Systems.2016. Vol. 147. Pp. 111-122.

CHAPTER-3

ENHANCING RICE GRAIN QUALITY THROUGH BREEDING

1. INTRODUCTION: AN OVERVIEW

The pursuit of enhancing grain quality in rice through strategic breeding represents a pivotal facet of agricultural research and development. As a staple food for a significant portion of the global population, rice not only serves as a primary source of sustenance but also holds cultural and economic significance. Recognizing the multifaceted importance of rice, the focus on breeding for grain quality improvement becomes imperative to address the evolving demands of consumers and market preferences. This endeavor involves a nuanced approach that integrates scientific advancements in genetics, genomics, and breeding techniques to optimize traits related to nutritional content, milling quality, and overall consumer acceptability.

Rice (Oryza sativa L.) is ranked as first in human food crop for more than half of the world's population [1] and economically imperative food crop with nutritional diversification that helps in poverty alleviation. With the intensifications of diverse food demands and living standards of global populations, rice grain appearance and grain quality have become a primary concern for rice breeders [2]. Rice grain quality characters has been grouped into two classes: (i) grain appearance and (ii) cooking and eating qualities. The grain appearance includes kernel length, kernel breadth, kernel length-breadth ratio, endosperm translucency and cooking and eating qualities includes cooked kernel elongation, amylose content, gelatinization temperature (measured as alkali spreading value), aroma and volume expansion. In addition processing qualities, hulling percentage, milling percentage and head rice recovery are important for market value. Most of the quality traits are complex trait governed by quantitative inheritance.

Generally, Japanese people prefer short and sticky rice, while Indians prefer aromatic basmati rice which elongates when cooked. Indian basmati rice and Thai jasmine rice are highly priced due to their distinctive aroma when cooked and some European countries have led consumers to prefer better quality rice. Due to consumer's demand for better rice quality, rice breeders are striving to develop rice varieties with improved qualities that meet local demand. Basmati varieties has superior quality traits *viz.*

Superfine slender grains, fine cooking quality, pleasant aroma, and lengthwise elongation during cooking and fetched with premium price in the local and global market [3]. The physicochemical and cooking characteristics are the good indicators of grain quality [4, 5]. Most of the consumers prefer rice with soft to medium gel consistency, intermediate amylose content and gelatinization temperature.

Kernel elongation after cooking is an important character of fine rice and most of the rice consumers prefer length-wise elongation. Rice kernels absorb water during cooking and increase in length through swelling of kernels. Breadth-wise increase is not desirable, whereas length-wise increase without increase in girth or a crack on the kernel is a desirable characteristic of high quality rice. In northeast India, cultivated varieties having high kernel elongation might also have originated from this region. The elongation genes might have moved towards both eastern and western parts from the periphery of the centre of origin. Some varieties showed marked elongation in eastern countries such as Burma and Thailand. However varieties in western region with Punjab showed good kernel elongation.

2. MEASUREMENT OF KERNEL ELONGATION RATIO

Cooked kernel elongation can be measured by three ways *viz.*, grain elongation ratio, proportionate change, grain elongation index. Grain elongation ratio is defined as the ratio of the length of cooked rice (L_2) to the length of milled rice (L_1). Kernel elongation ratio indicates the proportionate change of rice grain after cooking. Sood and Siddiq [6] followed following formula to measure kernel elongation

Proportionate change=LF/BF-L0/B0L0-B0E1

Where, L_F – kernel length after cooking, B_F – kernel breadth after cooking, L_0 – kernel length before cooking, B_F – kernel breadth before cooking. Grain elongation ratio is a better index of cooking quality than grain elongation index (A) or proportionate change [7].

A=L2/W2L1/W1E2

Where, L_2 – cooked kernel length, W_2 – cooked kernel breadth, L_1 - milled rice length, W_2 - milled rice breadth.

Grain elongation index is a more reliable estimate of kernel elongation during cooking [8].

Elongation index=BGSGSE3

Grain elongation index values were less precise and less sensitive than elongation ratio. It needs the additional measurement of grain width, use of kernel L/B ratio instead of length offered no advantage of sensitivity since, grain expanded both length and width wise during cooking [9, 10].

Cooked rice elongation=After cooking length–Before cooking length Before cooking length×100E4

3. ASSAY OF COOKED RICE ELONGATION

Ten selected grains were soaked in distilled water for 30 min which were then placed between two pieces of wet filter paper in a petri dish filled with an appropriate amount of water. The dish was then placed in a covered container and the grains were cooked by steaming over boiling water for 10 min and simmering for 10 min (with power off). The cooked rice grain was transferred onto a piece of dry filter paper at the bottom of a fresh petri dish, which was placed in a desiccator with a constant temperature (19°C). Then the CRL was measured [11].

Ten randomly chosen intact polished grains were boiled in a water bath for 7 min after soaking in 7.5 ml of distilled water for 20 min. The cooked rice grains were then placed on absorbent paper for 15 min, and the boiled grain length (BGL, mm) and width (BGW, mm) were measured using the same scanner. The treatments were repeated twice, and the average grain length and width before and after cooking were calculated for estimation of elongation index.

The length of 10 whole rice kernels after cooking was measured by using the micro-scale, and the average kernel length determined. Kernel elongation ratio was calculated by dividing the average length of cooked kernel by the average length of the raw (uncooked) rice.

An inter-laboratory collaborative test was conducted for measurement of grain elongation of milled rice during cooking in 19 laboratories in 16 countries. Details of individual laboratory methods for measurement of grain elongation given below [9].

Collaborator	Methods
2	10–20 grains soaked for 30 min in 20 ml water in 25 × 100 mm test tubes and tubes placed directly in 100°C bath for 15 min
4	Aggregate length and width of 20 raw and cooked grains measured in graduated grooved board. Rice (2 g) placed in boiling water bath without pre-soaking and cooled until centre becomes translucent, determined by sampling every minute from 10 min onwards
6	Fifty grains (not presoaked) boiled for 15 min and sampled at 2 min intervals until 90% of the grains had translucent centers, after which 10grains are measured
7	Rice grains (2 g) boiled for 22 min without presoaking.
8	Milled rice (7 g) is washed, soaked for 30 min and a duplicate 7 g sample soaked for 60 min in 12·6 g (1·8 volumes) water in aluminum pudding cup and cooked in a rice cooker with water (200 ml) for 32 min (at least 20 min at >99°C). Cooked rice is allowed to stand 15 min before measuring 10grains in the middle layer with caliper gauge
11	No presoaking. Grains measured directly with microscope fitted with an eyepiece graticule
14	15 min cooking time in boiling water
15	Aggregate length and width of 10 random raw and cooked grains measured
16	28 min heating of rice (10 g) in water (50 ml) in 250 ml beaker using hot plate

4. MECHANISM OF KERNEL ELONGATION

Although the mechanism of kernel elongation of rice is not clear. Water uptake might be the only trait showed positive and significant influence on kernel elongation. The anatomical study of the endosperm of elongating and non elongating varieties showed variation in the shape and arrangement of cells. Elongation pattern might be influenced by either size and arrangement of cells in endosperm or structural arrangement of starch molecules within cells and in non elongating rice, long belt of radially arranged cells might restrict linear expansion of cells while cells in the intermittent region may tend to expand towards the periphery leads to breadthwise elongation [12]. Aging also influences the kernel elongation [13].

Both genetic and environment factors influence the kernel elongation. The ambient temperature of about 25°C during day time and 21°C during night at the time of ripening is a favourable condition for maximum elongation [14]. Pre-soaking before cooking also play major role in kernel elongation [15].

Grain width and length, endosperm cell size and starch granule structure and arrangement are influenced the cooked kernel elongation. No specific mechanism to determine kernel elongation, but it might be due to physical and chemical phenomenon occurred after cooking [16].

Soaking of grains upto 30 min increases kernel elongation during cooking and leads to grain cracking. Elongating and non-elongating rices showed no difference in the composition of endosperm cell walls.

In basmati rice, equidistant pentagonal or hexagonal cells arranged in honeycomb fashion in the elongating variety and in breadth wise swelling types, long, rectangular and arranged radially in columns extending from the centre to the periphery. Microscopic examination of endosperm sections of the experimental materials revealed differences between elongating and non elongating groups of rice in respect of shape and arrangement of cells as represented by Sood et al. in Figure 1

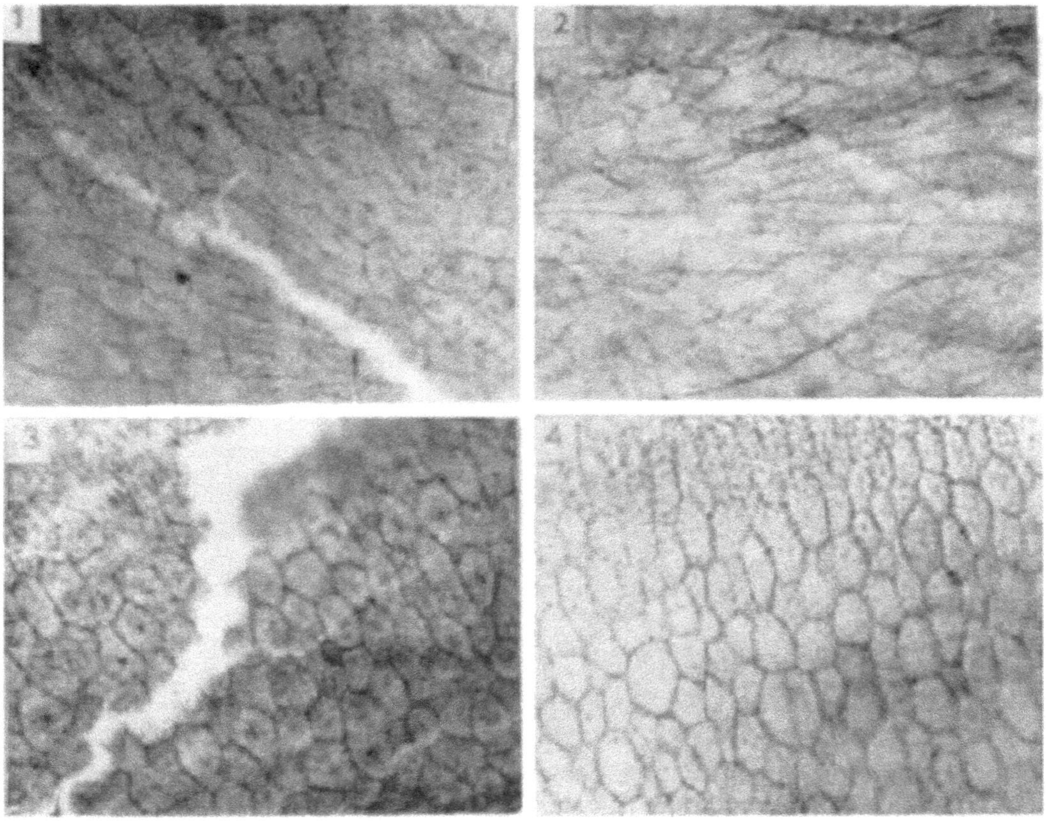

Figure 1. Anatomical study of elongating and non elongating rice kernels. (1, 2) Transverse sections of elongating and non elongating types (3, 4) Longitudinal sections of elongating and non elongating types. Source: Sood et al.

Internal kernel cracks were higher in aged rice kernels than the non-aged rice kernels in three popular Malaysian rice cultivars (Mahsuri, Mahsuri Mutant and Puteri). The number of cracks in kernel during cooking increased at higher temperature (110°C) compared to lower temperature (90°C). Kernel aging and temperature positively associated with internal physical structure of the rice kernel. From the above study, it is concluded that ageing and internal kernel cracks might influence the cooked kernel elongation [17].

5. GENETICS OF KERNEL ELONGATION

Kernel lengthwise increase without increase in girth is a desirable characteristic in high quality premium rice. Bao *et al.,* [18] reported that in *indica* rice variety, cooked kernel elongation was governed mainly by major gene with environmental interactions. Genetic improvement of any characters depends on efficiency of selection among the progenies which has different genetic value. Gene actions are reported to be associated with breeding value. To estimate the breeding value of progeny by studying the additive and dominance effects and their interactions which would benefit breeders to select the most appropriate breeding approaches to develop the superior variety.

Inheritance pattern of cooked kernel elongation ratio is controlled by single gene and influenced by some modifiers [13]. Kernel elongation governed by additive [19] or non-additive [20, 21, 22, 23, 24, 25, 26, 27, 28] or both additive and non additive gene action [29]. Hence genetics of kernel elongation has not been elucidated due to its complex nature and inconsistent pattern [30].

Elongation ratio and elongation index are highly heritable [31, 32]. This trait under polygenic control, involving epistasis, additivity, and genotype × environment interactions [33]. Three crosses were made among Mahsuri mutant, Mahsuri and 9192 to study the segregation pattern of cooked kernel elongation. The frequency distribution of kernel elongation ration of the population showed bimodal curve. However, in crosses of Mahsuri mutant and mahsuri (and its reciprocal crosses) the bimodal was skewed towards lower kernel elongation. Hence this character might be influenced by one or two loci [13].

The parental lines did not showed significant differences in kernel elongation ratio, however the progeny exhibited much larger range of variation due to transgressive segregation. It is difficult to conclude about inheritance pattern of kernel elongation in rice since the segregation pattern is not stable in different crosses and very little information is available on about inheritance pattern of kernel elongation of rice [34].

Genetic models used to analyse the cooked rice elongation traits of indica rice in two environments and identified that rice grain quality was mainly controlled by genotype, major gene effects and environment interactions. Milled and cooked rice length exhibited higher narrow sense heritability's than cooked rice elongation which indicate that early selection would be possible for milled and cooked rice length but not for cooked rice elongation [33].

6. ASSOCIATION OF KERNEL ELONGATION WITH OTHER QUALITY TRAITS

Waxy gene allele involved in determination of cooked rice elongation. The closely linked marker of waxy locus will facilitate the replacement of poor quality parent alleles [11] and waxy gene region control the cooked rice grain quality traits viz., elongation, length, width, width expansion ad water absorption [35]. Elongation index showed positive significant association with most of the grain quality traits which might be due to starch deformation caused by soaking or boiling.

Water absorption showed positive association with cooked rice elongation [36]. Aroma showed positive correlation with kernel elongation and both are highly influenced by environment. Elongation of rice can be influenced by both the l/b ratio and the amylose contents [37]. Kernel elongation is a physical phenomenon and influenced by gelatinization temperature [38]. Grain elongation has been associated with intermediate amylose and low gelatinization temperature.

Gene or quantitative trait loci for aroma and cooked kernel elongation were linked and present on chromosome 8.

7. QTL ASSOCIATED WITH COOKED KERNEL ELONGATION RATIO

A QTL was identified on chromosome 8 for kernel elongation [39]. Rani [40] found that a functional marker targeting an SNP in the GS3 is associated with kernel elongation. Tian *et al.* [36] detected 3, 2, and 2 QTLs for water absorption, volume expansion and cooked rice elongation, respectively in a DH population.

Amarawathi *et al.* [41] identified a QTL (*elr11-1*) for linear elongation ratio in chromosome 11 in the marker interval of RM1812 – RM209. Mallikarjuna *et al.* [42] reported that 1 QTL on chromosome 3 in *Oryza nivara* × Swarna derived backcross populations in the marker interval RM55–RM520. Two QTLs were identified for KLAC and both were derived from *O. nivara*, and these were located on chromosomes 5 and 12. Dewei *et al.* [43] identified 12 QTLs for rice elongation traits were detected on chromosomes 3, 4, 6, 8, 9, 10, and 11, among which two QTLs for MRL were located on chromosome 3, one QTL for MRL on chromosome 8, four QTLs for CRL on chromosome 3, 6, 8, and 9, and five QTLs for CRE on chromosome 4, 6, 9, 10, and 11.

Acga *et al.* [44] identified two QTLs for grain elongation on chromosome 2, designated *qGE-2-1* and *qGE-2-2*. The *qGE-2-1* mapped to the interval RM53-RM174, while *qGE-2-2* was mapped to the marker interval RM525-RM6, One QTL for grain elongation was previously reported on chromosome 2 in the marker interval R2510 - RM211 in the study conducted by Ge *et al.* [35]. Chen [45] reported that RM44 is associated with kernel elongation. Sathyasheela [46] reported that RM 209 was associated with LER. Liu *et al.* [47] detected three CRE QTLs on chromosome 4, 5, and 12, respectively, and the *qCRE-4* on chromosome 4 near *qER-4* was detected in this study. Li *et al.* [48] mapped a CRE QTL on chromosome 3, with the favorable allele obtained from the African rice *O. glaberrima*. Using a RIL population, Wang *et al.* [10] identified four CRE QTLs on chromosome 3, 6, 7, and 8, respectively.

8. APPLICATIONS OF CRISPR/CAS9 FOR RICE GRAIN QUALITY IMPROVEMENT

Rice grain quality improvement through targeted genome editing is a fast, sustainable and cost effective approach. Conventional plant breeding methods depends on naturally existing germplasm variations. The introgression through backcrossing requires much time and screening of large population by marker assisted selection requires more energy. The reverse genetic approaches enhance the speed of plant breeding through targeted genome modifications [49] (Figure 2).

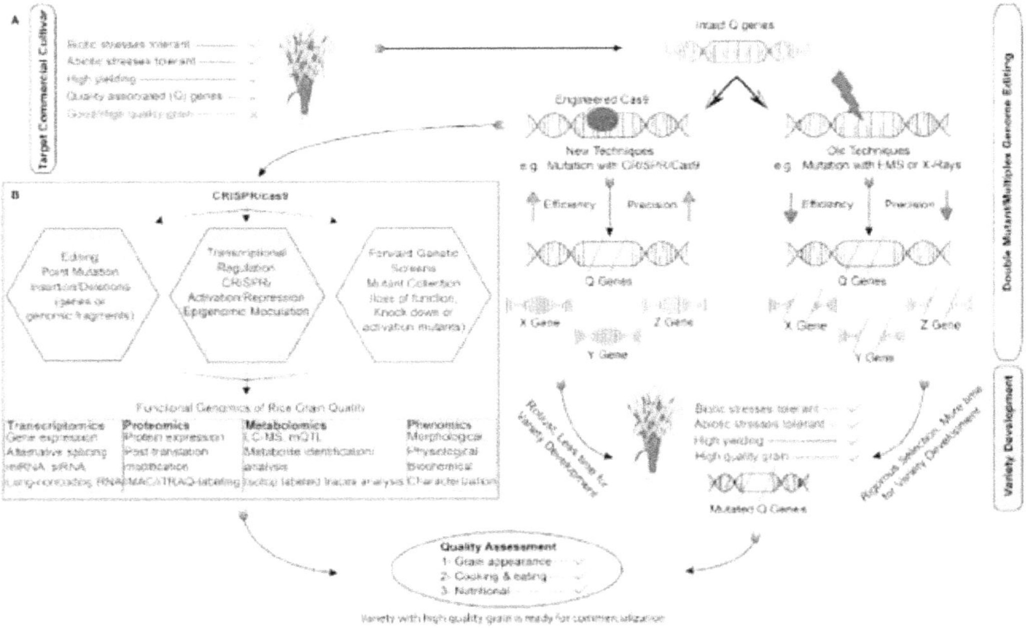

Figure 2. An illustration of rice grain quality improvement through the CRISPR/Cas9 system.

9. CONCISE SUMMARY

Premium quality rice grain is the demand of a growing population with better living standards. Presently, the CRISPR/Cas9 system has all genome editing capabilities, e.g., knock-in, knockout, knockdown, and expression activation. This system has tremendous untapped potential, has formed an ever-expanding genetic toolbox for plant biologists to investigate functional genomics, and is a helping hand for breeders to integrate important genes into the genomes of important crops. The successful application of CRISPR/Cas9 for tissue engineering and human stem cell modification has led to further developments in the field of precise genome editing. The ability to target multiple genes via multiplexed genome editing strategies can facilitate pathway-level research to engineer complex multigenic rice grain quality attributes. Previously, few studies have been conducted that are related to targeted mutagenesis for rice grain quality improvement. The pathways of rice grain quality are not well understood, and they can be investigated for the genetic mechanisms controlling quality attributes. The development of novel regulatory components from naturally existing peripherals (genes, promoters, cis-regulatory elements, small RNAs, and epigenetic modifications) can facilitate the engineering of regulatory pathways for different elements of rice grain quality. The rapid shift of research toward the utilization of CRISPR/Cas9 systems for targeted mutagenesis could be a promising approach for overcome barriers to breeding improved quality rice.

AT A GLANCE

Oryza sativa holds a unique position among domesticated crop species as it is one of the most important staple foods globally. Without rice, the day will not be fulfilled in most of the Asian countries. Requirement of rice for consumption is anticipated from 450 million tons in 2011 to about 490 million tons in 2020 and to around 650 million tons by 2050 globally. To meet the food demands, it has been estimated that 40 per cent more rice is needed to be produced by 2050 for the ever increasing population. Increasing incidences of both biotic and abiotic stresses under changing climate are the major constraints in rice production to meet the rapidly escalating population. Crop improvement in rice will not be completed lacking of grain quality analysis. Rice grain quality embraces storage, milling, market quality, cooking and eating quality and nutritive quality of grain. Demand for high quality rice has increased globally in recent years and continues to trend upward due to the taste preferences. Since, consumer demand in Asia and all over the world are diverse due to varied demographics and culture, defining uniform attributes to grain quality becomes more challenging. The Middle Eastern consumers highly prefer long grain, well milled rice with strong aroma while European consumers prefer long grain non aromatic rices. In Asia, Chinese consumers prefer semi-aromatic rice to pure aromatic rice. Cooked kernel elongation is the most important quality traits, which differentiate the highly valued basmati rice from other rice types. Kernel elongation after cooking is an important character of fine rice and the most rice consumers prefer lengthwise elongation.

REFERENCES

1. Itani T, Masahiks T, Eiks A, Toshroh H. Distribution of amylose, nitrogen and minerals in rice kernel with various characteristics. Journal of Agricultural and Food Chemistry. 2002; 50(19): 5326-5332

2. Leila Bazrkar-Khatibani, Barat-Ali Fakheri, Maryam Hosseini-Chaleshtori, Anumalla Mahender, Nafiseh Mahdinejad and Jauhar Ali. Genetic Mapping and Validation of Quantitative Trait Loci (QTL) for the Grain Appearance and Quality Traits in Rice (Oryza sativa L.) by Using Recombinant Inbred Line (RIL) Population. International Journal of Genomics. 2019;13
3. Kaur Shubhneet, Parmjit S, Panesar, Manab BB. Studies on evaluation of grain quality attributes of some basmati and non-basmati rice cultivars. Journal of Food Quality. 2011; 34: 435-441
4. Bhattacharya KR, Sowbhagya CM. Water uptake by rice during cooking. Cereal Sci Today,1971; 16:420-424
5. Sidhu JS, Gill MS, Bains GS. Milling of paddy in relation to yield and quality of different Indian varieties. J Agric.Food. Chem., 1975; 23:1183-1185
6. Sood BC, Siddiq EA. Study on component quality attributes of basmati rice. Z P flanzenschutz. 1980; 84:294-301
7. Pilaiyar P. Quality characteristics of Tamil Nadu rices. Madras Agric J., 1988; 75:307-317
8. Kumar MS, Genetic analysis of height and cooking quality traits in basmati rice. MSc thesis, IARI, New Delhi (1989)
9. Juliano BO, Perez CM. Results of a collaborative test on the measurement of grain elongation of milled rice during cooking. J Cereal Sci.,1984; 2:281-292
10. Wang XS, Zhu HB, Jin GL, Liu HL, Wu WR, Zhu J. Genome scale identification and analysis of LEA genes. Plant Science, 2007; 172: 414-420
11. Yang D, Zhang Y, Zhu Z, Chen T, Zhao Q, Yao S, Ling Zhao, Zhu W, Wang C. Substitutional mapping the cooked rice elongation by using chromosome segment substitution lines in rice, 2013; 4(13)
12. Sood BC, Siddiq EA, Zaman FU, Mechanism of kernel elongation in rice. Indian J Genet., 1979; 39:457-460
13. Faruq Golam, Mohamad O, Hadjim K, Meisner CA. Inheritance of Kernel Elongation in Rice. Int. J. Agri. Biol., 2004; 6(5)
14. Dela Cruz N, Kumar I, Kaushik RP, Khush GS. Effect of temperature during grain development on the performance and stability of cooking quality components of rice. J Breed., 1989; 39: 299-306
15. Vikas Kumar, Rastogi NK, Patel NB, Swarup Nanda Mandal. Inheritance of kernel length and kernel elongation in basmati rice (Oryza sativa L.). Electronic Journal of Plant Breeding. 2018; 9 (2) : 429-432
16. Singh VP. 2002. Pusa 1121: A Rice Line with Exceptionally High Cooked Kernel Elongation and Basmati Quality. IRRN 27. 1: 25-26

17. Golam F, Prodhan H Zakaria. Kernel elongation in rice. J. Sci Food Agric, 2012;
18. Bao JS, Zheng XW, Xia YW, He P, Shu QY, Lu X, Chen Y, Zhu LH. QTL mapping for the paste viscosity characteristics in rice (Oryza sativa L.). Theoretical and Applied Genetics. 2000;100 (2):280-284
19. Umadevi M, Veerabadhiran P, Manonmani S, Shanmugasundaram P. Physico-chemical and cooking characteristics of rice genotypes. Electronic Journal of Plant Breeding. 2010; 1 (2):114-123
20. Asfaliza R, Rafii MY, Saleh G, Omar O, Puteh A. Combining ability and heritability of selected rice varieties for grain quality traits. Australian Journal of Crop Science 2012; 6 (12):1718
21. Gonyanayak, Sreedhar M, Surender Raju Ch, Vanisree S. Heterosis and gene action studies involving aromatic lines for grain quality traits in rice. International Journal of Life Sciences 2016; 4 (4):517-528
22. Krishna L, Chandra Mohan Y, Sudheer Kumar S. Combining ability analysis for grain yield and quality characters in aromatic rice (Oryza sativa L.). Indian Journal ofChemicl Studies. 2018; 6 (5):3185-3189
23. Malini N, Sundaram T, Hari Ramakrishnan S. Combining ability for quality traits in rice (Oryza sativa L.). Plant Archives. 2014; 14:469-473
24. Mohammad, Naseer, Surendra Singh, Ekhlaque Ahmad and Salil Tewari. Assessment of combining ability for productivity and quality traits in aromatic rice (Oryza sativa L.). International of J. Bio-Res. Stress Manag., 2016;7 (4):525-532
25. Roy SK, Senapati BK. Combining ability analysis for grain yield and quality characters in rice (Oryza sativa L.). Indian journal of agricultural sciences.2012; 82 (4):293-303
26. Rukmini Devi, Satish Chandra B, Hari Y, Venkanna V. Heterosis and inbreeding studies for agronomic and quality traits in rice (Oryza sativa L.). Agricultural Science Digest-A Research Journal. 2018; 38 (4):313-318
27. Santha S, Vaithilingam R, Karthikeyan A, Jayaraj T. Combining ability analysis and gene action of grain quality traits in rice (Oryza sativa L.) using Line x Tester analysis. Journal of Applied and Natural Science. 2017; 9 (2):1236-1255
28. Sreenivas G, Cheralu C, Rukmini Devi K, Gopala Krishna Murthy K. Combining ability analysis for grain quality traits in rice (Oryza sativa L.). Environment and Ecology 2015; 33 (1):186-191
29. Mahalingam L, Nadarajan N. Genetic analysis of grain yield and milling quality characters of two line rice hybrids. International Journal of Plant Sciences 2010; 5 (2):630-635

30. Siwaret Arikit, SamartWanchana, Srisawat Khanthong, Chatree Saensuk, TripopThianthavon, ApichartVanavichit and TheerayutToojinda. QTL-seq identifies cooked grain elongation QTLs near soluble starch synthase and starch branching enzymes in rice (Oryza sativa L.). Scientific reports. 2019;9:8328

31. Hussain AA, Maurya DM, Vaish CP. Studies on quality status of indigenous upland rice (Oryza sativa L.). Indian J. Genet. 1987; 41: 12-17

32. Vivekanandan P, Giridharan S. Combining ability for grain traits in rice. Madras Agric J., 1995; 84:129-132

33. Bao JS, Cai YZ, Corke H. Prediction of rice starch quality parameters by near-infrared reflectance spectroscopy. J Food Sci., 2001; 66:936-939

34. Yellari A, Singh R, Singh AK, Singh VP, Trilochan M, Tilak RS, et al. Mapping of quantitative trait loci for basmati quality traits in rice (Oryza sativa L.). Mol Breeding. 2007; 21:49-65

35. Ge XJ, Xing YZ, Xu CG, He YQ. QTL analysis of cooked rice grain elongation, volume expansion and water absorption using a recombinant inbred population. Plant Breed. 2005; 124: 121-126

36. Tian R, Jiang GH, Shen LH, Wang LQ, He Q. Mapping quantitative loci underlying cooking and eating quality of rice using a DH population. Mol. Breed. 2005; 15: 117-124

37. Singh RK. Heterosis breeding in aromatic rice (Oryza sativa L.) for yield and quality characters. The Indian Journal of Genetics and Plant Breeding. 2005; 65 (3):176-179

38. Sood BG, Siddiq EA, Zamman FU. Genetic analysis of kernel elongation in rice. Indian J. Genet. 1983; 43: 40-43

39. Ahn, S.N., C. N. Bollich, A.M. McClung and S. D. Tanksley. 1993. RFLP analysis of genomic regions associated with cooked-kernel elongation in rice. Theor Appl Genet. 87: 27-32

40. Rani N.S., A.K.P. Sivaranjani, M.S. Madhav, R.M. Sundaram, G.S.V. Prasad, S. Srikanth, M. Pandey, K. Suneetha and I. Sudharshan 2011. Identification of molecular markers for cooking quality traits of rice. Indian J Genet Plant Breed. 71: 129-138

41. Amarawathi Y, R. Singh R, Singh AK, Singh VP, Mohapatra T, Sharma TR, Singh N. K. Mapping of quantitative trait loci for basmati quality traits in rice (Oryza sativa L.). Mol Breed. 2008; 21:49-65

42. Mallikarjuna SBP, Kaladhar K, Shobha rani N, Prasad GSV, Viraktamath BC, Ashok reddy G, Sarla N. QTL Analysis for Grain Quality Traits in 2 BC2F2 Populations Derived from Crosses between Oryza sativa cv Swarna and 2 Accessions of Oryza nivara. Journal of Heredity. 2012; 103(3):442-52

43. Dewei Y, Zhang Y, Zhu Z, Chen T, Zhao Q, Yao S, Zhao L, Lin J, Zhu W, Wang C. Substitutional Mapping the Cooked Rice Elongation by Using Chromosome Segment Substitution Lines in Rice. Mol. Pl. Breed. 2013; 4(13):107-115

44. Acga C, Ismail I, Osman M, Hashim H. Mapping of quantitative trait loci for aroma, amylose content and cooked grain elongation traits in rice. Plant Omics Journal. 2014;7(3): 152-157

45. Chen L. Theory and Practice in Two-Line Hybrid Rice. Shanghai Sci. Tech. Press, 2001; Shanghai. 136-184

46. Sathyasheela KRV. 2009. Genetic Analysis of Grain Quality Traits in Rice Using Segregating Population of CB 04110 X KDML 105. Ph.D., Thesis. Tamil Nadu Agricultural University, Coimbatore

47. Liu QQ, Li QF, Cai XL, Wang HM, Tang SZ, H.X. Yu, Z.Y. Wang and M.H. Gu. 2006. Molecualr marker-assisted selection for improved cooking and eating quality of two elite parents of hybrid rice. Crop Sci 46: 2354-2360

48. Li J, Thomson M, McCouch SR. Fine mapping of a grain-weight quantitative trait locus in the pericentrometric region of rice chromosome 3. Genetics. 2004; 168:2187-2195

49. Sajid Fiaz, Shakeel Ahmad, Mehmood Ali Noor, Xiukang Wang, Afifa Younas, Aamir Riaz, Adeel Riaz and Fahad Ali. Applications of the CRISPR/Cas9 System for Rice Grain Quality Improvement: Perspectives and Opportunities. International joural of molecular sciences. 2019; 20:888

CHAPTER-4

DEVELOPING STRATEGIES FOR ENHANCING OMEGA-3 FATTY ACID LEVELS IN SOYBEAN THROUGH CONVENTIONAL BREEDING, GENETIC MAPPING, AND GENOMICS

1. INTRODUCTION

In the pursuit of healthier dietary options and the continuous evolution of agricultural practices, the focus on enhancing the nutritional profile of crops has become imperative. One such endeavor involves the development of strategies aimed at elevating the levels of omega-3 fatty acids in soybeans. This multifaceted approach integrates conventional breeding techniques, genetic mapping, and genomics to meticulously engineer soybean varieties with heightened omega-3 fatty acid content. This holistic strategy not only underscores the significance of conventional methods but also leverages cutting-edge genomic technologies to expedite and refine the breeding process. The amalgamation of these diverse methodologies reflects a commitment to addressing nutritional challenges through innovative agricultural advancements, thereby contributing to the cultivation of more health-conscious and nutritionally enriched soybean varieties.

Soybean [*Glycine max* (L.) Merr.], an important oil crop, accounted for 28% of the total vegetable oil consumption in the world in 2019 [1]. The last 7 years have seen 22.5% increases in soybean oil consumption worldwide (Figure 1) and soybean oil production is expected to rise in the future. Commodity soybean seed contains 20% oil, 40% protein, and 12% soluble carbohydrates on a dry weight basis [2]. Fatty acid compositions of food and oil have received considerable attention in the past few decades for human nutrition and health concerns. Hence, improving the FA composition of the soybean oil is crucial to reduce the risks of associated coronary heart and other diseases.

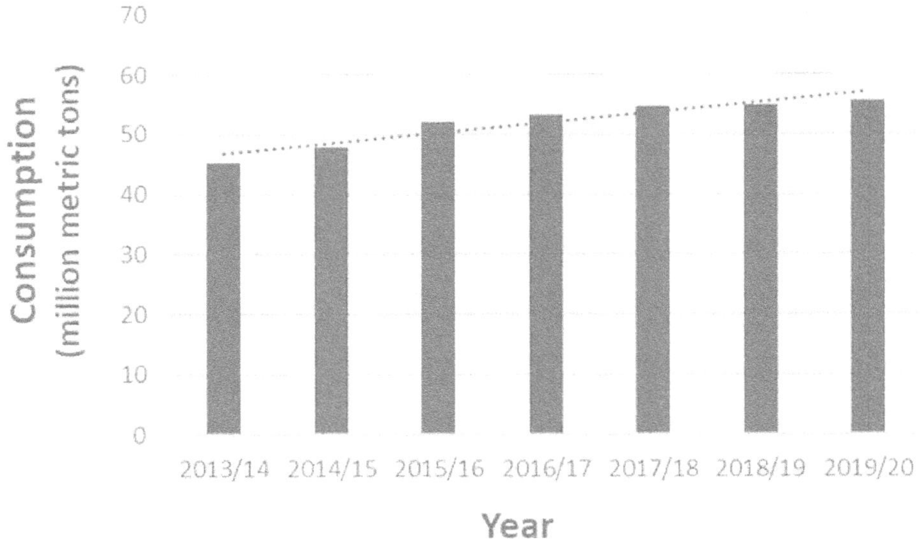

Figure 1. Soybean oil consumption worldwide from 2013/14 to 2019/20 (in million metric tons). Retrieved from https://www.statista.com/statistics.

1.1 Fatty acid composition of important oil crops

Oil and fatty acids are essential elements for the development and growth of all the creatures. These elements are the structural components of the cellular membrane, as well as play a pivotal role in storing energy and involved in the cellular signaling processes. Known natural resources of oil and fatty acids are plants, animals, and oleaginous microorganisms. Oil and fatty acids played a crucial role in human life in several ways as food and fuel resource, mostly as a nutritional element of the diet. Edible oilseed crops (palm, olive, rapeseed, canola, sunflower, and soybean) primarily contain five fatty acids: palmitic acid (16:0), stearic acid (18:0), oleic acid (18:1), linoleic acid (ω-6, 18:2), and α-linolenic acid (ALA, ω-3, 18:3) in their seed oil. The fatty acid composition varies with the oilseed plant species as given in Table 1. Fatty acids accumulated in plants as a form of triacylglyceride, which consists of three fatty acids linked to glycerol as a backbone through ester bond [4, 5]. The triacylglyceride is a key source of renewable energy in the form of reduced carbon used as food, feedstock, and fuel.

In the plant fatty acids, basic biosynthetic pathways are well established but FA/lipid operating between the plastid and the endoplasmic reticulum remains to be determined [6, 7]. The fatty acid biosynthesis takes place in the chloroplast stroma of leaves and proplastids of seeds by *de novo* and further incorporation of triacylglyceride occurs in the endoplasmic reticulum [8]. Besides, poly-unsaturated fatty acids (PUFA) synthesis in higher plants takes place by both prokaryotic and eukaryotic enzymatic pathways [7, 9]. These pathways are regulated through diverse sets of genes.

Table 1. Fatty acid composition of major oilseed crops.

Crops	Saturated fatty acids (Palmitic and stearic acid) (%)	Unsaturated fatty acids (%)		
		Oleic acid	Linoleic acid	α-linolenic acid
Canola	7.4	61.8	18.6	9.1
Coconut	82.5	6.0	—	—
Corn	12.9	27.3	58.0	1.0
Cottonseed	25.9	19.0	54.0	1.0
Flaxseed	8.0	19.0	17.0	53
Olive	13.8	71.3	9.8	0.7
Palm	49.3	40.0	9.1	0.2
Peanut	20.3	46.5	31.4	—
Safflower	7.5	75.2	12.8	—
Sesame	—	36.2	44.2	0.3
Sunflower	—	25.4	59.9	0.1
Soybean	15.6	22.6	51.0	7.0

Data retrieved from Kim [3].

1.2 Polyunsaturated fatty acids in soybean

Soybean seed primarily contains two saturated fatty acids, which are palmitic acid, and stearic acid, and three unsaturated fatty acids, which are oleic acid, linoleic acid, and ALA. The relative ratio of these fatty acids commonly found in cultivated soybean are, 11% of palmitic acid, 4% of stearic acid, 23% of oleic acid, 55% of linoleic acid, and 8% of ALA [10]. However, wide variation for fatty acids content has been reported in several studies [11, 12, 13]. Commonly, the proportion ratio of fatty acids in soybean is influenced by the genotypes as well as the environmental factors which are very crucial to determine the entire quality of the oil.

Linoleic acid and ALA are the PUFA, also termed essential fatty acids (EFA), present in the soybeans. Due to the presence of two or more double bonds between the carbons of fatty acid chains, PUFA are distinguished from saturated and monounsaturated fatty acids. The EFA is primarily classified into two forms, ω-6 and ω-3, which are metabolically interlinked but functionally diverse. ω-6 and ω-3 comprise long chains of carbon atoms with a carboxyl group at one end of the chain and a methyl group at the other end.

ω-3 fatty acids have a carbon–carbon double bond located between the third and fourth carbon atoms from the methyl end of the chain. ω-3 fatty acids exhibit *cis-*

trans isomerism with its extension to *E-Z* configuration [14]. ALA, eicosapentaenoic acid (EPA), and docosahexaenoic acid (DHA) are the three important ω-3 fatty acids. The carbon backbone of these contains 18 carbon atoms, 20, and 22 for ALA, EPA, and DHA, respectively. Their structures are shown in Figure 2. ALA is the precursor of DHA and EPA, which are essential for the growth and development of the brain and retina in humans [15].

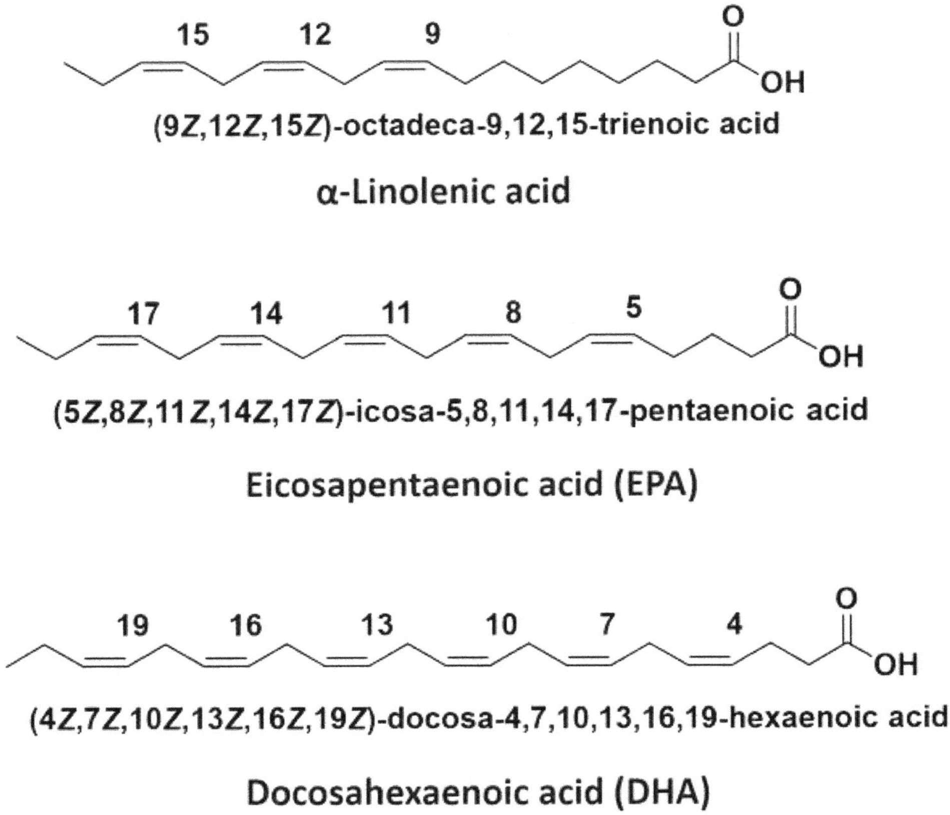

Figure 2. Major three ω-3 polyunsaturated fatty acid structure: Three unsaturated fatty acids are shown with the cis configuration of the double bonds with a methyl end and a carboxyl (acidic) end represented with commonly followed nomenclature numerical scheme.

1.3 Sources of ω-3 fatty acids

Seafood products such as fish are the main sources of ω-3 fatty acids (ALA, EPA, and DHA). However, they are not a routine part of the traditional diet in many countries. The ω-3 fatty acid is abundantly available in nature and found in most of the oilseed crops. The ALA is also synthesized in the plants found in green leafy vegetables, and in the seeds of flax, as rapeseeds (*Brassica campestris*), chia (*Salvia hispanica*), perilla (*Perilla frutescens*), walnut (*Juglans sinensis*), and soybean. The ω-3 is dietary EFA for humans; however, because of the absence of delta-12 and delta-15 desaturase enzymes, humans and other animals are unable to synthesize ω-6 and ω-3 fatty acids. Therefore, these EFAs need to acquire through diet or dietary supplements [16].

1.4 Importance of ω-3 fatty acid and ω-6/ω-3 ratio

Several studies reported the nutritional and health benefits of ω-3 in humans [17]. Besides, ω-3 fatty acids are known for therapeutic uses and to offer protection against numerous diseases [15]. Thus, the nutritional value of ω-3 fatty acids is now widely accepted. Earlier diets comprised meat, plants, eggs, fish, nuts, and berries, which contained substantial amounts of ω-3 fatty acid [17, 18]. With the changes in dietary habits, consumption of ω-6 fatty acid was enhanced, which consequently reduced the level of ω-3 fatty acids in human. Thus, the contemporary diets are now comprised of a high intake of saturated and ω-6 fatty acids, decreased ω-3 fatty acid intake, and an overuse of salt and refined sugar [19]. These dietary changes led to diets with an undesirable ω-6/ω-3 ratio up to 20:1 [20]. Ultimately, an altered ratio of ω-6/ω-3 is considered unhealthy and reported to be the prevalent cause of prothrombotic and proinflammatory diseases, such as atherosclerosis, obesity, and diabetes [16, 21, 22, 23]. Several studies have reported a positive correlation between lower ω-6/ω-3 ratio and reduced risks of cardiovascular disease (CVD), cancer, including breast, colon, prostate, liver, and pancreatic cancers, inflammation, favor apoptosis and exert antiproliferative effects cancers [24].

The balanced ω-6/ω-3 ratio is an important determinant in decreasing the risk for CVD [17]. Increased intake of linoleic acid is known to interfere with the incorporation of EPA and DHA (which have the most potent inflammatory effects) in cell membrane lipids, and causes platelet aggregation and oxidation of low-density lipoprotein. Intake of ω-3 fatty acids may help in preventing the development of CVD as well as other associated diseases. Therefore, it is highly essential to increase the intake of ω-3 fatty acids and to reduce the consumption of ω-6 fatty acids. It has been estimated that the present Western diets have a ω-6/ω-3 ratio of 15-20:1, which is highly imbalanced. Several studies in animals, such as *Caenorhabdtis elegans,* rats, mice, and pigs have shown the importance of balanced ω-6/ω-3 ratio [25, 26, 27]. Experimental studies suggested the changes in mucosal inflammation and reduced the patient suffering from an increased ratio of ω-6/ω-3. The combination of rheumatoid arthritis drug treatment and an adequate ω-6/ω-3 ratio has been suggested to cause significant changes in inflammatory markers [24]. Such studies have provided evidence for the need to have a balanced ratio of ω-6/ω-3 (1:1 to 2:1) [20, 24]. Besides, ω-6 and ω-3 fatty acids are involved differentially in the cellular process. It was reported that ω-6 fatty acids increase triacylglyceride content in the cell through altering membrane permeability; whereas, ω-3 fatty acids lower fat deposition in adipose tissues by suppressing lipogenic enzymes and increasing β-oxidation [28].

2. BREEDING GOALS FOR IMPROVEMENTS IN THE Ω-3 FATTY ACID CONTENT OF SOYBEAN SEEDS

ALA (ω-3) in soybean oil is unstable and has undesirable flavors. Due to the presence of a double bond at the 12th carbon in the fatty acid hydrocarbon chain, the ALA is oxidized easily, which causes unwanted odor and off-flavors [29, 30, 31]. Ultimately, this reduces the functional quality of fry food or soy food items [31, 32]. Hence, ALA content is negatively associated with the stability and shelf life of soybean oil. To improve the shelf life, stability, desirable flavor, and palatability, soybean oil is chemically hydrogenated, which leads to the formation of *trans* fats. The *trans* fatty acids are linked to coronary heart diseases, and hence it is desirable to reduce the consumption of foods containing *trans* fats. Nearly 13 million Americans are reported to suffer from coronary heart disease, and over 500,000 die annually from causes related to coronary heart disease. For these reasons, the addition of *trans* fat labels to food nutrition labels directly under the line for saturated fat was started in 2006. This resulted in a change in the narrative of food industries to promote and explore substitutions for hydrogenated soybean oils. In the past few decades, soybean geneticists and breeders have shown that oil with reduced *trans*-fat content can be obtained by decreasing the content of ω-3, and ω-6 fatty acids through conventional and modern breeding approaches. Soybean oils derived from cultivars with reduced ω-3 fatty acid content are shown to increase stability, lower hydrogenation, and comprise reduced *trans*-fat levels [29, 33, 34]. This shifted the breeder's focus to develop soybeans with reduced ω-3 fatty acid content [35, 36]. In this section, the genetic basis of ALA (ω-3) content and breeding approaches fro altering ALA content are discussed.

2.1 Genetic basis of PUFA in soybean

The biosynthesis of PUFAs in soybean involves a variety of pathways, which are catalyzed by a complex series of desaturation and elongation steps [37]. Fatty acid desaturases introduce double bonds into the hydrocarbon chains of fatty acids to produce unsaturated fatty acids [38]. The delta-12 fatty acid desaturase-2 enzyme (*FAD2*) catalyzes the conversion of oleic acid to linoleic (ω-6) in the developing soybean seeds [39]. The microsomal ω-3 fatty acid desaturases (*FAD3*) catalyze the transformation of linoleic into ALA [36]. Thus, the genes coding for these fatty acid desaturases may act together to control the ALA content in soybean.

Two identical copies of FAD2 enzymes (FAD2-1, and FAD2-2) have been identified in the soybean. Five *FAD2* gene family members (two *FAD2-1* members: *FAD2-1A* and *FAD2-1B*, and three *FAD2-2* members: *FAD2-2A*, *FAD2-2B*, and *FAD2-2C*) are present in the soybean genome [39, 40]. Through syntenic, phylogenetic, and in silico analysis, Lakhssassi et al. [41] revealed two additional members of the *FAD2* gene family: *FAD2-2D* and *FAD2-2E*, positioned on chromosomes 9 and 15, respectively. Of these

FAD-2 genes, *FAD2-1A* is highly expressed in developing soybean seeds [41]. The chromosomal locations and gene model names of these genes are given in Table 2. Mutations in one or more of these genes have been utilized to alter the fatty acid content of the soybean seeds.

Table 2. Chromosomal location and gene model information of *FAD2* and *FAD3* genes in the soybean genome.

Gene family	Gene/paralogs	Gene model (Wm82.a1.v1)	Gene model (Wm82.a2.v1)	Chromosome (Linkage group)
FAD2	FAD2-1A	Glyma10g42470	Glyma.10g278000	10 (O)
	FAD2-1B	Glyma20g24530	Glyma.20g111000	20 (I)
	FAD2-2A	Glyma19g32930	Glyma.19g147300	19 (L)
	FAD2-2B	Glyma19g32940	Glyma.19g147400	19 (L)
	FAD2-2C	Glyma03g30070	Glyma.03g144500	3 (N)
	FAD2-2D	Glyma09g17170	Glyma.09g111900	9 (K)
	FAD2-2E	Glyma15g23200	Glyma.15g195200	15 (E)
FAD3	FAD3A	Glyma14g37350	Glyma.14g194300	14 (B2)
	FAD3B	Glyma02g39230	Glyma.02g227200	2 (D1b)
	FAD3C	Glyma18g06950	Glyma.18g062000	18 (G)
	FAD3D	Glyma11g27190	Glyma.11g174100	11 (B1)

The genetic basis ω-3 fatty acid trait in soybean has been identified based on the experimental study of gene information from the model plant *Arabidopsis thaliana* (L.) Heynh. and screening of lower ω-3 fatty acid mutant soybean lines. Arabidopsis contains a single gene encoding a microsomal ω-3 fatty acid desaturase and two chloroplast targeted enzymes (FAD7 and FAD8). At least three independent loci (*fan, fan2*, and *fan3*), influencing the ALA content have been identified [42]. Genes underlying all the three loci have been identified as homologous genes of *FAD3* [43]. In soybeans, the level of ALA is controlled by three *FAD3* genes: *FAD3A* (*Glyma.14g194300* for Wm82.a2.v1 assembly), *FAD3B* (*Glyma.02g227200*), and *FAD3C* (*Glyma.18g062000)*. Of these, *FAD3A* has been reported to show consistent high expression in developing seed, and hence have the greatest effect in controlling the accumulation of ALA contents [42, 44]. These three loci have a greater effect on ALA concentration, as combining mutant alleles of these genes resulted in soybean oil having ~1% ALA [45, 46, 47, 48].

The full-length genomic DNA sequences for *FAD3A, FAD3B, FAD3C*, and *FAD3D* genes were found to share 78 to 95% similarity, and have similar structure, and contain eight exons [49]. These eight exons of *FAD3* genes are highly conserved in soybean and correspond to the sizes of exons in Arabidopsis. However, significant variation is found among the introns of *FAD3* genes. Based on the structure and similarity, the four *FAD3* genes could be separated into two groups: *FAD3A/FAD3B*, and *FAD3C/FAD3D.*

2.2 Reducing ω-3 fatty acid content to avoid the need for chemical hydrogenation

Initial breeding efforts were made by the USDA-ARS in 1952 to identify soybean germplasm with lower ω-3 content to replace chemical hydrogenation of soybean oil [50]. During that period, the cultivars with lower ω-3 levels were identified but no cultivar having <4% ω-3 were found. In 1981, USDA-ARS and North Carolina State University collaborated and developed the line N79-2245 having a reduced ω-3 content of 4.2% by recurrent selection approach [32]. The major cultivars/lines with low ω-3 content developed using conventional breeding, germplasm screening, mutation breeding, and recurrent selection [44, 51, 52, 53] are given in Table 3.

Table 3. List of soybean mutants and germplasm lines with low levels of α-linolenic acid in the seed oil.

Lines/Cultivars	Selection Type	α-linolenic acid (%)	Reference
M5	Mutant	<4.6	[54]
N79-2245	Recurrent selection	4.2	[32]
C1640	Mutant	3.4	[55, 56]
A5	Mutant	4	[57]
A23	Mutant	5.6	[58]
IL-8	Mutant	4.5	[54]
M-24	Mutant	4.5	[59]
RG10	Mutant	< 2.5	[60]
PI123440	Germplasm	< 4.0	[61, 62, 63]
PI361088B	Germplasm	4	[61, 62, 63]
A16	Mutant	< 2.5	[58]
A17	Mutant	< 2.5	[58]
MOLL	Recurrent selection	< 3.0	[64, 65]
LOLL	Recurrent selection	< 3.0	[65, 66]
A29	Mutant	3.0	[67, 68]
CX1512–44	Mutant	3.0	[46]
J18	Mutant	3.0	[43]
PE1690	Mutant	3.7	[51, 58]
	Mutant	4.0	[69]

Lines/Cultivars	Selection Type	α-linolenic acid (%)	Reference
RCAT 0716L	Mutant	3.0	[45]
MS382	Mutant	<4.0	[65]
19,457	Mutant	3.9	[70]
18,777	Mutant	4.0	[70]
21,249	Mutant	4.5	[70]

The natural accessions reported with the low ω-3 level in the USDA germplasm is known as PI 123440 and PI 361088B with allelic variant at *fan* locus. This mutation was reported as allelic or identical to the initial single recessive allele derived from the C1640 genotype [57, 61, 62, 63]. Burton et al. [71] used the PI 123440 as a parent source to develop a low ω- trait known as Soyola and Satelite.

Through the EMS and X-ray mutagenesis approach, several mutants were previously reported for the lower ω-3 fatty acid content ranging from <2.5% to 5.6% that are linked with the *Fan* loci such as C1640 (*fan*), A5 (*fan*), A23 (*fan2*), KL-8 (*fanx*), M-5 (*fan*), M-24 (*fanxa*) and RG10 (*fan-b*) [54, 58, 60, 64, 66, 67, 72]. Besides, mutants A16, A17, A29, MOLL, and LOLL with reduced ω-3 acid content showed allelic variation at *fan* loci [58, 64, 66, 67]. The RG10 line was developed from the mutagenesis of 4% ω-3 line C1640 [55, 60]. Several studies used the RG10 line to develop the novel lines (*GmFAD3aabbcc*) with low ω-3 fatty acid content and also used for mapping and validating the quantitative trait loci (QTLs) and *FAD3* genes [45, 49, 73]. The EMS mutant line PE1690 with the reduced ω-3 fatty acid was reported to have a single base mutation in the *FAD3A* gene, resulting in the desaturase enzyme being nonfunctional [51]. Recently, Held et al. [69] identified a novel mutant allele of the *FAD3C* gene in a screen of a N-nitroso-N-methylurea (NMU)-mutagenized population. This allele resulted in 2 to 3% reduction in ω-3 FA levels.

2.3 Increasing ω-3 fatty acid content for improving ω-6/ω-3 ratio in soybean

Soybean production focuses on providing high protein meals for livestock and the manufacture of vegetable oils in both Western and Asian countries, while soybean has traditionally been used as a staple food in many Asian countries [2, 74]. The consumption of soy foods has been increasing in North America, following the recognition of the health benefits of soy foods.

Since the shortage of resources in cultivated soybean with elevated ALA content [75], researchers tried to find suitable genetic resources to develop new cultivars with high ALA concentrations in soybean breeding programs. Wild soybean can be a possible resource to achieve the goal to increase ALA concentration because those soybeans have an average of 15% ALA concentration, which is almost twice the ALA concentration present in the cultivated soybean [76]. Cultivated soybeans have an ω-6/ ω-3 ratio of 6–7:1, whereas wild soybeans have an ω-6/ ω-3 ratio of 3–4:1, which has better health benefits [76, 77, 78]. Thus, wild soybean can be exploited as a genetic resource to develop soybean lines with high ALA concentrations, although exploiting wild soybeans in breeding programs is challenging due to their poor agronomic traits. Several studies reported soybean lines with elevated ALA from wild soybean using conventional breeding methods. Asekova et al. [77] reported that three recombinant inbred lines with elevated ALA concentrations from an interspecific cross between *G. max* and *G. soja* were stable for the accumulation of ALA across the environments. Also, since *G. soja* as donor plant was backcrossed with three different cultivars, new genotypes with elevated ALA concentration and agronomically similar to the cultivated soybeans have been developed [78]. These soybean lines developed by classical breeding could be exploited as genetic resources for the development of novel soybean cultivars with high levels of ALA concentration, which could be sources of ω-3 fatty acids.

To date, there have been few genetic mapping studies with high ALA concentration in soybean. Shibata et al. [79] identified four QTLs controlling ALA concentration in the wild soybean accession Hidaka 4. Also, Ha et al. [80] identified nine putative QTLs controlling ALA concentration in a wild soybean accession PI 483463. According to these studies, high ALA concentrations in wild soybean were controlled by multiple QTLs. Besides, Pantalone et al. [81] suggested that high ALA concentration in wild soybean was controlled by a different set of desaturase alleles from cultivated soybean. Recently, the application of gamma-ray irradiation has generated new mutant soybeans with a high level of ALA concentration [82]. They concluded that the phenotype of high ALA concentration in these mutant lines was related to *FAD3* gene expression levels, although they observed no direct relationship between elevated gene expression level and gene sequence variations. Taken together, we assume that increased expression levels of *FAD3* genes during seed development may be associated with the gene expression regulators.

Since *FAD2* genes play an important role in regulating ω-6/ω-3 ratio in soybean, *FAD2* mutant alleles were found to increase in oleic acid and decrease in linoleic acid contents [40, 83, 84]. Populations segregating for *FAD2–1A* and *FAD2–1B*mutant alleles have been investigated for increases in oleic acid content [40]. By combining either of a mutant allele in M23 (the deletion of the *FAD2–1A* gene) or 17D (*FAD2–1A* S117N) with either of a missense mutant allele in *FAD2-1B* from PI 283327 (*FAD2-1B* P137R) or PI 567189A (*FAD2-1B* I143T), soybean genotypes with 77.3–82.2% oleic acid content were developed [84]. The ω-6/ω-3 ratio in these lines ranged from 0.6 to 1.3. Similarly, progenies containing *FAD2-1A* allele from the 17D lines, and *FAD2-1B* allele from S08-14788 were found to show an ω-6/ ω-3 ratio in the range of 0.62-0.97 [85]. These soybean genotypes had high oleic acid content (~80%) and lower ω-6/ω-3 ratio, but the overall ω-3 acid content (~5%) was also very low [86]. Kulkarni et al. [87] suggested the genetic improvement of the system to increase ALA concentration with a balanced ω-6/ω-3 ratio. Soybean containing either of the *FAD2-1* mutant alleles with ALA-related alleles from wild soybean reduced the seed ω-6/ω-3 ratio as well as increased ω-3 fatty acid concentration. Among *FAD2* genes, soybean genotypes with a mutant allele of the *FAD2-1A* gene had higher oleic acid and ALA content in soybean oil than one with *FAD2-1B* mutant allele. Further genetic improvements in the FA biosynthetic pathways were made by combining mutant alleles of either of *FAD2-1A* or *FAD2-1B* genes with alleles governing ALA in wild soybeans to develop soybean genotype with lower ω-6 and higher ω-3, resulting in low ω-6/ ω-3 ratio (Figure 3; [87]). Similar genetic improvements involving new sources of ALA-controlling alleles from the wild soybeans can guide development of soybeans with a balanced ω-6/ω-3 ratio in their seed oils.

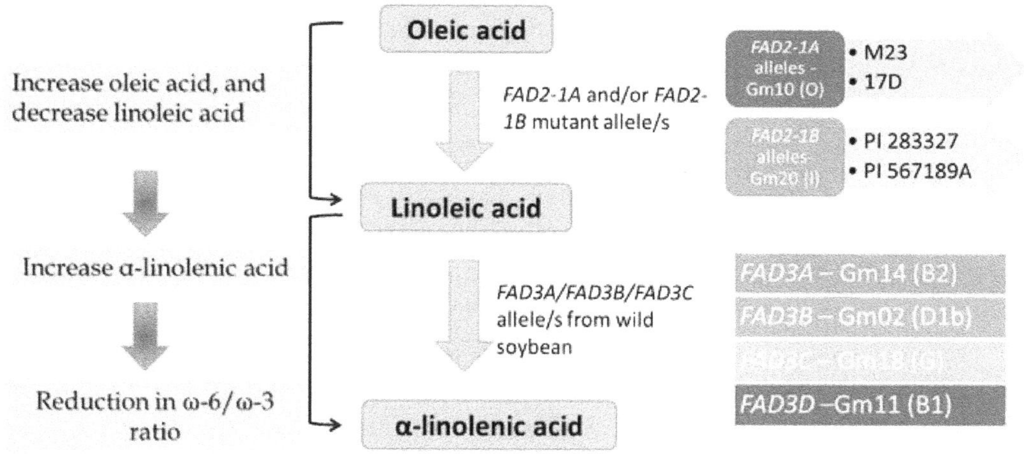

Figure 3. Schematic representation of the genetic improvements in the steps that result in significant reductions in ω-6/ω-3 ratio in soybean. The mutant alleles of FAD2-1A or FAD2-1B increase the oleic and reduce linoleic acid content, whereas, various alleles of FAD3 cause increases in the α-linolenic acid contents. These improvements significantly alter the ω-6/ω-3 ratio in soybean seed oil.

3. BIOTECHNOLOGICAL APPROACHES FOR IMPROVING THE FATTY ACID COMPOSITION

3.1 Transgenic soybeans with improved fatty acid profile

Soybean is widely recognized as a dual-use crop because of its high protein and oil content [29], and several loci controlling both the traits have been identified. The negative correlation between these two traits [88] pose a challenge in genetic improvement programs. Introducing a transgene that can specifically modulate one pathway without disrupting the other can be useful to overcome the linkage between oil and protein. Several transgenic approaches have been tried to improve seed oil content in oilseed crops, In Arabidopsis, transcription factor gene, *WRI1*, and metabolic enzyme, acetyl-CoA carboxylase have been targeted [89, 90]. In soybean, Lardizabal et al. [91] first reported the development of a transgenic soy crop with increased oil that shows no major impact on protein content or yield. They achieved an increase in oil by 1.5% (by weight) in the mature seed by expressing a codon-optimized version of a diacylglycerol acyltransferase (DGAT)-2A from the soil fungus *Umbelopsisramanniana* in soybean seed during development. Later, increased oil content of soybean seeds by an average of 3% was also reported with the use of an improved variant of soybean type 1 DGAT [92].

In recent years, RNA interference (RNAi) has gained significant attention due to its success for efficient metabolic engineering across the plant species. RNAi uses small interfering RNAs (siRNAs) to mediate the degradation of mRNA to regulate the expression of a desired plant gene. Using this approach, Flores et al. [93] showed that silencing of *GmFAD3* by siRNA caused a reduction in the ALA contents in *fad3*-mutant. A similar approach was used by Wagner et al. [94] for simultaneous suppression of soybean *FAD2* and *fatty acyl-ACP thioesterase (FATB)* genes to produce soybean seed with low-saturated, low-polyunsaturated oil phenotype.

Many studies in the recent past have demonstrated the role of *GmFAD2* family members in metabolically engineered oilseed plants. Using antisense RNA mediated posttranscriptional gene silencing approach, Zhang et al. [95] were successful in inhibiting the expression of *Gmfad2-1b* to develop transgenic soybean lines with increased oleic acid contents up to 51.7%. To simultaneously elevate stearic acid and reduce PUFA content in soybean, Park et al. [96] introduced the mangosteen (*Garcinia mangostana*) stearoyl-ACP thioesterase into soybean and subsequently stacked it with a soybean event that is down-regulated in both palmitoyl-ACP thioesterase activity and Δ12 fatty acid desaturase activity in a seed-specific fashion. This approach generated soybeans with a seed lipid phenotype of approximately 11–19% stearic acid and approximately 70% oleic acid. Recently, the introduction of the *PfFAD3-1* gene from Lesquerella (*Physariafendleri*) into soybean resulted in an increase in the ALA content up to 42% in the seeds of T_2 homozygous plants [97].

It is important to note that the transgenes expressing RNAi constructs are subject to variation in transgene expression, and hence a large number of events need to be screened to select the candidate providing stable expression. They also need to go through the regulation process, which is not only expensive but also time-consuming. Nevertheless, these approaches are expected to guide further improvement in the fatty acid composition without largely affecting the other traits, mainly the protein content and yield.

3.2 Targeted mutagenesis to improve ω-3 fatty acid contents

Targeted genome engineering (also known as genome editing) using designed nucleases has emerged as an alternative to conventional plant breeding and transgenic means to improve crop plants [98]. The discovery of sequence-specific nucleases (SSNs) such as TAL effector nucleases (TALENs) and CRISPR (clustered regularly interspaced short palindromic repeats)/Cas 9, made it possible to introduce targeted knockout mutations within gene/s of interest [99, 100]. These SSNs make DNA double-stranded breaks at defined genomic loci, which are subsequently repaired by two main DNA repair pathways, which result in frameshift mutations that often create genetic knockouts. Such knockout lines have been generated across the plant species, making genome editing an emerging tool for trait improvement.

Using genome editing approach, Haun et al. [101] engineered TALENs to recognize and cleave conserved DNA sequences in *FAD2-1A* and *FAD2-1B* genes. In the plants that carried homozygous mutations in both *FAD2-1A* and *FAD2-1B* genes, oleic acid was increased from 20% to 80% and linoleic acid was reduced from 50% to 4%. Further reduction in the linoleic acid (up to 2.5%) was achieved by stacking mutations within *FAD2-1A* and *FAD2-1B* with mutations in *FAD3A* [44]. Such an approach of TALENs-mediated targeted mutagenesis of *FAD2* was found to be effective in the development of the high oleic peanut (*Arachis hypogaea* L.) varieties [102]. The low to average mutagenic frequency by TALENs has been observed in the genome editing studies done so far in peanut and soybeans. The efficiency of genome editing can further be enhanced by using the CRISPR/Cas system.

In recent years, the CRISPR/Cas9 system has revolutionized functional genomics due to its simplicity, efficiency, cost-effectiveness, and versatility [103]. The CRISPR system has two components: a nuclear-localized CRISPR-associated (Cas) 9 protein and a guide RNA (gRNA). Cas9 is a large protein containing two nuclease domains, whereas the gRNA is a synthetic 100 nucleotide RNA molecule, of which the first ~20 nucleotides are the targeting site, and the 3′ end forms a hairpin structure that interacts with the Cas9 protein [104]. The Cas9 and the gRNA interact to identify DNA sequences complementary to the gRNA and generate a DNA double-strand break, which, after a repair result in genomic insertion or deletion (indel) mutations.

In plants, the CRISPR-Cas9 system has been effectively used in many species such as Arabidopsis thaliana, Nicotiana benthamiana, rice, tobacco, sorghum, wheat, and maize [105]. In soybean, CRISPR/Cas9-mediated genome editing has been successful in targeting DNA mutations in genes for soybean hairy roots and flowering [106, 107, 108, 109], plant architecture and yield [110], plant height [111], and seed storage protein genes [112]. However, researchers have started using this system to improve fatty acid composition. Do et al. [113], designed two gRNAs to guide Cas9 to simultaneously cleave two sites, spaced 1Kb apart, within the second exons of GmFAD2–1A and GmFAD2–1B to yield a high-oleic, low-linoleic, and low-ALA phenotype in soybean. In this study, dramatic increases in oleic acid content to >80%, and decreases of 1.3–1.7% in linoleic acid were observed in the T1 seeds derived from CRISPR-edited plants homozygous for both GmFAD2 genes. In a similar study, increases in oleic acid from 17.10% to 73.50%, and decreases in the linoleic acid content from 62.91% to 12.23% have been reported by inserting mutations in GmFAD2–1A and GmFAD2–2A soybean fatty acid desaturase mutants based on CRISPR/Cas9 Technology [114]. Overall, these studies demonstrated the CRISPR/Cas9 system as a rapid and highly efficient method to simultaneously edit homeologous soybean genes to facilitate gene discovery and breeding programs.

4. CONCISE SUMMARY

Altering the ω-6 and ω-3 fatty acid profile of the soybean seed/oil has been an important goal for soybean breeders. While low-ALA oils are better-suited for vegetable oil, genotypes with high ALA can be suited in food products that use whole soybeans in various fermented/non-fermented recipes. Therefore, breeding strategies according to the specific requirements are required. For these reasons, three major breeding strategies need be considered to achieve improvement in ω-3 fatty acid content in soybean. 1. To reduce ω-3 fatty acid for soybean oil, which is being achieved with the use of available several mutant lines with reduced ALA concentration in breeding programs. 2. To increase ω-3 fatty acid for soybean foods, which can be achieved by finding new alleles in wild soybeans, and introgressing such alleles in desired cultivars. However, there are many difficulties in this breeding process. Generating mutants with increased ω-3 fatty acid could be very crucial in achieving this goal. Wild soybeans [76] and some mutants [82] have relatively higher ω-3 fatty acid; however, there is still a lack of clarity and research information on the genes that regulate FAD3 genes. Therefore, studies investigating the regulators controlling ω-3 fatty acid in soybean need to be carried out. 3. To increase ω-3 fatty acid along with decreasing ω-6/ω-3 ratio, which can be achieved by combining mutant alleles of either of FAD2-1A or FAD2-1B genes with alleles (genes) governing elevated ALA in wild and cultivated soybeans. The success of these three strategies rely on the availability of genetic and genomic information governing ALA content, which at the moment is limited. Hence, an integrated approach (Figure 4) comprising genetic dissection, breeding, and biotechnological approaches is necessary to develop soybeans with desired fatty acid profile.

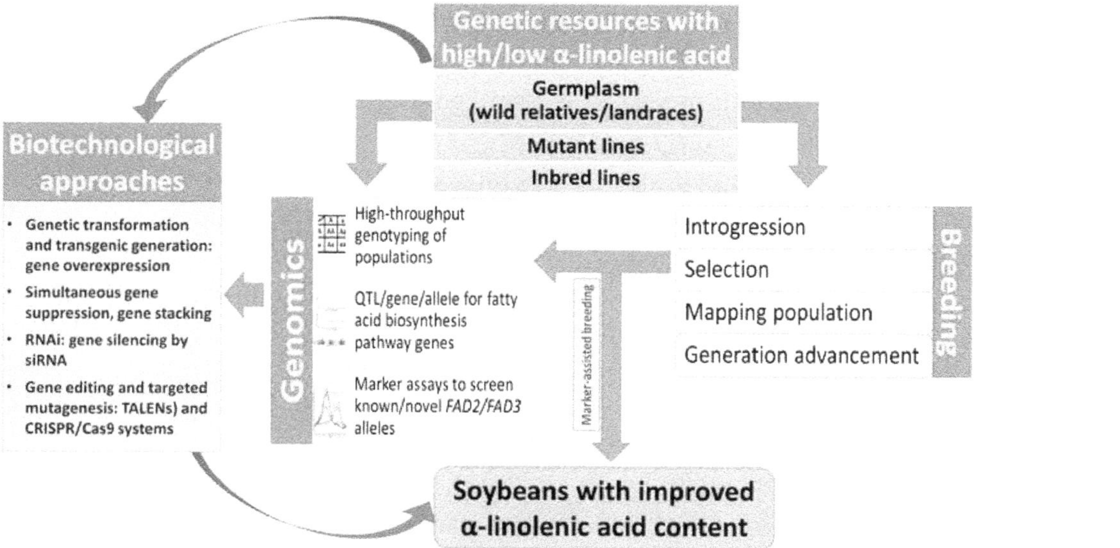

Figure 4. Schematic representation of the integrated approach involving genomic, biotechnological, and breeding approaches for the improving α-linolenic acid in soybean cultivars.

In last two decades, advances in the genomic and DNA sequencing technologies facilitated the genetic discovery of fatty acid biosynthesis in soybean and other oilseed crops [115]. It is now feasible to screen a large germplasm and mutant collections in quick time using high-density genotyping platforms (such as Axiom SoyaSNP array; [116]), and use the data for genetic and association mapping. Several wild and cultivated soybean genotypes with varied seed fatty acid contents are already known and have been used to develop improved cultivars. Also, many artificial mutant lines have been used in developing segregating mapping populations to identify novel alleles, for which genotyping assays have been developed and used for introgression of desired fatty acid trait in a soybean cultivar. Besides, the recent success of gene-editing technologies in targeting selected sites in the genes regulating fatty acid composition traits has shown the potential to selectively insert mutations in target genes. TALENs, and CRISPR/Cas9 has shown a great potential in soybean for many agronomic traits, and need to be exploited for improving the seed fatty acid composition.

AT A GLANCE

Plant-derived omega (ω)-3 polyunsaturated fatty acid is an essential fatty acid in human and animal diets and is a precursor of eicosapentaenoic acid and docosahexaenoic acid, which exists as α-linolenic acid (ALA, ω-3) in plant oil. Several epidemiological studies have revealed the health benefits of regular consumption of ω-3 fatty acid-containing diets. Soybean [Glycine max (L.) Merr.] is one of the major oil crops in the world and has around 8% ALA (ω-3) in seed oil. Soybean-derived ω-3 can be potential alternative sources of ω-3 fatty acids for populations living in countries with high risks of inadequate ω-3 intake. Therefore, increasing ω-3 concentration became an important goal in soybean breeding. Conversely, higher content of ω-3 fatty acids makes seed oil rancid, necessitating chemical hydrogenation, which generates trans fats. Since trans fats have been associated with the heart and other diseases, demand for soybeans with reduced ALA content is growing. In this book chapter, we described the importance of ω-3 fatty acid and consumption of diets with balanced ω-6/ω-3 ratio and discussed breeding and biotechnological means (and integrated approaches) for altering the ω-3 fatty acid content to avoid the need for chemical hydrogenation as well as to improve the ω-6/ω-3 ratio.

REFERENCES

1. SoyStats. The American Soybean Association. http://www.soystats.com. 2020
2. Liu K. Chemistry and nutritional value of soybean components. Soybeans: Springer; 1997. p. 25-113
3. Kim M. Identification of a new candidate gene controlling □-linolenic acid concentration in soybean. Kyungpook National University, Daegu; 2019
4. Drexler H, Spiekermann P, Meyer A, Domergue F, Zank T, Sperling P, et al. Metabolic engineering of fatty acids for breeding of new oilseed crops: Strategies, problems and first results. Journal of Plant Physiology. 2003;160(7):779-802
5. Voelker T, Kinney AJ. Variations in the biosynthesis of seed-storage lipids. Lipids. 2001;42:358-365
6. Guerra DJ, Holbrook L. Proceedings: world conference on biotechnology for the fats and oils industry. American Oil Chemists' Society; 1988. p. 364
7. He M, Qin CX, Wang X, Ding NZ. Plant Unsaturated Fatty Acids: Biosynthesis and Regulation. Frontiers in Plant Science. 2020;11(April):1-13
8. Ohlrogge J, Browse J. Lipid Biosynthesis. plant metabolism and biotechnology. 1995;7(July):957-970
9. Browse J, Warwick N, Somerville C, Slack C. Fluxes through the prokaryotic and eukaryotic pathways of lipid synthesis in the '16: 3' plant Arabidopsis thaliana. Biochemical Journal. 1986;235(1):25-31
10. Fehr WR. Breeding for modified fatty acid composition in soybean. Crop Science. 2007;47:S-72-S-87

11. Fernandez-Martinez J, Del Rio M, De Haro A. Survey of safflower (*Carthamus tinctorius* L.) germplasm for variants in fatty acid composition and other seed characters. Euphytica. 1993;69(1-2):115-122

12. Johnson R, Bergman J, Flynn C. Oil and meal characteristics of core and non-core safflower accessions from the USDA collection. Genetic Resources and Crop Evolution. 1999;46(6):611-618

13. Rebetzke GJ, Burton JW, Carter Jr TE, Wilson RF. Changes in agronomic and seed characteristics with selection for reduced palmitic acid content in soybean. Crop Science. 1998;38(2):297-302

14. Chatgilialoglu C, Ferreri C, Melchiorre M, Sansone A, Torreggiani A. Lipid geometrical isomerism: From chemistry to biology and diagnostics. Chemical Reviews. 2014;114(1):255-284

15. Wang J, Maldonado MA. The ubiquitin-proteasome system and its role in inflammatory and autoimmune diseases. Cellular & Molecular Immunology. 2006;3(4):255-261

16. Simopoulos AP. An increase in the omega-6/omega-3 fatty acid ratio increases the risk for obesity. Nutrients. 2016;8(3):1-17

17. Simopoulos AP. Omega-3 fatty acids in wild plants, nuts and seeds. Asia Pacific Journal of Clinical Nutrition. 2002;11(SUPPL. 6)

18. Crawford MA. Fatty acid ratios in free-living and domestic animals: Possible Implications for Atheroma. The Lancet. 1968;291(7556):1329-1333

19. Lee D-H, Kulkarni KP, Kim B-O, Seok YM, Song JT, Lee J-D. Comparative assessment of quality characteristics of Chungkookjang made from soybean seeds differing in oleic acid concentration. Journal of Functional Foods. 2019;52:529-536

20. Simopoulos AP. The importance of the omega-6/omega-3 fatty acid ratio in cardiovascular disease and other chronic diseases. Experimental Biology and Medicine. 2008;233(6):674-688

21. Del Gobbo LC, Imamura F, Aslibekyan S, Marklund M, Virtanen JK, Wennberg M, et al. ω-3 Polyunsaturated fatty acid biomarkers and coronary heart disease: Pooling project of 19 cohort studies. JAMA Internal Medicine. 2016;176(8):1155-1166

22. Mallick H, Franzosa EA, McLver LJ, Banerjee S, Sirota-Madi A, Kostic AD, et al. Predictive metabolomic profiling of microbial communities using amplicon or metagenomic sequences. Nature Communications. 2019;10(1):1-11

23. Shahidi F, Ambigaipalan P. Omega-3 polyunsaturated fatty acids and their health benefits. Annual Review of Food Science and Technology. 2018;9:345-381

24. Simopoulos AP. Evolutionary aspects of diet, the omega-6/omega-3 ratio and genetic variation: nutritional implications for chronic diseases. World Review of Nutrition and Dietetics. 2009;100:1-21

25. Kang J. Balance of omega-6/omega-3 essential fatty acids is important for health. World review of nutrition and dietetics. 2005;95(R):93

26. Kang ZB, Ge Y, Chen Z, Cluette-Brown J, Laposata M, Leaf A, et al. Adenoviral gene transfer of Caenorhabditis elegans n- 3 fatty acid desaturase optimizes fatty acid composition in mammalian cells. Proceedings of the National Academy of Sciences. 2001;98(7):4050-4054

27. Lai L, Kang JX, Li R, Wang J, Witt WT, Yong HY, et al. Generation of cloned transgenic pigs rich in omega-3 fatty acids. Nature Biotechnology. 2006;24(4):435-436

28. Ukropec J, Reseland JE, Gasperikova D, Demcakova E, Madsen L, Berge RK, et al. The hypotriglyceridemic effect of dietary n-3 FA is associated with increased β-oxidation and reduced leptin expression. Lipids. 2003;38(10):1023-1029

29. Clemente TE, Cahoon EB. Soybean oil: Genetic approaches for modification of functionality and total content. Plant physiology. 2009;151(3):1030-1040

30. Prabakaran M, Lee KJ, An Y, Kwon C, Kim S, Yang Y, et al. Changes in soybean (Glycine max L.) flour fatty-acid content based on storage temperature and duration. Molecules. 2018;23(10)

31. Reinprecht Y, Luk-Labey S-Y, Peter K. A versatile soybean recombinant inbred line population segregating for low linolenic acid and lipoxygenase nulls - molecular characterization and utility for soymilk and bioproduct production. Soybean - Molecular Aspects of Breeding. 2011

32. Wilson RF, Burton JW, Brim CA. Progress in the selection for altered fatty acid composition in soybeans. Crop Science. 1981;21(5):788-791

33. Liu HR, White PJ. Oxidative stability of soybean oils with altered fatty acid compositions. Journal of the American Oil Chemists Society. 1992;69(6):528-532

34. Mounts T, Warner K, List G, Kleiman R, Fehr W, Hammond E, et al. Effect of altered fatty acid composition on soybean oil stability. Journal of the American Oil Chemists' Society. 1988;65(4):624-628

35. Lee J-D, Bilyeu KD, Shannon JG. Genetics and breeding for modified fatty acid profile in soybean seed oil. Journal of Crop Science and Biotechnology. 2007;10:201-210

36. Yadav NS, Wierzbicki A, Aegerter M, Caster CS, Perez-Grau L, Kinney AJ, et al. Cloning of higher plant omega-3 fatty acid desaturases. Plant Physiology. 1993;103(2):467-476

37. Cardinal AJ. Molecular genetics and breeding for fatty acid manipulation in soybean. Plant Breeding Reviews. 2007;30:259-294

38. Singh SC, Sinha RP, Hader D-P. Role of lipids and fatty acids in stress tolerance in cyanobacteria. Acta protozoologica. 2002;41(4):297-308

39. Schlueter JA, Vasylenko-Sanders IF, Deshpande S, Yi J, Siegfried M, Roe BA, et al. The FAD2 gene family of soybean: insights into the structural and functional divergence of a paleopolyploid genome. Crop Science. 2007;47:S-14-S-26

40. Pham A-T, Lee J-D, Shannon JG, Bilyeu KD. Mutant alleles of FAD2-1A and FAD2-1B combine to produce soybeans with the high oleic acid seed oil trait. BMC Plant Biology. 2010;10(1):195

41. Lakhssassi N, Zhou Z, Liu S, Colantonio V, AbuGhazaleh A, Meksem K. Characterization of the FAD2 gene family in soybean reveals the limitations of gel-based TILLING in genes with high copy number. Frontiers in Plant Science. 2017;8:324

42. Bilyeu K, Palavalli L, Sleper D, Beuselinck P. Three microsomal omega-3 fatty-acid desaturase genes contribute to soybean linolenic acid levels. Crop Science. 2003;43(5):1833-1838

43. Anai T, Yamada T, Kinoshita T, Rahman SM, Takagi Y. Identification of corresponding genes for three low-α-linolenic acid mutants and elucidation of their contribution to fatty acid biosynthesis in soybean seed. Plant Science. 2005;168(6):1615-1623

44. Demorest ZL, Coffman A, Baltes NJ, Stoddard TJ, Clasen BM, Luo S, et al. Direct stacking of sequence-specific nuclease-induced mutations to produce high oleic and low linolenic soybean oil. BMC plant biology. 2016;16(1):1-8

45. Bilyeu K, Gillman JD, LeRoy AR. Novel FAD3 mutant allele combinations produce soybeans containing 1% linolenic acid in the seed oil. Crop Science. 2011;51(1):259-264

46. Bilyeu K, Palavalli L, Sleper D, Beuselinck P. Mutations in soybean microsomal omega-3 fatty acid desaturase genes reduce linolenic acid concentration in soybean seeds. Crop Science. 2005;45(5):1830-1836

47. Bilyeu K, Palavalli L, Sleper DA, Beuselinck P. Molecular genetic resources for development of 1% linolenic acid soybeans. Crop Science. 2006;46(5):1913-1918

48. Reinprecht Y, Luk-Labey SY, Larsen J, Poysa VW, Yu K, Rajcan I, et al. Molecular basis of the low linolenic acid trait in soybean EMS mutant line RG10. Plant Breeding. 2009;128(3):253-258

49. Reinprecht Y, Pauls KP. Microsomal omega-3 fatty acid desaturase genes in low linolenic acid soybean line RG10 and validation of major linolenic acid QTL. Frontiers in Genetics. 2016;7(MAR):1-16

50. White HB, Quackenbush FW, Probst AH. Occurrence and inheritance of linolenic and linoleic acids in soybean seeds. Journal of the American Oil Chemists Society. 1961;38(3):113-117

51. Kim M, Song JT, Bilyeu KD, Lee JD. A new low linolenic acid allele of GmFAD3A gene in soybean PE1690. Molecular Breeding. 2015;35(8):1-6

52. Lee J-D, Kim M, Kulkarni KP, Song JT. Agronomic traits and fatty acid composition of high–oleic acid cultivar Hosim. Plant Breeding and Biotechnology. 2018;6(1):44-50

53. Yamada T, Takagi K, Ishimoto M. Recent advances in soybean transformation and their application to molecular breeding and genomic analysis. Breeding Science. 2011;61(5):480-494

54. Rahman SM, Takagi Y, Kumamaru T. Low linolenate sources at the Fan locus in soybean lines M-5 and IL-8. Japanese Journal of Breeding. 1996;46(2):155-158

55. Wilcox J, Cavins J. Inheritance of low linolenic acid content of the seed oil of a mutant in Glycine max. Theoretical and Applied Genetics. 1985;71(1):74-78

56. Wilcox J, Cavins J, Nielsen N. Genetic alteration of soybean oil composition by a chemical mutagen. Journal of the American Oil Chemists' Society. 1984;61(1):97-100

57. Rennie BD, Tanner JW. Fatty acid composition of oil from soybean leaves grown at extreme temperatures. Journal of the American Oil Chemists Society. 1991;68(2):1622-1624

58. Fehr WR, Welke GA, Hammond EG, Duvick DN, Cianzio SR. Inheritance of reduced linolenic acid content in soybean genotypes A16 and A17. Crop Science. 1992;32(4):903-906

59. Rahman SM, Kinoshita T, Anai T, Arima S, Takagi Y. Genetic relationships of soybean mutants for different linolenic acid contents. Crop Science. 1998;38(3):702-706

60. Stijšin D, Luzzi BM, Ablett GR, Tanner JW. Inheritance of low linolenic acid level in the soybean line RG10. Crop Science. 1998;38(6):1441-1444

61. Howell RW, Brim CA, Rinne RW. The plant geneticist's contribution toward changing lipid and amino acid composition of soybeans. Journal of the American Oil Chemists' Society. 1972;49(1):30-32

62. Rennie B, Zilka J, Cramer M, Beversdorf W. Genetic analysis of low linolenic acid levels in the soybean line PI 361088B. Crop Science. 1988;28(4):655-657

63. Rennie BD, Tanner JW. Mapping a second fatty acid locus to soybean linkage group 17. Crop Science. 1989;29(4):1081-1083

64. Rahman S, Takagi Y. Inheritance of reduced linolenic acid content in soybean seed oil. Theoretical and Applied Genetics. 1997;94(3-4):299-302

65. Wasala SK, T. Kinoshita, S. Arima and Y. Takagi. Agronomic performances of low linolenic acid soybean mutant lines developed from cultivar Bay. Bulletin of Faculty of Agriculture, Saga University. 1997;82:29-36

66. Rahman SM, Kinoshita T, Anai T, Takagi Y. Combining ability in loci for high oleic and low linolenic acids in soybean. Crop Science. 2001;41(1):26-29

67. Ross AJ, Fehr WR, Welke GA, Cianzio SR. Agronomic and seed traits of 1%-linolenate soybean genotypes. Crop Science. 2000;40(2):383-386

68. Chae J-H, Dhakal KH, Asekova S, Song JT, Lee J-D. Variation of Fatty Acid Composition in Soybean 'Pungsannamul' Mutation Population from EMS Treatment. Current Research on Agriculture and Life Science. 2013;31(1):45-50

69. Held JP, Carrero-Colón M, Hudson KA. Combination of novel mutation in FAD3C and FAD3A for low linolenic acid soybean. Agrosystems, Geosciences & Environment. 2019;2(1):1-4

70. Thapa R, Carrero-Colón M, Addo-Quaye C, Held J, Dilkes B, Hudson KA. New alleles of FAD3A lower the linolenic acid content of soybean seeds. Crop Science. 2018;58(2):713-8

71. Burton J, Wilson R, Novitzky W, Carter T. Registration of 'Soyola' soybean. Crop Science. 2004;44(2):687-a-8

72. Hammond EG, Fehr WR. Oil Quality Improvement in Soybeans-Glycine max (L.) Merr. Fette SeifenAnstrichmittel. 1975;77(3):97-101

73. Reinprecht Y, Poysa VW, Yu K, Rajcan I, Ablett GR, Pauls KP. Seed and agronomic QTL in low linolenic acid, lipoxygenase-free soybean (Glycine max (L.) Merrill) germplasm. Genome. 2006;49(12):1510-1527

74. Watanabe T. Science of Tofu. Food Journal Co. Ltd, Kyoto, Japan. 1997

75. Dhakal KH, Lee J-D, Jeong Y-S, Kim H-S, Shannon JG, Hwang Y-H. Stability of linolenic acid in seed oil of soybean accessions with elevated linolenic acid concentration. The Journal of Food, Agriculture and Environment. 2013;11:80-85

76. Chae JH, Ha BK, Chung G, Park JE, Park E, Ko JM, et al. Identification of environmentally stable wild soybean genotypes with high alpha-linolenic acid concentration. Crop Science. 2015;55(4):1629-1636

77. Asekova S, Chae J-H, Ha B-K, Dhakal KH, Chung G, Shannon J, et al. Stability of elevated α-linolenic acid derived from wild soybean (Glycine soja Sieb. &Zucc.) across environments. Euphytica. 2014;195(3):409-418

78. Jo H, Kim M, Ali L, Tayade R, Jo D, Le DT, et al. Environmental stability of elevated α-linolenic acid derived from a wild soybean in three Asian countries. Agriculture. 2020;10(3):70

79. Shibata M, Takayama K, Ujiie A, Yamada T, Abe J, Kitamura K. Genetic relationship between lipid content and linolenic acid concentration in soybean seeds. Breeding Science. 2008;58(4):361-366

80. Ha B-K, Kim H-J, Velusamy V, Vuong TD, Nguyen HT, Shannon JG, et al. Identification of quantitative trait loci controlling linolenic acid concentration in PI483463 (Glycine soja). Theoretical and applied genetics. 2014;127(7):1501-1512

81. Pantalone V, Rebetzke G, Burton J, Wilson R. Genetic regulation of linolenic acid concentration in wild soybean Glycine soja accessions. Journal of the American Oil Chemists' Society. 1997;74(2):159-163

82. Hong MJ, Jang YE, Kim DG, Kim JM, Lee MK, Kim JB, et al. Selection of mutants with high linolenic acid contents and characterization of fatty acid desaturase 2 and 3 genes during seed development in soybean (Glycine max). Journal of the Science of Food and Agriculture. 2019;99(12):5384-5391

83. Hoshino T, Takagi Y, Anai T. Novel GmFAD2-1b mutant alleles created by reverse genetics induce marked elevation of oleic acid content in soybean seeds in combination with GmFAD2-1a mutant alleles. Breeding science. 2010;60(4):419-425

84. Pham A-T, Lee J-D, Shannon JG, Bilyeu KD. A novel FAD2-1A allele in a soybean plant introduction offers an alternate means to produce soybean seed oil with 85% oleic acid content. Theoretical and Applied Genetics. 2011;123(5):793-802

85. Kulkarni KP, Patil G, Valliyodan B, Vuong TD, Shannon JG, Nguyen HT, et al. Comparative genome analysis to identify SNPs associated with high oleic acid and elevated protein content in soybean. Genome. 2018a;61(3):217-222

86. Pham AT, Shannon JG, Bilyeu KD. Combinations of mutant FAD2 and FAD3 genes to produce high oleic acid and low linolenic acid soybean oil. Theoretical and Applied Genetics. 2012;125(3):503-515

87. Kulkarni KP, Kim M, Song JT, Bilyeu KD, Lee J-D. Genetic improvement of the fatty acid biosynthesis system to alter the ω-6/ω-3 ratio in the soybean seed. Journal of the American Oil Chemists' Society. 2017;94(11):1403-1410

88. Patil G, Mian R, Vuong T, Pantalone V, Song Q, Chen P, et al. Molecular mapping and genomics of soybean seed protein: a review and perspective for the future. Theoretical and Applied Genetics. 2017;130(10):1975-1991

89. Baud S, Mendoza MS, To A, Harscoët E, Lepiniec L, Dubreucq B. WRINKLED1 specifies the regulatory action of LEAFY COTYLEDON2 towards fatty acid metabolism during seed maturation in Arabidopsis. The Plant Journal. 2007;50(5):825-838

90. Cernac A, Benning C. WRINKLED1 encodes an AP2/EREB domain protein involved in the control of storage compound biosynthesis in Arabidopsis. The Plant Journal. 2004;40(4):575-585

91. Lardizabal K, Effertz R, Levering C, Mai J, Pedroso M, Jury T, et al. Expression of Umbelopsisramanniana DGAT2A in seed increases oil in soybean. Plant Physiology. 2008;148(1):89-96

92. Roesler K, Shen B, Bermudez E, Li C, Hunt J, Damude HG, et al. An improved variant of soybean type 1 diacylglycerol acyltransferase increases the oil content and decreases the soluble carbohydrate content of soybeans. Plant Physiology. 2016;171(2):878-893

93. Flores T, Karpova O, Su X, Zeng P, Bilyeu K, Sleper DA, et al. Silencing of GmFAD3 gene by siRNA leads to low α-linolenic acids (18: 3) of fad3-mutant phenotype in soybean [Glycine max (Merr.)]. Transgenic research. 2008;17(5):839-850

94. Wagner N, Mroczka A, Roberts PD, Schreckengost W, Voelker T. RNAi trigger fragment truncation attenuates soybean FAD2-1 transcript suppression and yields intermediate oil phenotypes. Plant Biotechnology Journal. 2011;9(7):723-728

95. Zhang L, Yang X-d, Zhang Y-y, Yang J, Qi G-x, Guo D-q, et al. Changes in oleic acid content of transgenic soybeans by antisense RNA mediated posttranscriptional gene silencing. International Journal of Genomics. 2014;2014

96. Park H, Graef G, Xu Y, Tenopir P, Clemente TE. Stacking of a stearoyl-ACP thioesterase with a dual-silenced palmitoyl-ACP thioesterase andΔ 12 fatty acid desaturase in transgenic soybean. Plant Biotechnology Journal. 2014;12(8):1035-1043

97. Yeom WW, Kim HJ, Lee K-R, Cho HS, Kim J-Y, Jung HW, et al. Increased production of α-linolenic acid in soybean seeds by overexpression of Lesquerella FAD3-1. Frontiers in Plant Science. 2020;10:1812

98. Belhaj K, Chaparro-Garcia A, Kamoun S, Nekrasov V. Plant genome editing made easy: targeted mutagenesis in model and crop plants using the CRISPR/Cas system. Plant Methods. 2013;9(1):1-10

99. Lee J, Chung JH, Kim HM, Kim DW, Kim H. Designed nucleases for targeted genome editing. Plant Biotechnology Journal. 2016;14(2):448-462

100. Malzahn A, Lowder L, Qi Y. Plant genome editing with TALEN and CRISPR. Cell & Bioscience. 2017;7(1):21

101. Haun W, Coffman A, Clasen BM, Demorest ZL, Lowy A, Ray E, et al. Improved soybean oil quality by targeted mutagenesis of the fatty acid desaturase 2 gene family. Plant Biotechnology Journal. 2014;12(7):934-940

102. Wen S, Liu H, Li X, Chen X, Hong Y, Li H, et al. TALEN-mediated targeted mutagenesis of fatty acid desaturase 2 (FAD2) in peanut (Arachis hypogaea L.) promotes the accumulation of oleic acid. Plant Molecular Biology. 2018;97(1-2):177-185

103. Chen K, Wang Y, Zhang R, Zhang H, Gao C. CRISPR/Cas genome editing and precision plant breeding in agriculture. Annual review of plant biology. 2019;70:667-697

104. Jinek M, Chylinski K, Fonfara I, Hauer M, Doudna JA, Charpentier E. A programmable dual-RNA–guided DNA endonuclease in adaptive bacterial immunity. science. 2012;337(6096):816-821

105. Sun X, Hu Z, Chen R, Jiang Q, Song G, Zhang H, et al. Targeted mutagenesis in soybean using the CRISPR-Cas9 system. Scientific Reports. 2015;5(1):1-10

106. Cai Y, Chen L, Liu X, Sun S, Wu C, Jiang B, et al. CRISPR/Cas9-mediated genome editing in soybean hairy roots. PLoS One. 2015;10(8):e0136064

107. Cai Y, Chen L, Sun S, Wu C, Yao W, Jiang B, et al. CRISPR/Cas9-mediated deletion of large genomic fragments in soybean. International journal of molecular sciences. 2018;19(12):3835

108. Han J, Guo B, Guo Y, Zhang B, Wang X, Qiu L. Creation of Early Flowering Germplasm of Soybean by CRISPR/Cas9 Technology. Frontiers in Plant Science. 2019;10:1446

109. Jacobs TB, LaFayette PR, Schmitz RJ, Parrott WA. Targeted genome modifications in soybean with CRISPR/Cas9. BMC biotechnology. 2015;15(1):1-10

110. Bao A, Chen H, Chen L, Chen S, Hao Q, Guo W, et al. CRISPR/Cas9-mediated targeted mutagenesis of GmSPL9 genes alters plant architecture in soybean. BMC plant biology. 2019;19(1):1-12

111. Cheng Q, Dong L, Su T, Li T, Gan Z, Nan H, et al. CRISPR/Cas9-mediated targeted mutagenesis of GmLHY genes alters plant height and internode length in soybean. BMC plant biology. 2019;19(1):1-11

112. Li C, Nguyen V, Liu J, Fu W, Chen C, Yu K, et al. Mutagenesis of seed storage protein genes in Soybean using CRISPR/Cas9. BMC Research Notes. 2019;12(1):176

113. Do PT, Nguyen CX, Bui HT, Tran LT, Stacey G, Gillman JD, et al. Demonstration of highly efficient dual gRNA CRISPR/Cas9 editing of the homeologous GmFAD2-1A and GmFAD2-1B genes to yield a high oleic, low linoleic and α-linolenic acid phenotype in soybean. BMC plant biology. 2019;19(1):311

114. Wu N, Lu Q, Wang P, Zhang Q, Zhang J, Qu J, et al. Construction and analysis of GmFAD2-1A and GmFAD2-2A soybean fatty acid desaturase mutants based on CRISPR/Cas9 technology. International Journal of Molecular Sciences. 2020;21(3):1104

115. Kulkarni KP, Tayade R, Asekova S, Song JT, Shannon JG, Lee J-D. Harnessing the potential of forage legumes, alfalfa, soybean, and cowpea for sustainable agriculture and global food security. Frontiers in Plant Science. 2018b;9:1314

116. Lee YG, Jeong N, Kim JH, Lee K, Kim KH, Pirani A, et al. Development, validation and genetic analysis of a large soybean SNP genotyping array. The Plant Journal. 2015;81(4):625-636

SECTION-2

BREEDING FOR STRESS CONDITIONS

CHAPTER-5

SMART BREEDING STRATEGIES TO CLIMATE RESILIENT AGRICULTURE

1. INTRODUCTION: AN OVERVIEW

In response to the escalating challenges posed by climate change on global agriculture, the imperative to develop innovative and adaptive strategies has become increasingly evident. Among these strategies, Smart Breeding emerges as a pivotal approach aimed at enhancing the resilience of crops to the dynamic and often unpredictable impacts of climate change. The amalgamation of advanced technologies, data analytics, and traditional breeding practices under the umbrella of Smart Breeding marks a paradigm shift in agricultural research and development. This transformative methodology seeks not only to address current agricultural challenges but also to proactively prepare crops for the evolving climate scenarios, thereby fostering sustainable and climate-resilient agriculture.

Land is shrinking but world population is increasing in a rapid phase, so, modern agricultural practice is struggling to meet the level of primary productivity required to feed approximately 10 billion people by 2050 [1]. From last few decades the adverse effects of climate change and higher CO_2 concentrations, the consequence of expected impacts on the water-use efficiency of dryland as well as irrigated crop production, potential effects on biosecurity, production, and quality of product through increased the frequency of introduced various abiotic (heat, salinity and drought) and biotic stresses (pests and diseases). In addition, climate change is also expected to cause losses of biodiversity, mainly in more marginal environments. Drought alone is expected to reduce crop productivity in half of the global arable land and it's estimated around 50% in the next five decades [2]. It has been predicted that, on average, global yields of major economic important crops will be reduced by the unfavorable climatic conditions in wheat (6.0%), rice (3.2%), maize (7.4%) and soybean (3.1%) for every degree celsius increase in global mean temperature [3].

Climate resilience is an ability of the plant/crop to survive and recover from the effects of climate change. Some important practices that may help to adapt the climate change are soil organic carbon build up or carbon sequestration, in-situ moisture conservation, residue incorporation instead of burning, water harvesting and recycling for supplemental irrigation, growing biotic and abiotic resistance/tolerant varieties, location specific agronomic and nutrient management and breeding for multiple traits of interest including quality.

Plant Breeding has always played a pivotal role in human history from revolutionizing agriculture to feed the ever-growing population. The key role of plant breeding in agriculture is to develop a genetically superior genotype/variety, which is suitable for a specific as well as general cultivation of particular environment towards higher production [4]. Realizing the importance of genomic resources to expedite the breeding programs, huge amount of genetic data related to genes and QTLs (Quantitative Trait Loci) are generated after the advent of molecular biology and biotechnology [5]. The progress in precise phenotyping and genotyping offers tremendous opportunities to develop crop varieties that are suit for better changing the climatic conditions, which ameliorate in boosting the plant breeding activities for developing climate resilient varieties/cultivars [6]. Hence, development of climate resilient varieties utilizing Smart breeding tools to ensures the food security in adverse climatic conditions.

2. EFFECT OF CLIMATE CHANGE ON AGRICULTURE AND FOOD PRODUCTION

The effect of climate on agriculture is related to variability's in local climates rather than in global climate patterns. The changes in the rainfall patterns, temperature, CO_2 level and greenhouse gases resulting in the frequency and severity of extreme events such as flooding, drought, hail, and hurricanes etc. are major hindrance in achieving the food security for ever increasing population [7].

According to Intergovernmental Panel on Climate Change (IPCC), global temperature may be rise from 1.7 to 4.8°C during the twenty-first century and precipitation pattern will also be altered [8]. In recent times, it has been reported that the Yangtze river basin in China has become hotter and it is expected that the temperature will increase up to 2°C by 2050 relative to 1950 [9], and also reduce the rice (41%) and maize (50%) production by the end of the 21st century. This shift in climate will affect the environment, including the soil ecology and thus has the potential to threaten food security through its adverse effects on soil properties and processes [10]. Additionally, the direct and indirect effects of climatic change would lead to alter the nutrient and their bioavailability in soils (Figure 1). The effect of climate changes on biotic and abiotic stresses have already reduced the global agricultural production from 1 to 5% during the past three decades [11].

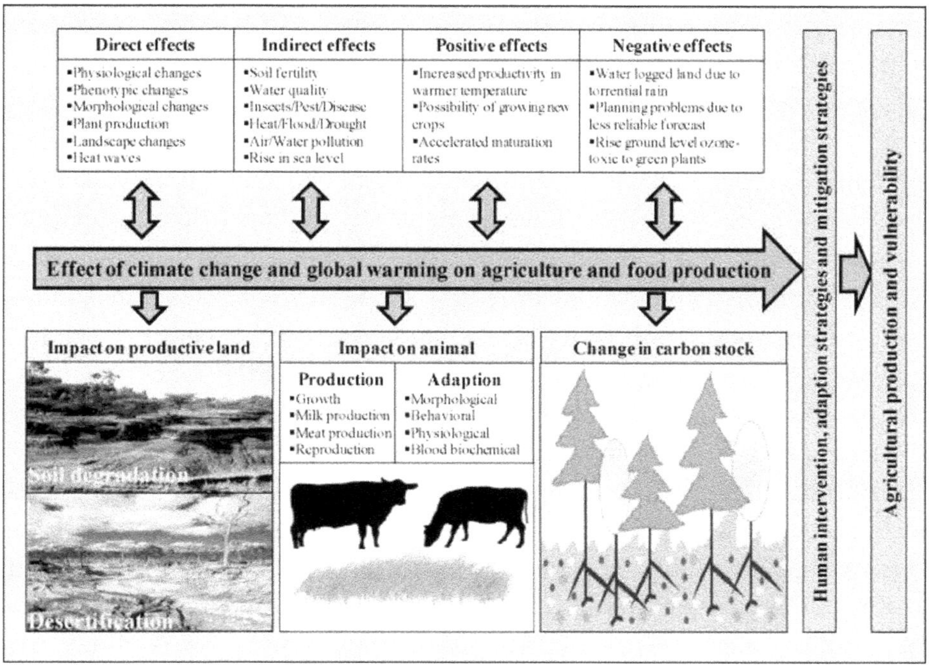

Figure 1. Adverse effects of climate change on agriculture and food production.

3. TECHNOLOGICAL APPROACHES TOWARDS CLIMATE RESILIENT AGRICULTURE

Some important practices that assist to adapt the climate changes for crop production including (i) Building resilience in soil (tillage management, avoid bare soil, fertilizer application after mandatory soil testing, increase soil carbon through organic manure, green manuring, crop rotation or intercropping with legume sequester carbon and biochar), (ii) Adapted cultivars and cropping systems (crop diversification, shallow-deep root and legume-cereal cropping system, improved early/short duration cultivars for tolerant against drought, heat and submergence capturing optimum yields despite climatic stresses), (iii) Rainwater harvesting and recycling (inter-row water harvesting, inter-plot water harvesting, in farm ponds and reservoirs and recycling), (iv) Farm machinery (chisel and para plow to opening the furrows which conserves rain water, laser leveler helps in increasing nutrient as well as water use efficiency), (v) Crop contingency plans (livestock and fishery interventions), (vi) Weather based agro advisories (automatic weather stations establishment at experimental farms and mini-weather observatories records for real time weather parameters such as rainfall, temperature and wind speed, which customized through agro advisories and improve weather literacy among the farmers).

Plant breeding procedures have been constantly evolving to meet the increasing food demand. The art of plant breeding has been practiced in various forms since the start of human civilization. In conventional plant breeding, development of a new cultivar take around 10–14 years and may even exceed this period based on the plant habit, reproductive cycle and complexity of traits involved. The rapid climate change necessitates the development of varieties in a shorter period to tackle with the unpredictable weather parameters. The concept of Smart breeding is an integration of conventional breeding strategies with advanced molecular, genomic and phenomic tools to efficiently and effectively breed the resilient crop cultivars with enhanced yield potential. New breeding approaches such as rapid generation advancement, doubled haploid (DH), marker assisted back crossing (MABC), marker assisted recurrent selection (MARS), genomic selection (GS) etc. have been used to help shorten the breeding cycle along with efficient screening for specific biotic and abiotic stresses. Biotechnology-based breeding technologies (marker-assisted breeding and genetic modifications) will be essential to assist and accelerate genetic gain, but their application requires additional investment in the understanding, genetic characterization and phenotyping for complex adaptive traits to be exploited for climate resilient breeding.

4. PRE-BREEDING AND CROP WILD RELATIVES (CWR): EXPLORING UNTAPPED HUB OF GENETIC DIVERSITY

Climate change leading to severe weather fluctuations would also lead to evolution of plant diseases and pests, exposing crops to higher biotic pressure in addition to abiotic stresses. To make crop adaptation feasible in the era of changing climate, there is indispensible need to breed the crop plants with diverse genetic backgrounds. In order to feed the mushrooming population, there is urgent need to use crop wild relatives for developing broader spectrum varieties to tackle various biotic and abiotic stresses. During the era of domestication, selection preferences lead to modern crops with narrow genetic background, resulting in limitation of environmental adaptation and breeding capacity using modern germplasm [12]. Wild relatives and ancestral species relatively possess broader adaptation to environment and climates ultimately higher potential in crop improvement.

Prebreeding activity is a bridge for linking the desirable traits of CWR to the modern cultivar development by providing breeders with wild genetic diversity in a more immediately usable form [13, 14]. Pre-breeding is an opportunity to introgression of desirable genes, from wild species (primary, secondary and tertiary gene pools) into elite breeding lines/cultivars/genotypes, to overcome the linkage drag (Figure 2). Almost all cultivated crop species were originally domesticated from wild plants by humans, due to domestication inherently reduced the genetic variation [15]. The genetic potential of wild relatives has been reported in different crops like rice, wheat, maize, potato, tomato, cotton, tobacco, sugarcane, chickpea and pigeonpea [16, 17, 18, 19, 20, 21].Genomics strategies have been widely utilized in staple crops for transferring major genes (i.e. disease resistance) from wild germplasm to elite cultivars [22]. It is well documented that application of molecular mapping and sequencing to could be useful to unlock the genetic potential of CWR [23]. So, crop wild relatives (CWRs) are good reservoir of untapped genetic diversity, which may not exist in the cultivated gene pool that can be used to improve the numerous trait of interest including resistance/tolerance against diseases, insect-pests, drought, salinity, cold, heat and good agronomic adaption with quality improvement.

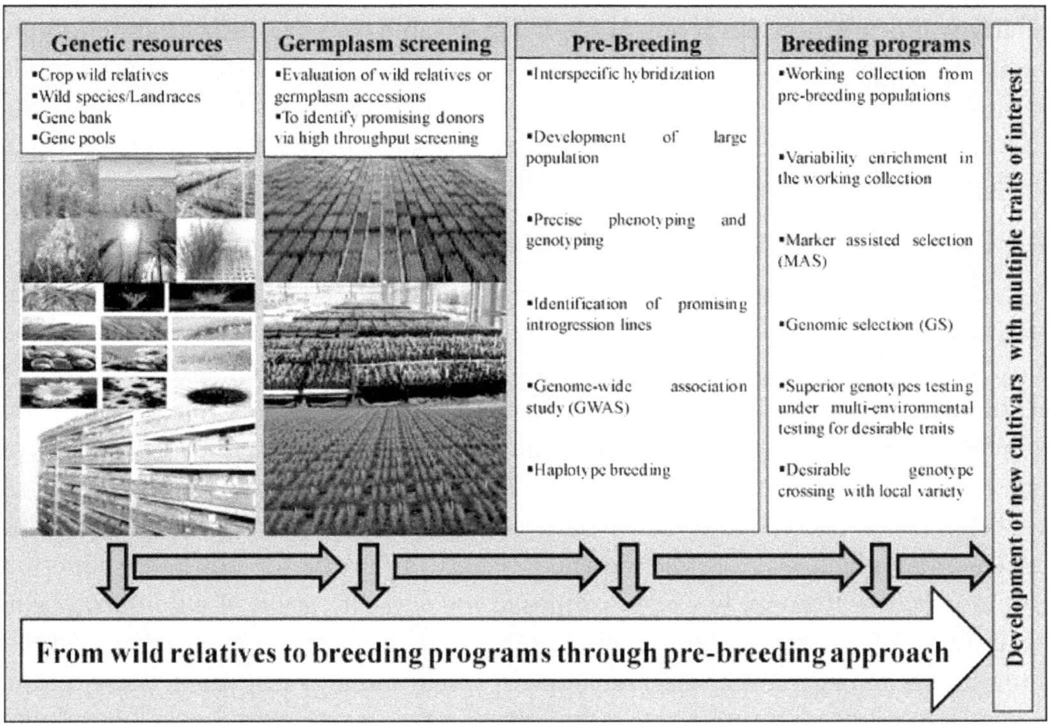

Figure 2. Untapped genetic resources/ CWRs towards the germplasm enhancement.

4.1 Exotic introgression into elite varieties

Wild species are used mainly for the introgression of disease and insect resistance into crops although drought, cold, heat and salinity tolerance have also been addressed in some staple crops. This is because most pathogens have faster adaptation to climate rendering cultivars vulnerable to novel deadly diseases [24]. The use of interspecific or intergeneric hybridization for disease resistance introgression is conventional one. Another potential technique to enhance genetic diversity and facilitate crop vigor with adaptation to different environmental niches is creating the polyploidy crops mimicking natural evolution through hybridization [25]. Enriched genes for biotic and abiotic stress resistance of CWR can be studied using comparative pool sequencing of genome assemblies, elucidating the potential genomic segments responsible for adaptation to different ecological niches. These have been explored in wild relatives of many crops including chickpea, barley and maize [26, 27, 28, 29].

To address the diversity within species, pan-genomics based on entire gene repository of a species can reveal the genetic variations such as structure variants (SVs) and single nucleotide polymorphism (SNPs) abundantly found in plants. One such example under SVs is presence/absence variants (PAVs) of Elicitin response (ELR) gene between wild and cultivated potato leads to resistance/susceptibility response to late blight disease [30]. Larger pan-genomes including both wild relatives and cultivars can acquire glut Of dispensable genes resulting in phenotypic variations; thereby easing out with characterization of the trait associated genomic variants [31]. To tackle the deadly rust diseases in wheat in the context of changing climate, several pan-genomic R genes have been successfully identified and cloned from wild diploid wheat *Aegilops tauschii* [32].

4.2 De novo domestication for future crops

Considering the risks of introducing foreign alleles into cultivars, other potential technique for developing climate-friendly crops is de novo domestication [33]. As most staple crops are grown majorly in the regions other than where they were originally domesticated with different climatic regimes. Nevertheless, their wild relatives and landraces exhibit better adaptation to local climate in the native regions. In the scenario of climatic change, there is chance to leverage this opportunity to use those underutilized or orphan crops e.g. rise in *Sinapis alba* (white mustard) acreage replacing the *B. napus* in Europe for biofuel production [34]. A pipeline strategy has been proposed for domestication of wild germplasm in some orphan crops such as quinoa [35]. In addition to direct planting of non-domesticated crop plants, relatively advance methodology of CRISPR/Cas9 boosts the wild germplasm domestication by editing of domesticated genes e.g. editing in wild tomatoes (*Solanum pimpinellifolium*) and ground cherry (*Physalis pruinosa*) mainly focused on flower improvement, plant architecture improvement, fruit size, fruit number and nutritional content [36, 37, 38]. It is evident from such a few successful introgressions of domesticated genes that use of wild germplasm in regular plant breeding is quite promising in countering the effects of climate change on agriculture and hence, food security.

4.3 Introgressiomics approach for adaptation to climate change

The actual potential of the CWR in plant breeding largely remains underexploited due to linkage drag and frequent breeding barriers with the crops. Introgressiomics approach allows mass scale development of plant material and populations with introgression lines from CWR into the genetic background of crops [39]. This pre-emptive breeding technique could be focused or unfocused depending upon the objective. Besides genetic analysis of traits present in CWR, MAS driven generation of chromosome substitution lines (CSL), introgression lines (IL) or MAGIC populations allow the development of genetically characterized elite material. Genomic tools like high throughput molecular markers facilitate the characterization and development of Introgressiomics populations, which can be easily incorporated into major breeding programs for coping with the accelerating environmental challenges.

4.4 Some other techniques for CWR use

After the introgression into domesticated background from CWR, populations such as backcross populations (BC), recombinant inbred lines (RILs), doubled haploids (DH), near isogenic lines (NILs), multiparent advance generation intercross (MAGIC) populations as well as nested association mapping (NAM) populations are developed to study the introgressed gene(s). After mapping their locations on to the genome and it genotypic validation with molecular markers, they are further deployed using Marker assisted selection (MAS). Systematic screening of the huge number of progenies with MAS enhances the efficiency of breeding program (van de Weil 2010). Desirable recombinants can be developed at early generations using larger populations e.g. using marker-assisted backcrossing (MABC), an important QTL was introduced into a new lowland rice background in just 2 rounds of backcrossing [40].

Genomic scans can also reveal candidate domestication and improvement loci as well as post-domestication introgression using CWR [41, 42] to be further harnessed in the scenario of climatic challenges. In case of CWR, high throughput sequencing offers a cheap and rapid way to deploy thousands to millions of markers for mapping purposes [43]. Reduced representation techniques as genotyping by sequencing (GBS) or even nimble exom capture have been exploited to this effect in several CWR species already [42, 44, 45]. These technologies offer rapid marker density as required for rapid fine mapping and can saturate mapping populations in terms of capturing all of the recombinants.

The availability of a reference genome sequence in CWR during recent times greatly boosts the use of high-throughput sequence data. Some large scale genomic sequencing and re-sequencing programs are well underway [27, 46] often with reduced representation methods. Whole genome shotgun sequencing (WGS) techniques can also be utilized to characterize CWR germplasm for climate resilience breeding in major staple crops. E.g. Rice having smaller genome size (430 mb) long with its wild relatives has been re-sequenced using WGS [47, 48, 49]. Already sequenced germplasm collections including Chickpea [50], Rice [48],

Soybean [51] and Wheat [52] etc. will provide insights into these diverse gene pools to be exploited in combating various biotic and abiotic challenges during this era of climate change. More recently, a massive scale genomic study of almost 80000 accessions from CIMMYT and ICARDA unraveled unprecedented amount of genetic diversity in 29 wheat species comprising cultivated wheats, CWRs and landraces to be exploited in wheat improvement for range of climate related plant traits [53].

Potentially revolutionary technology in modern plant breeding like genome editing has enabled scientists to alter genome of any organism with unprecedented precision without involvement of any foreign DNA [54]. CWR and their sequence information may serve as a reference library for all kind of diversity. This information on allelic diversity and its phenotype is a vital requirement for many genome editing approaches. In fact, these approaches will allow the use of this information from more distantly related, cross-incompatible CWR and domesticated species to be further utilized in crop improvement [55, 56].

5. BIOTECHNOLOGY: A TOOLKIT FOR CLIMATE RESILIENT AGRICULTURE

Considering the various direct and indirect impacts of climate change on food production and agriculture along with rapid deterioration of arable land and perplexity of rainfall patterns, all these factors triggering various abiotic stresses such as drought, heat stress and biotic stresses like pest and disease attacks, the sophisticated techniques laden biotechnology toolkit has potential to address these immense challenges of developing the stress tolerant food crop cultivars in this hour of need [57]. With population growing at rapid rate under threatening scenario of climate change, it is high time to shift resilience from conventional breeding along with fertilizers and pesticides to genomics-assisted crop improvement techniques in order to achieve more sustainable and efficient yield gains [58].

5.1 Marker assisted breeding

Recent advances in biotechnology tools have the potential to understand the function of genes/QTLs that govern the economic traits, and applying this information's to Smart breeding programs, leading to crop improvement. The advent of molecular markers such as Restriction fragment length polymorphism (RFLP), Rapid Amplified Polymorphic DNA (RAPD), Simple Sequence repeat (SSR), Kompetitive allele specific PCR (KASP), Cleaved amplified polymorphic sequence (CAPS) and especially Single Nucleotide polymorphism (SNP) have revolutionerized the field of plant genetics and facilitated molecular crop breeding [59].

The ultimate goal of crop breeding to develop super-varieties by assembling multiple desirable traits, such as yield related, superior quality, tolerance/resistance against biotic and abiotic stress and good environmental adaption. It is very challenging, difficult and time consuming to combine all traits in single genotypes by traditional breeding, so some alternates need to be compiling all important traits, into single varieties, can be done through marker assisted selection (MAS), which have become an integral component of genotypes/germplasm improvement. The potential benefits of using molecular markers linked to the genes/QTLs of interest in breeding programmes, which have shifted from phenotype-based (traditional breeding) to a combination of phenotype and genotype-based selection, are of great importance to the Smart breeding programme [60].

Breeding programme combine, with MAS strategies have major advantages compared to traditional phenotype-dependent breeding in terms of convenience and efficiency for transferring the genes/QTLs of interest to the plant genome [61]. Selection can be done selectively with the genotypes of molecular markers linked to the target traits, selection in off-season nurseries (reduce breeding cycle), making the technique more cost effective to grow for more generations per year (speed breeding), reduction of required population size because many lines can be discarded in earlier breeding generations after MAS. The most effective and usefulness of MAS approaches, for traits of simple inheritance (qualitative traits controlled by one or a few genes) have been well proven in many important crops [62].

Basically, two major MAS strategies are usually applied in breeding programme, (i) backcrossing for favorable alleles into elite germplasm, i.e. marker-assisted-backcrossing (MABC) and (ii) stacking multiple genes of different sources into elite breeding lines, i.e. marker-assisted gene pyramiding (MAGP). The success of MAS has depends to search the important QTLs for complex traits (controlled by minor genes), which account for a large proportion of phenotypic variation (major QTLs). Successful applications of MABC and MAGP for improving yield or yield component traits by using well characterized major QTLs/genes in important crops [63]. Successful implementation of MAS breeding in broad range of crops including barley, beans, cassava, chickpea, cowpea, groundnut, maize, potato, rice, sorghum, and wheat [64]. Genetic markers associated with agronomic traits can be introgressed into elite crop genetic backgrounds via marker assisted breeding (MAB). It allows stacking of desirable traits into elite varieties to make them better adapted to climatic changes.

5.2 DNA sequencing and advent of genomics assisted breeding

With plummeting cost and greater accessibility of high throughput genome sequencing technology, the breadth of genomic data is expanding rapidly. In order to capture diversity of specific gene families within a large group, DNA samples can preferentially be enriched before sequencing. This approach can be adopted to define genetic variation in disease resistance gene repositories in Solanaceae and Triticeae (RNA seq) [65] and gluten gene families I bread wheat (GlutEnSeq) [66].

Sanger sequencing to study plant genomes is unfeasible due to low throughput and high sequencing costs. In 2005, Roche released its revolutionary 454 pyrosequencing platform [67]. Subsequently, several sequencing platforms such as developed by Illumina, ABI, Life technologies, PacBio, Oxford Nanopore and Complete genomics were released commercially, changing the scenario of genome sequencing. Depending on chemistry, second generation sequencing (SGS) approaches are classified as ligation based approaches and synthesis based approaches [68]. To rectify the problems of assembling repetitive genomic regions, long read sequencing offers solution by producing reads spanning the repeat regions [69].

Rapid cost reduction in genome wide genotyping allows large scale assessment of crop species diversity to capture climate related traits. It leverages cheaper sequencing to identify up to millions of SNPs in plant population [70]. High SNP density approach like whole genome resequencing (WGR) & low SNP density approach like reduced representation sequencing (RRS) are majorly used approaches. However, high density genotyping assay "SNP chips" enable large scale genotyping using SNP specific oligonucleotide probes rather than direct sequencing.

The variants identified by genotyping by sequencing (GBS) can be used for conventional QTL analysis and modern approach like genome wide association studies (GWAS). GWAS exploits the past recombinations in a diverse association panels to identity genes lined to phenotypic traits [71]. SNP genotyping have been widely used in many crops including wheat [72] and Maize [73]. Extensive use of GWAS is resulting in our enhanced understanding of genetics of important climate specific traits viz. drought and heat tolerance. In light of reducing sequencing cost and expensive validation of candidate genes, use of WGR to further enhance resolution of mapping studies is likely to become routine task in future [70].

The availability of reference genome assembly rewards us with information about gene content, ability to associate the traits with specific genes with subsequent insights into related biophysical and biochemical roles of gene(s) in the expression of that particular trait [74]. Resequencing of diverse crop cultivars reveals the gene content variation and DNA sequence differences between allelic variants, while sequencing of expressed gene products provides information on where and when genes are functioning. Such information when integrated within breeding pipelines, offers promise to accelerate the development of climate smart crop varieties.

The recent explosion in genomic data is rapidly triggering a fundamental shift to genomic based breeding [75]. The ability to identify and genotype umpteen SNPs at ever reducing costs facilitated expansion of MAS in breeding to plethora of traits and across wider range of crops [76]. A major outcome of availability of high throughput genome wide markers is a move towards population based trait association and breeding i.e. NAM or MAGIC populations to ultimately enhance the trait mapping resolution by greatly increasing the number of recombinations in the population. After identification and validation of the candidate genes, there achieved the deeper understanding of biological mechanism underlying the trait, which can subsequently be improved through MAB or genetic alterations. Furthermore, precise understanding of the molecular basis of traits enables the engineering of novel alleles or mining of potentially desirable alleles from CWR, facilitating further enhancement of the trait.

6. GENOME EDITING: A REVOLUTIONARY TOOL IN BREEDERS' TOOLKIT

Genome editing has enabled breeders to precisely add or delete any DNA sequence in the genome and has shown enormous potential to revolutionize the crop improvement in this very decade [70, 77]. Some approaches like transcription activator-like effector nucleases (TALENs) and zinc finger nucleases (ZFNs) have been in the game for more than 2 decades. However, type II clustered regularly interspaced short palindromic repeat (CRISPR)/CRISPR-associated protein (Cas) system from *Streptococcus pyogenes* [78] developed in last decade has been most versatile tool in breeder's toolkit to introduce desirable or novel traits and accelerate development of climate smart crop varieties.

Usually, a custom-made guide RNA (gRNA) along with Cas9 nuclease is delivered into plant protoplast, where Cas9 produces double strand break (DSB) 3 bp upstream of the NGG motif (protospacer adjacent motif-PAM sequence) [78]. Cellular repair machinery through non-homologous end joining (NHEJ) can lead to frameshift mutation causing a knock-out. Otherwise, a donor DNA template can be provided for precise genetic knock-in through homologous recombination (HR). CRISPR/Cas9 was initially used to disrupt genes related to disease susceptibility in crops such as OsERF922 gene disruption in rice for blast resistance [79] and loss of function in susceptibility gene TaMLO for powdery mildew resistance in wheat [80]. Genome editing has also been used to tackle some abiotic stresses in staple crops like a promoter of a gene AGROS8 was replaced with a stronger one to impart drought tolerance in maize [81].

Due to changing climates, it may be quite beneficial for the farmers to have early maturing varieties, which enables plants to complete crucial developmental periods before the onset of a stress. It has been achieved by disrupting a flower repressing gene SP5G to develop early maturing tomato varieties [82]. For instance, developing climate rice to grow in diverse climates, generally desirable traits are cold, heat and drought tolerance at seedling and reproductive stages

[83]. Secondary characters like root and flag leaf traits can be useful to generate cultivars with improved drought and heat tolerance [84]. Here, CRISPR tools could prove to be of great value for exploration of the candidate genes from CWR (*O. officinalis*, *O. nivara* and *O. glaberrima*) for abiotic stress resistance [85].

Genome editing has also huge potential to accelerate the domestication of novel crops form CWR or minor crops with valuable traits for coping with extreme climatic events. This would allow the editing of key genes for domestication in potential new crops for rapid enhancement of currently limited gene pools to maximize the use of germplasm adapted to climate change. Also, multiplexing of CRISPR systems for simultaneous editing of multiple genetic loci can boost the speed and efficiency manifolds. However, there are a number of shortcomings in this approach including off target effects [86], low efficiency of HR, restrictive PAM sequences and regulatory concerns, which paved the way for advent of more sophisticated technologies like DNA free genome editing, base editing and prime editing.

6.1 DNA free genome editing (DFGE)

Conventional genome editing using recombinant DNA (rDNA) leads to random host genome integration and can generate undesirable genetic changes or DNA damage [87], along with concerns over genetically modified organism (GMO) regulations with introduction of foreign DNA [88]. DFGE takes care of such critical issues along with reduced risk of off-targets. Initially, it was successfully deployed in rice and tobacco with transfection of protoplast with CRISPR-Cas9 ribonucleoprotein (RNP) [89]. Also, a particle bombardment mediated DFGE approach has been developed in wheat and maize [90, 91].

6.2 Base editing

It is evident that a single base change can cause variation in the elite traits [92], so there required an efficient technique to cause precise and efficient point mutations in plants. CRISPR-Cas9 driven base editing is new approach which accurately transform one DNA base to another without repair template [93]. E.g. Cytidine deaminases convert cytosine (C) to uracil (U), which is treated as thymine (T) in subsequent DNA repair and replication, thus creating C•G to T•A substitution. It has been utilized in wheat, maize and tomato [94] and can be quite useful for gene functional analysis and therefore can assist breeding for better stress adapted varieties.

6.3 Prime editing

Another latest milestone in this genome engineering era called prime editing allows introduction of all known 12 base to base conversions in addition to mutations such as insertions and deletions using prime editing guide RNA (pegRNA) [95]. This promising approach opening up numerous possibilities for effectively targeting and modifying desirable genome sequences to accelerate functional genomics and introduction of genes for adaptation to diverse climates can boost breeding for climate smart crop varieties in near future [96].

In this rejuvenated plant mutagenesis breeding era, genome editing can be used in functional genomics for the identification of candidate genes for climate related agronomic, physiological and phonological traits, which can be exploited for crop improvement in adaptation to changing climate. Despite having enormous potential and real world applications of genome editing technologies, the regulatory and ethical concerns may limit it, as happened in a few European countries. In the nutshell, genome editing in complementation with conventional plant breeding can be adopted to develop and deploy climate smart crop varieties in the farmers' fields.

7. PHENOMICS AND ARTIFICIAL INTELLIGENCE (AI): SUPPLEMENTING THE GENETIC GAINS

Advances in phenomics and genomics have generated unprecedented amount of new data, enabling breeders to continuously pushing the crop yields on positive side [97]. Despite success in techniques like genomic selection (GS) in cereals and legumes, lack of predictive accuracy for many complex traits (yield) have revealed their inability to adequately model all relevant factors inherent to such traits due to complexity of the interactions between genetic and environmental components of phenotypic variation [98]. Several mapping studies have shown that such complex traits are controlled by minor genes (polygenes) with small but cumulative effect, hence go undetected while analyzing them in smaller population size.

Relationship between genotype and phenotype is not always linear and small changes on one hierarchical level may have bigger impact on other levels. Many statistical models therefore fail to accurately delineate the non-linear relationships. Additionally, epistatic interactions are hard to detect while mapping genotype to phenotype with linear models due to low power and sheer computational demand [99]. With continuously falling cost of genome sequencing, advent of innovative genetic assays to explore missing heritability and genetic regulation, breeders have access to wide range of high-throughput sensors and imaging techniques for spectrum of traits and field conditions.

Omics technologies (genomics, transcriptomics, proteomics, metabolomics, phenomics, epigenomics and microbiomics) together with approaches to gather information about climate and field environment conditions have become routine in breeding programs now a days. However, ability to accurately predict &

select best lines for the specific environment relies on our ability to model these immensely complex systems from web of genomic and phenomic data at hand e.g. multiomics big data. Integrating with phenomics and genomics, AI technologies by assisting with big data, can boost up the development of climate resilient crop varieties with enhanced yield potential and stability and improved tolerance to expected simultaneous environmental stresses (abiotic and biotic).

7.1 Field phenomics

Accelerated plant breeding for climate resilience is critically dependent upon high resolution, high throughput, field level phenotyping that can effectively screen among better performing breeding lines within larger population across multiple environments [100]. With advent of novel sensors (unmanned air vehicle-UAV), high resolution imagery and new platforms for wide range of traits and conditions, phenomics has been elevating the collection of more phenotypic data over the past decade [101, 102]. High throughput phenotyping (HTP) allows the screening for plant architectural traits and early detection of desirable genotypes. It enables accurate, automated and repeatable measurements for agronomic traits (seedling vigor, flowering time, flower counts, biomass and grain yield, height and leaf erectness, canopy structure) as well as physiological traits (photosynthesis, disease and stress tolerance). HTP methods such as RGB imaging, 3-D scanning, thermal and hyper spectral sensing and fluorescence imaging have been successfully utilized to identify, quantify and monitor plant diseases [103].

By coupling GWAS with high throughput phenotyping facilities, phenomics can be adopted as novel tool for studying plant genetics and genomic characterization enhancing the crop breeding efficiency in era of climate change [104]. Recently, deep learning (DL) has been extensively used to analyze and interpret more phenomic big data, especially for advancing plant image analysis and environmental stress phenotyping [105].

7.2 Next gen based GS

Genomic selection as been extensively used breeding approach for climate resilience in agriculture in last decade, especially for complex polygenic traits. It involves prediction models developed by estimating the combined effect of all existing markers simultaneously on a desirable phenotype. Highly accurate prediction can result into enhanced levels of yields by shortening the breeding cycles. Omics layers (gene expression, metabolite concentration and epistatic signals) can be better predictors of phenotype than SNPs alone due to their molecular proximity to the phenotype. Many such omics layers that explain trait variation have not been made available to the statistical models lowering down its efficacy. Several approaches such as mixed effect linear models and Bayesian models to select only most important predictive SNPs are majorly used.

From the prospective of breeding, by accessing the rich set of omics and environmental data lying between plant genotype and its phenotype, superior and refined impact can be achieved on desirable phenotype. Next gen AI holds promise for GS as acquisition of large scale genomics and phenomics data in addition to molecular layers between them such as transcriptomics, proteomics and epigenomics will facilitate a period, where AI models can identify and explain the complex biological interactions [99].

Next gen AI will surely require knowledge and rationality of breeders as well as farmers to evaluate the efficacy of outcomes. In coming times, agriculture will rely on Next Gen AI methods for making decisions and recommendations from big data (highly heterogeneous and complex) that are representative of environment and system biology based understanding of the behavioral response of plants.

8. SPEED BREEDING: AN ACCELERATION TO CROP IMPROVEMENT

The current pace of yield increase in staple crops like wheat, rice and maize is insufficient to meet the future demand in the wake of climate change [106]. A major limiting factor in plant breeding is the longer generation times of the crops, typically allowing 1–2 generations in a year. Several 'speeding breeding' protocols, using extended photoperiods and controlled temperatures have enabled breeders to harvest up to 6 generations per year by reducing the generation time by more than half [107]. Such protocols have been reported in several important crops such as spring wheat (*Triticum aestivum*) [108], barley (*Hordeum vulgare*) [109], chickpea (*Cicer arietinum*), rice (*Oryza sativa*) [110] and canola (*Brassica napus*).

Speed breeding can potentially accelerate the discovery and use of allelic diversity in landraces as well as in CWR to be further used in developing climate resilient crop varieties. One such example is recent discovery of new sources of leaf rust resistance after screening of the Vavilov wheat collection using speed breeding along with gene specific molecular markers [111].

Interestingly, speed breeding can also be integrated with advanced technique like gene editing to precisely alter the plant genes for better coping with various biotic and abiotic stresses in threatening climatic changes. In traditional CRISPR gene editing, the sgRNA directs Cas9 enzymes to cut target sequence. 'CRISPR-ready' genotypes containing heterologous Cas9 gene can be created. For instance, a transformant harboring a Cas9 transgene can be used a donor to create a stock of elite inbred lines using speed marker-assisted backcrossing. Such an integrated system like ExpressEdit could circumvent the bottlenecks of in vitro manipulation of plant materials also making gene editing fast-tracking [1]. Integration of both the techniques without tissue culture/foreign DNA requires handful of technological breakthroughs with the desirable outcomes being allelic modification, these would bypass genetically modified organism (GMO) label. It has been widely reported that single or multiplex edits can be obtained [112] and could be implemented with some tissue culture free techniques like CRISPR-Cas9 ribonucleoprotein (RNP) complexes in wheat [91] and maize [90].

Genomic selection (GS) unlike MAS uses genome-wide DNA markers in order to predict the genetic gain of breeding individuals for complex traits such as yield [113]. The effect of large number of genetic variants for such a complex traits is captured through linkage disequilibrium (LD) with the genome-wide markers (SNPs), effects of which are determined in large training populations (lines in which marker genotype and trait are measured). Since speed breeding can substantially lowers down the generation periods, it can maximize the benefits by applying genomic selection at every generation to select parents for next generation. Modern genotyping techniques such as rAmpSeq may considerably reduce the genotyping cost for genomic selection [114]. When combined with speed breeding protocol, the approach for stacking of best haplotypes (ones with desirable resistance alleles/ desirable edits) could be used rapidly to develop new cultivars [1] with improved performance across multiple traits like coping with adverse climatic variations or any pathogen/insect attack.

Re-domestication of crop plants for capturing the desirable alleles for climate resilience can be sped up by linking it with speed breeding. Re-creation of the polyploids such as groundnut (*Arachis hypogea*) and banana (*Musa spp.*) can be benefitted by such approach. Speed breeding could accelerate re-domestication at multiple selection steps after crossing of diploids followed by colchicine application [115]. Ultimately, it will provide access to novel plant traits for developing cultivars of these crops exhibiting disease resistance and stress adaptation. Also, Gene editing and targeted mutagenesis coupled with speed breeding could prove to be more efficient to create healthier foods by biofortification. For instance, the increased content of vitamin B9 in rice and antinutritional glucosinolates from *Brassica* seeds etc. [1].

Combining all these tools with speed breeding approach would provide rapid access to desirable alleles and novel variation present in CWR and would accelerate the breeding pipelines to develop more climate resilient varieties (Table 1).

Table 1.Utilization of smart breeding tools and techniques for crop improvement.

Crop species	Target trait/Improved trait	Technology/ Technique used	Reference
Rice	Submergence tolerance	MAB	[116]
Rice	Grain number, dense erect panicles and larger grain size	CRISPR/Cas9	[117]
Rice	Maintenance of heterosis	CRISPR/Cas9	[118, 119]
Wheat	Heat tolerance	GWAS	[120]
Wheat	Leaf rust, fusarium head blight and stripe rust resistance	Speed breeding	[121, 122, 123, 124]
Wheat	Powdery mildew-resistant	CRISPR/Cas9	[80]
Finger millet	Salt tolerance	RNA sequencing	[125]
Sorghum	Low and high nitrogen conditions	RNA sequencing	[126]
Sugarcane	Drought and chilling resistance	CRISPR/Cas9	[127]
Maize	Kernel row number	RNA sequencing	[128]
Maize	High amylopectin content	CRISPR/Cas9	[129]
Cotton	Salt and drought tolerance	GWAS	[130]
Soybean	Salt and drought tolerance	CRISPR/Cas9	[131, 132]
Soybean	Salt tolerance	RNA sequencing	[133]
Chickpea	Drought, salinity, cold and heavy metal stress resistance	RNA sequencing	[134]
Lentil	Seedling drought stress resistance	RNA sequencing	[135]
Tomato	High temperature stress responsiveness	GWAS	[136]
Tomato	Powdery mildew-resistant	CRISPR/Cas9	[137]
Tomato	Longer internodes and lighter green leaves with smoother margins	TALEN	[138]
Tomato	Short (hairy) roots with stunted meristematic, altered branching and increased yield	CRISPR/Cas9	[139, 140]
Tomato	Fruits never turn red, altered firmness	CRISPR/Cas9	[141]
Broccoli	Dwarf phenotype	CRISPR/Cas9	[142]
Watermelon	Albino phenotype	CRISPR/Cas9	[143]
Potato	Reduced steroidal glycoalkaloids in leaves and Undetectable level of reducing sugar in tubers	TALEN	[144, 145]

Crop species	Target trait/Improved trait	Technology/ Technique used	Reference
Mushroom	Reduced browning	CRISPR/Cas9	[146]
Banana	Cold and salt resistance	CRISPR/Cas9	[147]
Coconut	Root wilt disease	CRISPR/Cas9	[148]
Papaya	Drought, heat and cold resistance	CRISPR/Cas9	[149]
Apple	Albino phenotype and Blight resistance	CRISPR/Cas9	[150, 151]

9. FUTURE PROSPECTS

In the face of ongoing and projected climate change, including higher temperatures and more erratic climate events across extensive regions over the globe, breeding of crop plants with enhanced yield potential and improved resilience to such environments is crucial for global food security. Improved plant varieties that can withstand diseases and pests with efficient use of fewer resources, exhibiting stable yields amidst stressful climate in near future could only help to achieve the goal of climate resilient agriculture. In order to be able to make contribution in climatic resilience, research attention is indispensable for currently underutilized crop species. The concept of smart breeding largely depends upon generating large breeding populations, efficient high throughput phenotyping, big data management tools and downstream molecular techniques to tackle the vulnerability of crop plants to changing climate (Figure 3). The efficient preservation and conservation of plant genetic resources is also a pre requisite for climate smart breeding. Strategies for capturing the novel variation may include the state of the art tools such as gene editing to directly introduce novel alleles found in wild plants into domesticated crop varieties. Generating new crop cultivars with the capability to tolerate multiple stresses can be achieved with increasing information on their basal physiological and genetic mechanisms. The technological improvements in phenotypic and genotypic analysis, as well as the biotechnological and digital revolution could definitely pave the way for developing and deployment of climate smart varieties in coming times.

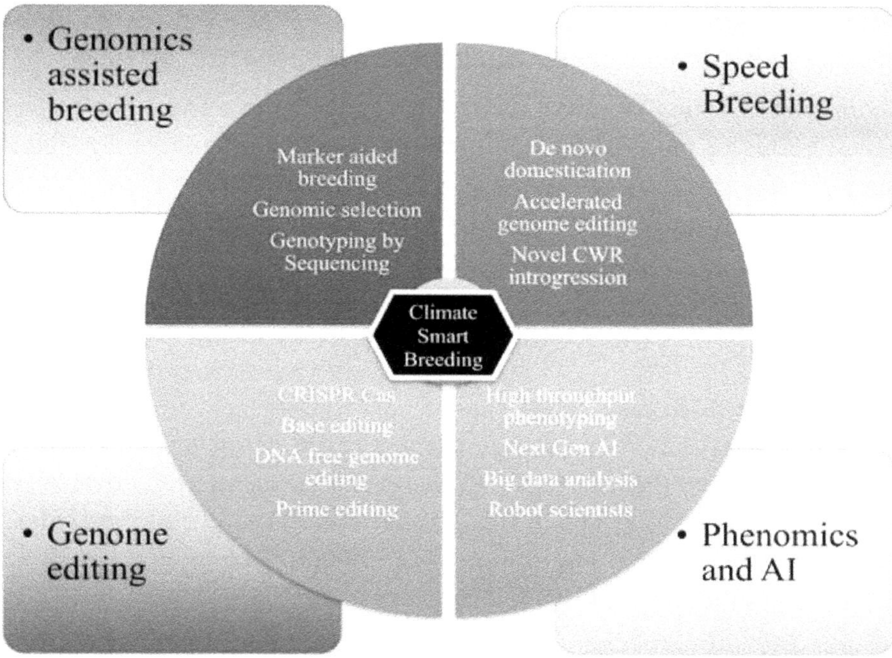

Figure 3.Compilation of state-of-the-art genomic, phenomic and computational tools comprising smart breeding approach for climatic resilience in agriculture.

AT A GLANCE

Human society is at a turning point of its time as climate change is becoming more and more real and inevitable. From rising temperature, which undermines the food production, to melting glaciers, causing disastrous flooding and erosion, the global repercussions of climate change are unprecedented. Plant breeding has always played a pivotal role in human history by revolutionizing agriculture to feed the ever-growing population. It can rescue humankind from imminent threats to agriculture posed by weather fluctuations, rapidly evolving pests and limiting resources. Unlocking the repository of genetic diversity and extensive utilization of wild germplasm invariably is imperative to every crop improvement program. But recent advancements in genomics, high throughput phenomics, sequencing and breeding methodologies along with state-of-the-art genome-editing tools in integration with artificial intelligence open up new doors for accelerated climate-resilient crop improvement. Therefore, holistic smart breeding approaches can be promising way out to tackle climate change and develop better-adapted crop varieties.

REFERENCES

1. Hickey LT, Hafeez AN, Robinson H, Jackson SA, Leal-Bertioli SC, et al. Breeding crops to feed 10 billion. Nature Biotechnology. 2019;37: 744-754

2. Dhankher OP, Foyer CH. Climate resilient crops for improving global food security and safety. Plant, Cell and Environment. 2018;41:877-884
3. Zhao C, Liu B, Piao S, Wang X, Lobell DB, et al. Temperature increase reduces global yields of major crops in four independent estimates. Proceedings of the National Academy of Sciences. 2017; 114:9326-9331
4. Falconer DS, Mackay TFC. Introduction to Quantitative Genetics. Essex, England: Longman. 1996. 1529-1536 p
5. Wang JK. Simulation modeling in plant breeding: Principles and applications. Agricultural Sciences in China. 2007;6:908-921
6. Gobu R, Shiv A, Kumar AC, Basavaraj PS, Harish D, et al. Accelerated crop breeding towards development of climate resilient varieties. In: Srinivasarao C, Srinivas T, Rao RVS, Rao NS, Vinayagam SS, Krishnan P, editiors. Climate change and Indian agriculture: challenges and adaptation strategies. ICAR-National Academy of Agricultural Research Management, Hyderabad, Telangana, India. 2020. 49-69 p
7. Leisner CP. Climate change impacts on food security-focus on perennial cropping systems and nutritional value. Plant Science. 2020;293:110412
8. IPCC (2014) Climate change: impacts, adaptation, and vulnerability. part A: global and sectoral aspects. Contribution of working group II to the fifth assessment report of the Inter governmental Panel on Climate Change. Cambridge University Press. 2014. 32 p
9. Ming X. World wide fund for nature: Yangtze river basin climate change vulnerability and adaptation report. WWF-China Program Office, Beijing. 2009
10. Karmakar R, Das I, Dutta D, Rakshit A (2016) Potential effects of climate change on soil properties: a review. Science International. 2016;4:51-73
11. Newbery F, Qi A, Fitt BD. Modelling impacts of climate change on arable crop diseases: progress, challenges and applications. Current Opinion in Plant Biology. 2016;32:101-109
12. Smykal P, Nelson MN, Berger JD, Von Wettberg EJ. The impact of genetic changes during crop domestication. Agronomy. 2018;8:119
13. Valkoun JJ. Wheat pre-breeding using wild progenitors. Euphytica. 2001;119:17-23
14. Sharma S, Upadhyaya HD, Varshney RK, Gowda CL. Pre-breeding for diversification of primary gene pool and genetic enhancement of grain legumes. Frontiers in Plant Science. 2013;4:309

15. Ladizinsky G. Founder effect in crop-plant evolution. Economic Botany. 1985;39:191-199

16. Hawkes JG The importance of wild germplasm in plant breeding. Euphytica. 1977;26:615-621

17. Stalker HT. Utilization of wild species for crop improvement. Advances in Agronomy. 1980;33:111-147

18. Plucknett DL, Smith NJ, Williams JT, Murthi Anishetty N. Gene Banks and the world's food Princeton University press. Princeton, NJ. 1987

19. Brar DS, Khush GS. Wide hybridization for rice improvements: Alien gene transfer and molecular characterization of introgression. In: Jones MP, Dingkhun M, Johnson DE, Fagade SO, editors. Interspecific Hybridization: Progress and Prospect. WARDA, Bouake. 1997. 21-29 p

20. Singh S, Gumber RK, Joshi N, Singh K. Introgression from wild Cicer reticulatum to cultivated chickpea for productivity and disease resistance. Plant Breeding. 2005;124:477-480

21. Singh G, Singh I, Taggar GK, Rani U, Sharma P, Gupta M, Singh S. Introgression of productivity enhancing traits, resistance to pod borer and Phytopthora stem blight from Cajanus scarabaeoides to cultivated pigeonpea. Physiology and Molecular Biology of Plants. 2020; 26:1399-1410

22. Rana K, Atri C, Gupta M, Akhatar J, Sandhu PS, et al. Mapping resistance responses to Sclerotinia infestation in introgression lines of Brassica juncea carrying genomic segments from wild Brassicaceae B. fruticulosa. Scientific Reports. 2017;7:5904

23. Tanksley SD, McCouch SR. Seed banks and molecular maps: unlocking genetic potential from the wild. Science. 1997;277:1063-1066

24. Garrett KA, Dendy SP, Frank EE, Rouse MN, Travers SE. Climate change effects on plant disease: genomes to ecosystems. Annual Review of Phytopathology. 2006;44:489-509

25. Mason AS, Batley J. Creating new interspecific hybrid and polyploid crops. Trends in Biotechnology. 2015;33:436-441

26. Von Wettberg EJ, Chang PL, Başdemir F, Carrasquila-Garcia N, Korbu LB, et al. Ecology and genomics of an important crop wild relative as a prelude to agricultural innovation. Nature Communications. 2018;9:649

27. Brozynska M, Furtado A, Henry RJ. Genomics of crop wild relatives: expanding the gene pool for crop improvement. Plant Biotechnology Journal. 2016;14:1070-1085

28. Fustier MA, Brandenburg JT, Boitard S, Lapeyronnie J, Eguiarte LE, et al. Signatures of local adaptation in lowland and highland teosintes from whole-genome sequencing of pooled samples. Molecular Ecology. 2017;26:2738-2756

29. Liu M, Li Y, Ma Y, Zhao Q, Stiller J, et al. The draft genome of a wild barley genotype reveals its enrichment in genes related to biotic and abiotic stresses compared to cultivated barley. Plant Biotechnology Journal. 2020;18:443-456

30. Du J, Verzaux E, Chaparro-Garcia A, Bijsterbosch G, et al. Elicitin recognition confers enhanced resistance to Phytophthora infestans in potato. Nature Plants. 2015;1:1-5

31. Tao Y, Zhao X, Mace E, Henry R, Jordan D. Exploring and exploiting pan-genomics for crop improvement. Molecular Plant. 2019;12:156-169

32. Arora S, Steuernagl B, Gaurav K, Chandramohan S, Long Y et al. Resistance gene cloning from a wild crop relative by sequence capture and association genetics. Nature Biotechnology. 2019;37:139-143

33. Fernie AR, Yan J. De novo domestication: an alternative route toward new crops for the future. Molecular Plant. 2019;12:615-631

34. Jaime R, Alcantara JM, Manzaneda AJ, Rey PJ. Climate change decreases suitable areas for rapeseed cultivation in Europe but provides new opportunities for white mustard as an alternative oilseed for biofuel production. PloS One. 2018;13:e0207124

35. DeHaan LR, Van Tassel DL, Anderson JA, Asselin SR, Barnes R, et al. A pipeline strategy for grain crop domestication. Crop Science. 2016;56:917-930

36. Zsögön A, Čermák T, Naves ER, Notini MM, Edel KH, et al. De novo domestication of wild tomato using genome editing. Nature Biotechnology. 2018;36:1211-1216

37. Li T, Yang X, Yu Y, Si X, Zhai X, Zhang H, Dong W, Gao C, Xu C. Domestication of wild tomato is accelerated by genome editing. Nature Biotechnology. 2018;36:1160-1163

38. Lemmon ZH, Reem NT, Dalrymple J, Soyk S, Swartwood KE, et al. Rapid improvement of domestication traits in an orphan crop by genome editing. Nature Plants. 2018;4:766-770

39. Prohens J, Gramazio P, Plazas M, Dempewolf H, Kilian B, et al. Introgressiomics: a new approach for using crop wild relatives in breeding for adaptation to climate change. Euphytica. 2017;213:158

40. Iftekharuddaula KM, Newaz MA, Salam MA, Ahmed HU, Mahbub MA, et al. Rapid and high-precision marker assisted backcrossing to introgress the SUB1 QTL into BR11, the rainfed lowland rice mega variety of Bangladesh. Euphytica. 2011;178:83-97

41. Lin T, Zhu G, Zhang J, Xu X, Yu Q, et al. Genomic analyses provide insights into the history of tomato breeding. Nature Genetics. 2014;46:1220-1226

42. Russell J, Mascher M, Dawson IK, Kyriakidis S, Calixto C, et al. Exome sequencing of geographically diverse barley landraces and wild relatives gives insights into environmental adaptation. Nature Genetics. 2016;48:1024-1030

43. Kilian B, Graner A. NGS technologies for analyzing germplasm diversity in genebanks. Briefings in functional genomics. 2012;11:38-50

44. He J, Zhao X, Laroche A, Lu ZX, Liu H, Li Z. Genotyping-by-sequencing (GBS), an ultimate marker-assisted selection (MAS) tool to accelerate plant breeding. Frontiers in Plant Science. 2014;5:484

45. Wendler N, Mascher M, Himmelbach A, Johnston P, Pickering R, Stein N. bulbosum to go: a toolbox to utilize Hordeum vulgare/bulbosum introgressions for breeding and beyond. Molecular Plant. 2015;8:1507-1519

46. The 100 Tomato Genome Sequencing Consortium, Aflitos S, Schijlen E, de Jong H, de Ridder D, Smit S, et al. Exploring genetic variation in the tomato (Solanum section Lycopersicon) clade by whole-genome sequencing. The Plant Journal. 2014;80:136-148

47. Xu X, Liu X, Ge S, Jensen JD, Hu F, et al. Resequencing 50 accessions of cultivated and wild rice yields markers for identifying agronomically important genes. Nature Biotechnology. 2012;30: 105-111

48. Li JY, Wang J, Zeigler RS, Hirochika H, Guiderdoni E, et al. The 3,000 rice genomes project: new opportunities and challenges for future rice research. Giga Science. 2014;3:8

49. Alexandrov N, Tai S, Wang W, Mansueto L, Palis K, Fuentes RR, et al. SNP-Seek database of SNPs derived from 3000 rice genomes. Nucleic Acids Research. 2015;43:1023-1027

50. Varshney RK. Exciting journey of 10 years from genomes to fields and markets: some success stories of genomics-assisted breeding in chickpea, pigeonpea and groundnut. Plant Science. 2016;242: 98-107

51. Song Q, Hyten DL, Jia G, Quigley CV, Fickus EW, et al. Fingerprinting soybean germplasm and its utility in genomic research. G3: Genes, Genomes, Genetics. 2015;5:1999-2006

52. Sehgal D, Vikram P, Sansaloni CP, Ortiz C, Saint Pierre C, et al. Exploring and mobilizing the gene bank biodiversity for wheat improvement. PloS One. 2015;10:e0132112

53. Sansaloni C, Franco J, Santos B, Percival-Alwyn L, Singh S, et al. Diversity analysis of 80,000 wheat accessions reveals consequences and opportunities of selection footprints. Nature Communications. 2020;11:4752

54. Voytas DF. Plant genome engineering with sequence-specific nucleases. Annual Review of Plant Biology. 2013;64:327-350

55. Nogue F, Mara K, Collonnier C, Casacuberta JM. Genome engineering and plant breeding: Impact on trait discovery and development. Plant Cell Reports. 2016;35:1475-1486

56. Spindel JE, McCouch SR. When more is better: how data sharing would accelerate genomic selection of crop plants. New Phytologist. 2016;212:814-826

57. Espeland EK, Kettenring KM. Strategic plant choices can alleviate climate change impacts: A review. Journal of Environmental Management. 2018;222:316-324

58. Scheben A, Yuan Y, Edwards D. Advances in genomics for adapting crops to climate change. Current Plant Biology. 2016;6:2-10

59. Nadeem MA, Nawaz MA, Shahid MQ, Doğan Y, Comertpay G, et al. DNA molecular markers in plant breeding: current status and recent advancements in genomic selection and genome editing. Biotechnology and Biotechnological Equipment. 2018;32:261-285

60. Tester M, Langridge P. Breeding technologies to increase crop production in a changing world. Science. 2010;327:818-822

61. Babu R, Nair SK, Prasanna BM, Gupta HS. Integrating marker assisted selection in crop breeding-prospects and challenges. Current Science. 2004;87:14

62. Collard BC, Mackill DJ. Marker-assisted selection: an approach for precision plant breeding in the twenty-first century. Philosophical Transactions of the Royal Society B: Biological Sciences. 2008; 363:557-572

63. Uga Y, Sugimoto K, Ogawa S, Rane J, Ishitani M, et al. Control of root system architecture by DEEPER ROOTING 1 increases rice yield under drought conditions. Nature Genetics. 2013; 45:1097-1102

64. Cobb JN, Biswas PS, Platten JD. Back to the future: revisiting MAS as a tool for modern plant breeding. Theoretical and Applied Genetics. 2019;132:647-667

65. Steuernagel B, Periyannan SK, Hernández-Pinzón I, Witek K, Rouse MN, et al. Rapid cloning of disease-resistance genes in plants using mutagenesis and sequence capture. Nature Biotechnology. 2016;34:652-655

66. Jouanin A, Borm T, Boyd LA, Cockram J, Leigh F, et al. Development of the GlutEnSeq capture system for sequencing gluten gene families in hexaploid bread wheat with deletions or mutations induced by γ-irradiation or CRISPR/Cas9. Journal of Cereal Science. 2019;88:157-166

67. Margulies M, Egholm M, Altman WE, Attiya S, Bader JS, et al. Genome sequencing in microfabricated high-density picolitre reactors. Nature. 2005;437:376-380

68. Goodwin S, McPherson JD, McCombie WR. Coming of age: ten years of next-generation sequencing technologies. Nature Reviews Genetics. 2016;17:333-351

69. Yuan Y, Bayer PE, Batley J, Edwards D. Improvements in genomic technologies: application to crop genomics. Trends in Biotechnology. 2017;35:547-558

70. Scheben A, Batley J, Edwards D. Genotyping-by-sequencing approaches to characterize crop genomes: choosing the right tool for the right application. Plant Biotechnology Journal. 2017;15:149-161

71. Yuan Y, Scheben A, Batley J, Edwards D. Using genomics to adapt crops to climate change. In: Sustainable Solutions for Food Security. Springer. 2019. 91-109 p

72. Winfield MO, Allen AM, Burridge AJ, Barker GL, Benbow HR, et al. High-density SNP genotyping array for hexaploid wheat and its secondary and tertiary gene pool. Plant Biotechnology Journal. 2016;14:1195-1206

73. Li H, Peng Z, Yang X, Wang W, Fu J, et al. Genome-wide association study dissects the genetic architecture of oil biosynthesis in maize kernels. Nature genetics. 2013;45:43-50

74. Edwards D. The impact of genomics technology on adapting plants to climate change. In: Plant genomics and climate change. Springer. 2016. 173-178 p

75. Hamblin MT, Buckler ES, Jannink JL. Population genetics of genomics-based crop improvement methods. Trends in Genetics. 2011;27:98-106

76. Varshney RK, Terauchi R, McCouch SR. Harvesting the promising fruits of genomics: applying genome sequencing technologies to crop breeding. PLoS Biol. 2014;12:e1001883

77. Belhaj K, Chaparro-Garcia A, Kamoun S, Patron NJ, Nekrasov V. Editing plant genomes with CRISPR/Cas9. Current Opinion in Biotechnology. 2015;32:76-84

78. Jinek M, Chylinski K, Fonfara I, Hauer M, Doudna JA, Charpentier E. A programmable dual-RNA- guided DNA endonuclease in adaptive bacterial immunity. Science. 2012;337:816-821

79. Wang F, Wang C, Liu P, Lei C, Hao W, et al. Enhanced rice blast resistance by CRISPR/Cas9-targeted mutagenesis of the ERF transcription factor gene OsERF922. PloS One. 2016;11:e0154027

80. Wang Y, Cheng X, Shan Q, Zhang Y, Liu J, et al. Simultaneous editing of three homoeoalleles in hexaploid bread wheat confers heritable resistance to powdery mildew. Nature Biotechnology. 2014;32:947-951

81. Shi J, Gao H, Wang H, Lafitte HR, Archibald RL, et al. ARGOS 8 variants generated by CRISPR-Cas9 improve maize grain yield under field drought stress conditions. Plant BiotechnologyJournal. 2017;15:207-216

82. Soyk S, Müller NA, Park SJ, Schmalenbach I, Jiang K, et al. Variation in the flowering gene SELF PRUNING 5G promotes day-neutrality and early yield in tomato. Nature Genetics. 2017;49:162-168

83. Biswal AK, Mangrauthia SK, Reddy MR, Yugandhar P. CRISPR mediated genome engineering to develop climate smart rice: Challenges and opportunities. Seminars in Cell and Developmental Biology. 2019;96:100-106

84. Biswal AK, Kohli A. Cereal flag leaf adaptations for grain yield under drought: knowledge status and gaps. Molecular Breeding. 2013;31:749-766

85. Feng F, Xu X, Du X, Tong H, Luo L, Mei H. Assessment of drought resistance among wild rice accessions using a protocol based on single-tiller propagation and PVC-tube cultivation. Australian Journal of Crop Science. 2012;6:1204

86. Jiang W, Zhou H, Bi H, Fromm M, Yang B, Weeks DP. Demonstration of CRISPR/Cas9/sgRNA-mediated targeted gene modification in Arabidopsis, tobacco, sorghum and rice. Nucleic Acids Research. 2013;41:188

87. Kim S, Kim D, Cho SW, Kim J, Kim JS. Highly efficient RNA-guided genome editing in human cells via delivery of purified Cas9 ribonucleoproteins. Genome Research. 2014;24:1012-1019

88. Jones HD. Regulatory uncertainty over genome editing. Nature Plants. 2015;1:14011

89. Woo JW, Kim J, Kwon SI, Corvalán C, Cho SW, et al. DNA-free genome editing in plants with preassembled CRISPR-Cas9 ribonucleoproteins. Nature biotechnology. 2015;33:1162-1164

90. Svitashev S, Schwartz C, Lenderts B, Young JK, Cigan AM. Genome editing in maize directed by CRISPR–Cas9 ribonucleoprotein complexes. Nature Communications. 2016;7:13274

91. Liang Z, Chen K, Li T, Zhang Y, Wang Y, et al. Efficient DNA-free genome editing of bread wheat using CRISPR/Cas9 ribonucleoprotein complexes. Nature communications. 2017;8:14261

92. Henikoff S, Comai L. Single-nucleotide mutations for plant functional genomics. Annual Review of Plant Biology. 2003;54:375-401

93. Komor AC, Kim YB, Packer MS, Zuris JA, Liu DR. Programmable editing of a target base in genomic DNA without double-stranded DNA cleavage. Nature. 2016;533:420-424

94. Zhang Y, Gao C. Recent advances in DNA-free editing and precise base editing in plants. Emerging Topics in Life Sciences. 2017;1:161-168

95. Anzalone AV, Randolph PB, Davis JR, Sousa AA, Koblan LW, et al. Search-and-replace genome editing without double-strand breaks or donor DNA. Nature. 2019;576:149-157

96. Marzec M, Hensel G. Prime Editing: Game Changer for Modifying Plant Genomes. Trends in Plant Science. 2020;25:722-724

97. Brown TB, Cheng R, Sirault XR, Rungrat T, Murray KD, et al. Trait Capture: genomic and environment modelling of plant phenomic data. Current Opinion in Plant Biology. 2014;18:73-79

98. Voss-Fels KP, Cooper M, Hayes BJ. Accelerating crop genetic gains with genomic selection. Theoretical and Applied Genetics. 2019;132:669-686

99. Harfouche AL, Jacobson DA, Kainer D, Romero JC, Harfouche AH, et al. Accelerating climate resilient plant breeding by applying next-generation artificial intelligence. Trends in Biotechnology. 2019;37:1217-1235

100. Fahlgren N, Gehan MA, Baxter I. Lights, camera, action: high-throughput plant phenotyping is ready for a close-up. Current Opinion in Plant Biology. 2015;24:93-99

101. Kyratzis AC, Skarlatos DP, Menexes GC, Vamvakousis VF, Katsiotis A. Assessment of vegetation indices derived by UAV imagery for durum wheat phenotyping under a water limited and heat stressed mediterranean environment. Frontiers in Plant Science. 2017;8:1114

102. Atkinson JA, Pound MP, Bennett MJ, Wells DM. Uncovering the hidden half of plants using new advances in root phenotyping. Current Opinion in Biotechnology. 2019;55:1-8

103. Shakoor N, Lee S, Mockler TC. High throughput phenotyping to accelerate crop breeding and monitoring of diseases in the field. Current Opinion in Plant Biology. 2017;38:184-192

104. Yang W, Guo Z, Huang C, Duan L, Chen G, et al. Combining high-throughput phenotyping and genome-wide association studies to reveal natural genetic variation in rice. Nature Communications. 2014;5:5087

105. Singh AK, Ganapathysubramanian B, Sarkar S, Singh A. Deep learning for plant stress phenotyping: trends and future perspectives. Trends in Plant Science. 2018;23:883-898

106. Ray DK, Ramankutty N, Mueller ND, West PC, Foley JA. Recent patterns of crop yield growth and stagnation. Nature Communications. 2012;3:1293

107. Pfeiffer NE. Microchemical and morphological studies of effect of light on plants. Botanical Gazette. 1926;81:173-195

108. Watson A, Ghosh S, Williams MJ, Cuddy WS, Simmonds J, et al. Speed breeding is a powerful tool to accelerate crop research and breeding. Nature Plants. 2018;4:23-29

109. Hickey LT, Germán SE, Pereyra SA, Diaz JE, Ziems LA, et al. Speed breeding for multiple disease resistance in barley. Euphytica. 2017;213:64

110. Collard BC, Beredo JC, Lenaerts B, Mendoza R, Santelices R et al. Revisiting rice breeding methods–evaluating the use of rapid generation advance (RGA) for routine rice breeding. Plant Production Science. 2017;20:337-352

111. Riaz A, Athiyannan N, Periyannan S, Afanasenko O, Mitrofanova O, et al. Mining Vavilov's treasure chest of wheat diversity for adult plant resistance to Puccinia triticina. Plant Disease. 2017;101:317-323

112. Zhang Z, Mao Y, Ha S, Liu W, Botella JR, Zhu JK. A multiplex CRISPR/Cas9 platform for fast and efficient editing of multiple genes in Arabidopsis. Plant Cell Reports. 2016;35:1519-1533

113. Meuwissen THE, Hayes BJ, Goddard ME. Prediction of total genetic value using genome-wide dense marker maps. Genetics. 2001;157:1819-1829

114. Buckler ES, Ilut DC, Wang X, Kretzschmar T, Gore M, Mitchell SE. rAmpSeq: Using repetitive sequences for robust genotyping. BioRxiv. 2016;1:096628

115. O'Connor DJ, Wright GC, Dieters MJ, George DL, Hunter MN, et al. Development and application of speed breeding technologies in a commercial peanut breeding program. Peanut Science. 2013; 40:107-114

116. Septiningsih EM, Pamplona AM, Sanchez DL, Neeraja CN, Vergara GV, et al. Development of submergence-tolerant rice cultivars: the Sub1 locus and beyond. Annals of Botany. 2009;103:151-160

117. Li M, Li X, Zhou Z, Wu P, Fang M, et al. Reassessment of the four yield-related genes Gn1a, DEP1, GS3, and IPA1 in rice using a CRISPR/Cas9 system. Frontiers in Plant Science. 2016;7:377

118. Khanday I, Skinner D, Yang B, Mercier R, Sundaresan V. A male-expressed rice embryogenic trigger redirected for asexual propagation through seeds. Nature. 2019;565:91-95

119. Wang C, Liu Q, Shen Y, Hua Y, Wang J, et al. Clonal seeds from hybrid rice by simultaneous genome engineering of meiosis and fertilization genes. Nature biotechnology. 2019;37:283-286

120. Paliwal R, Röder MS, Kumar U, Srivastava JP, Joshi AK. QTL mapping of terminal heat tolerance in hexaploid wheat (T. aestivum L.). Theoretical and Applied Genetics. 2012;125:561-575

121. Alahmad S, Dinglasan E, Leung KM, Riaz A, Derbal N, et al. Speed breeding for multiple quantitative traits in durum wheat. Plant Methods. 2018;14:36

122. Dinglasan E, Godwin ID, Mortlock MY, Hickey LT. Resistance to yellow spot in wheat grown under accelerated growth conditions. Euphytica. 2016;209:693-707

123. Riaz A, Periyannan S, Aitken E, Hickey L. A rapid phenotyping method for adult plant resistance to leaf rust in wheat. Plant Methods. 2016;12:1-10

124. Hickey LT, Wilkinson PM, Knight CR, Godwin ID, Kravchuk OY, et al. Rapid phenotyping for adult-plant resistance to stripe rust in wheat. Plant Breeding. 2012;131:54-61

125. Rahman H, Jagadeeshselvam N, Valarmathi R, Sachin B, Sasikala R, et al. Transcriptome analysis of salinity responsiveness in contrasting genotypes of finger millet (Eleusine coracana L.) through RNA-sequencing. Plant Molecular Biology. 2014;85:485-503

126. Gelli M, Mitchell SE, Liu K, Clemente TE, Weeks DP, et al. Mapping QTLs and association of differentially expressed gene transcripts for multiple agronomic traits under different nitrogen levels in sorghum. BMC Plant Biology. 2016;16:16

127. Chen Y, Ma J, Zhang X, Yang Y, Zhou D, et al. A novel non-specific lipid transfer protein gene from sugarcane (NsLTPs), obviously responded to abiotic stresses and signaling molecules of SA and MeJA. Sugar Tech. 2017;19:17-25

128. Jiang Q, Tang D, Hu C, Qu J, Liu J. Combining meta-QTL with RNA-seq data to identify candidate genes of kernel row number trait in maize. Maydica. 2018;61:9

129. Waltz E. CRISPR-edited crops free to enter market, skip regulation. Nature Biotechnology. 2016a;34:582

130. Yuan Y, Xing H, Zeng W, Xu J, Mao L, et al. Genome-wide association and differential expression analysis of salt tolerance in Gossypium hirsutum L at the germination stage. BMC Plant Biology. 2019;19:394

131. Cardi T, Batelli G, Nicolia A. Opportunities for genome editing in vegetable crops. Emerging Topics in Life Sciences. 2017;1:193-207

132. Curtin SJ, Xiong Y, Michno JM, Campbell BW, Stec AO, et al. Crispr/cas9 and talen s generate heritable mutations for genes involved in small rna processing of Glycine max and Medicago truncatula. Plant Biotechnology Journal. 2018;16:1125-1137

133. Qi X, Li MW, Xie M, Liu X, Ni M, et al. Identification of a novel salt tolerance gene in wild soybean by whole-genome sequencing. Nature Communications. 2014;5:4340

134. Abdelrahman M, Jogaiah S, Burritt DJ, Tran LS. Legume genetic resources and transcriptome dynamics under abiotic stress conditions. Plant Cell and Environment. 2018;41:1972-1983

135. Singh D, Singh CK, Taunk J, Tomar RS, Chaturvedi AK, et al. Transcriptome analysis of lentil (Lens culinaris Medikus) in response to seedling drought stress. BMC genomics. 2017;18:206

136. Wen J, Jiang F, Weng Y, Sun M, Shi X, et al. Identification of heat-tolerance QTLs and high-temperature stress-responsive genes through conventional QTL mapping, QTL-seq and RNA-seq in tomato. BMC Plant Biology. 2019;19:398

137. Nekrasov V, Wang C, Win J, Lanz C, Weigel D, Kamoun S. Rapid generation of a transgene-free powdery mildew resistant tomato by genome deletion. Scientific Reports. 2017;7:482

138. Lor VS, Starker CG, Voytas DF, Weiss D, Olszewski NE. Targeted mutagenesis of the tomato PROCERA gene using transcription activator-like effector nucleases. Plant Physiology. 2014; 166:1288-1291

139. Ron M, Kajala K, Pauluzzi G, Wang D, Reynoso MA, et al. Hairy root transformation using Agrobacterium rhizogenes as a tool for exploring cell type-specific gene expression and function using tomato as a model. Plant Physiology. 2014;166:455-469

140. Soyk S, Lemmon ZH, Oved M, Fisher J, Liberatore KL, et al. Bypassing negative epistasis on yield in tomato imposed by a domestication gene. Cell. 2017;169:1142-1155

141. Ito Y, Nishizawa-Yokoi A, Endo M, Mikami M, Toki S. CRISPR/Cas9-mediated mutagenesis of the RIN locus that regulates tomato fruit ripening. Biochemical and Biophysical Research Communications. 2015;467:76-82

142. Lawrenson T, Shorinola O, Stacey N, Li C, Østergaard L, et al. Induction of targeted, heritable mutations in barley and Brassica oleracea using RNA-guided Cas9 nuclease. Genome Biology. 2015;16:258

143. Tian S, Jiang L, Gao Q, Zhang J, Zong M, et al. Efficient CRISPR/Cas9-based gene knockout in watermelon. Plant Cell Reports. 2017;36:399-406

144. Sawai S, Ohyama K, Yasumoto S, Seki H, Sakuma T, et al. Sterol side chain reductase 2 is a key enzyme in the biosynthesis of cholesterol, the common precursor of toxic steroidal glycoalkaloids in potato. The Plant Cell. 2014;26:3763-3774

145. Clasen BM, Stoddard TJ, Luo S, Demorest ZL, Li J, et al. Improving cold storage and processing traits in potato through targeted gene knockout. Plant Biotechnology Journal. 2016;14:169-176

146. Waltz E. Gene-edited CRISPR mushroom escapes US regulation. Nature News 2016b;532:293

147. Miao H, Sun P, Liu Q, Miao Y, Liu J, et al. Genome-wide analyses of SWEET family proteins reveal involvement in fruit development and abiotic/biotic stress responses in banana. Scientific Reports. 2017;7:3536

148. Verma SK, Jasrotia RS, Iquebal MA, Jaiswal S, Angadi UB, Rai A, Kumar D. Deciphering genes associated with root wilt disease of coconut and development of its transcriptomic database (CnTDB). Physiological and Molecular Plant Pathology. 2017;100:255-263

149. Arroyo-Herrera A, Figueroa-Yánez L, Castano E, Santamaría J, Pereira-Santana A, et al. A novel Dreb2-type gene from C arica papaya confers tolerance under abiotic stress. Plant Cell, Tissue and Organ Culture. 2016;125:119-133

150. Malnoy M, Viola R, Jung MH, Koo OJ, Kim S, et al. DNA-free genetically edited grapevine and apple protoplast using CRISPR/Cas9 ribonucleoproteins. Frontiers in Plant Science. 2016;7:1904

151. Osakabe Y, Liang Z, Ren C, Nishitani C, Osakabe K, et al. CRISPR–Cas9-mediated genome editing in apple and grapevine. Nature Protocols. 2018;13:2844-2863

CHAPTER-6

PLANT BREEDING AND MICROBIOME INTERACTIONS

1. INTRODUCTION: AN OVERVIEW

The intersection of plant breeding and microbiome research represents a frontier in agricultural science that holds promise for transforming the way we approach crop improvement. As we delve into the intricate relationships between plants and their associated microorganisms, it becomes increasingly clear that the plant microbiome plays a pivotal role in shaping plant health, growth, and resilience. The synergy between traditional plant breeding methods and a deeper understanding of the plant microbiome opens new avenues for enhancing crop productivity, disease resistance, and environmental sustainability.

Since plants colonized the land, they have developed mechanisms to respond to changing environmental conditions and to settle in extreme habitats. Recent studies indicate that several plant species require associations with microorganisms to tolerate stress and to survive [1]. The human contribution to plant breeding has not only been the development of new breeding methods, but also the acceleration of progress in the evolution of crops.

On the other hand, in recent years the interest in the use of rhizobacteria that promote plant growth has increased. The beneficial effects of these microorganisms involve the ability to act as phytohormones or biofertilizers, increasing the yield of many important crops. Ecological factors such as temperature and nutritional conditions of the soil affect the behavior of microorganisms; inoculation has a better stimulating effect on plant growth in nutrient-deficient soil conditions than in fertile soils [2].

Although most plants lack the adaptive capacity under stress conditions, this ability seems to be associated with certain microorganisms, which suggests asking the question Through what mechanisms can microorganisms and plants adapt to stress conditions? Can plant species improve their tolerance to environmental stresses when associated with certain microorganisms? The answer to these questions could change our concepts about plant breeding and could lead us to new routes towards sustainability.

If food production is to increase by 50% in the next 40 years in a scenario of scarce resources and climate change, it will require a considerable investment in capital, time, and effort. A major component of the solution will have to start from the improvement in agricultural technologies, to produce sufficient and safe food that meets the needs and preferences of the human population, without affecting the sustainability of the natural environment.

From a conceptual perspective, the effects of microorganisms on plants have long been grouped under the idea of "promoting and regulating plant growth." However, the microbiota associated with plants influences multiple regulatory cascades of plants that together define their phenotype. In addition, the effect of the modified phenotype will depend on the context, as a function of the abiotic and biotic environmental parameters, giving rise to new phenotypes through the joint modification of genomic information and the microbiota associated with plants [3].

The plant microbiome not only helps plants survive in the ecosystem but also offers critical genetic variability, hitherto little used as a strategy by plant breeders, who have traditionally exploited only the genetic variability of the host plant to develop improved varieties of high yield or with tolerance to diseases, pests and abiotic stress [4]. In the words of Walters [5], resistance induced by microorganisms has the potential to revolutionize disease control in crops, but it remains an unconventional type of crop management. This is the subject of intense research at present.

2. CROP DOMESTICATION AND MICROBIAL DIVERSITY

Domestication refers to the selection and artificial reproduction of wild species to obtain cultivated variants that thrive in man-made niches and that meet human or industrial requirements [6]. Several genotypic and phenotypic forms of domestication have been described in crops and animals, however, domestication is not exclusive to higher organisms.

Before domestication, the wild ancestors of cultivated plants evolved in association with a wide set of microorganisms and insects, with which they participated in pathogenic, predatory, commensal, and mutualistic interactions [7]; most of the species of insects and microorganisms associated with crops in their centers of origin remain to be described [8].

Domestication has transformed hundreds of wild plant species into productive crops for human use. However, cultivation practices and intense artificial selection for yield can come at a hidden cost: disrupting interactions between plants and beneficial microbiota. To improve agricultural production and sustainability, research must develop breeding methods to optimize symbiotic results in crop species [9]. Microbial diversity has also been shaped by the emergence of new, highly specific, man-made environments such as food and beverage fermentations [6]. The domestication of plants is now recognized as a major driver of microbial diversity associated with plants [10]. Among other traits, domestication has changed root architecture, exudation, or defense responses that could have modified plant microbiota, as explained by Martínez-Romero et al. [11]. The authors present the comparison of reported data on the microbiota from cereals and legumes and their ancestors, showing that different bacteria were found in domesticated and wild plant microbiomes.

To date, a few hundred genes and loci have been identified by classical genetic and association mapping as targets of domestication and post domestication divergence. However, only a few of these have been characterized, and for even fewer is the role of the wild-type allele in natural populations understood [12].

According to Pérez-Jaramillo et al. [13], there is an impact of the domestication of crops on soil management, phenotypes, physiology, and the diversity of rhizobacteria associated with crops; the domestication process may have had an adverse impact on the composition and functions of the associated microbiota. In this regard, Martín-Robles [14] states that colonization by mycorrhizae is lower and the infection rate by nematodes is higher in the roots of plants that grow in soils previously cultivated by domesticated plants. Furthermore, domesticated plants showed lower mycorrhizal colonization and higher nematode infection rates than their wild progenitors.

Berg and Raaijmakers [10] argue that seeds are the place of residence of various microbial communities whose composition is determined by the genotype of the plant, the environment, and management practices. To what extent and how domestication affects the microbiomes in plant seeds is still little studied and the microbiomes present in the seeds of wild relatives of modern cultivars should be used to conserve and restore beneficial microorganisms to induce tolerance to biotic and abiotic stress.

The results of Shi et al. [15] indicate that the rhizosphere fungal communities in rice are more influenced than the bacterial communities by the domestication of crops; a more intense relationship has also been verified for fungi and bacteria in cultivated species than in their wild relatives. According to the aforementioned study, wild rice and soybean varieties have a greater abundance of beneficial symbionts and a lower abundance of pathogens compared to cultivated varieties; the domestication process may have a more pronounced effect on fungal communities than on bacterial communities, affecting the microbial relationship in the rhizosphere of these crops.

Leff et al. [16] indicate that the communities of associated fungi in the sunflower (*Helianthus annuus*) are more strongly influenced by the genetic factors of the host plant than the bacterial communities, a finding that could influence the strategies to optimize the use of microbial communities to improve crop yield, suggesting that there would be a vertical transmission of fungi from seeds to adult plants.

Analyzing the evolution of *Phaseolus vulgaris*, Pérez-Jaramillo et al. [17] observed a gradual decrease in the relative abundance of Bacteroidetes, mainly Chitinophagaceae and Cytophagaceae, and an increase in the relative abundance of Actinobacteria and Proteobacteria, in particular Nocardiaceae and Rhizobiaceae, establishing a link between the domestication of the common bean, the specific morphological traits in the root and the community of rhizobacteria associated with this species.

Given the planting system as a monoculture, maize can be seen as a crop responsible for shaping the agricultural environment for the species that cohabit with it [18]. Brisson et al. [19] argue that the analysis of the co-occurrence network in the case of corn revealed that the microbial co-occurrence patterns in the rhizosphere of pure lines of corn were significantly more similar to those of teosinte (an ancestor of corn) than modern hybrids. These results suggest that advances in the development of maize hybrids have had a significant impact on the microbial communities of the rhizosphere and on the assembly of their interaction networks. Wagner et al. [20] found that interactions with soil microorganisms are important for the expression of heterosis in corn and Zambonin et al. [21], found no significant interaction between corn hybrids and inoculation with *Azospirillum* sp. for the variables studied including grain yield, and the specificity between maize hybrids and inoculation was not verified.

According to Walters et al. [22], some local varieties of corn grown under traditional agricultural practices with little or no fertilizer could have developed strategies to improve grain yield under conditions of low nitrogen content in the soil and in these varieties of corn, 29% - 82% of the assimilated nitrogen was derived from the atmospheric form N_2. Rangel-Lucio et al. [23], found a degree of affinity or effect of the homologous strain of Azospirillum obtained from traditional H-28 and Chalqueño maize, and the re-inoculation in these same varieties and its subsequent recognition of the bacterial strain in modern varieties.

The composition of the metabolites in the rhizosphere of wheat is associated with differences between the genotypes of the domestication groups of this species, determined by a high heritability in some of these metabolites. In general, domestication and reproduction have had important effects on the exudates in the wheat rhizosphere, suggesting the adaptive nature of these changes [24]. Furthermore, the prominent role of neutral processes in the assembly of the domesticated wheat microbiota has been revealed and it has been proposed that domestication has relaxed the selective processes in the wheat microbiota [25].

Cultivated tomatoes (*Solanum lycopersicum*) are more likely to have negative feedback between plants and soil than wild parents according to Carrillo et al. [26], which could partially explain its sensitivity to monoculture in agricultural soils.

3. THE MICROBIOME: A SECOND GENOME FOR PLANTS

The biome is characterized by multiple complex interactions between plants and the associated microbiota, that is, endophytes with different functions, including pathogenic microorganisms, and the environment. Hardoim [27] maintains that the phenotype of a plant is not only determined by its response to the environment, but also by the associated microbiota, the response of the microbiota to the environment, and the complex interactions between members of the ecosystem.

According to Corbin et al. [28], understanding what makes a plant a suitable host for its microbiota is essential to take advantage of the plant-microorganism complex in improving crops. Identifying the genes that allow plants to regulate the assembly of microbiota in their roots is essential for future breeding programs aimed at sustainably improving productivity and product quality.

Tosi et al. [29] have reviewed current strategies for the manipulation of the plant microbiome and classify them as (i) introduction and engineering of microbiomes, (ii) reproduction and engineering of the plant-host relationship, and (iii) selecting agricultural practices that improve soil and plant-associated microbial communities.

Wei and Jousset [3] propose an alternative framework to produce new phenotypes by modifying genomic information and the microbiota associated with plants, thanks to a novel technology that would allow the transmission of the endophytic microbiota to the next generation of plants. The authors indicate that more studies are still needed to implement reproduction at the holobiont level, possibly due to the limited vertical transmission of microorganisms. Even if the bacteria could be transmitted reliably in F1, they would disappear in subsequent generations during the selection process to achieve the desired phenotype.

Sessitsch and Mitter [30] consider that plants could be improved by breeding methods, in relation to greater efficiency in their interaction with microorganisms. While in recent decades, crops have been improved and selected for higher yield and resistance to pests and diseases, it is anticipated that efficient interaction with certain beneficial microorganisms will be an additional factor in plant reproduction. New agricultural practices may include microbiome reproduction, and engineering of specific microbiomes, for example, through strategic soil amendments in which the selective addition of plant exudates can maintain beneficial microorganisms, or through the direct application of microbial consortia. as probiotics [31, 32].

4. SUSTAINABLE AGRICULTURE AND THE PLANT MICROBIOME

Sustainable agricultural practices are a response to the multifaceted problems that have originated from the prolonged and indiscriminate use of chemicals to improve crop production for many decades, for this reason, the search for eco-friendly options to replace chemical fertilizers and pesticides has been accentuated [33].

According to Sessitsch and Mitter [30] in many parts of the world where low-income agriculture is a common situation and improved germplasm or agricultural practices are poorly available, make better use of the functions of plants microbiomes. It will especially support agricultural production under these conditions and will promote the bioeconomy in less developed countries that use microbial inoculants and that can establish strain collections for local environments.

The functioning of the ecosystem is largely governed by the microbial dynamics of the soil; many global ecosystems are in various states of vulnerability, evidenced by erosion, low productivity, and poor water quality caused by intensive agricultural activity and continued use of land resources. Microorganisms in agricultural soil are known to exert profound influences on soil fertility status, particularly with regard to nutrient availability, as well as suppression of plant diseases [34].

There is evidence that soil biodiversity confers stability to stress and disturbance [35], but the mechanism is not yet fully understood; seems to depend on the type of stress and disturbance or a combination of both. Alternatively, the structure of the soil biotic community can play a role in the resilience of the agro-ecosystem; however, possible explanations for this require further investigation.

Many bacterial strains with growth-promoting activity have been reported belonging to the genera Azoarcus, Azospirillum, Azotobacter, Arthrobacter, Bacillus, Clostridium, Enterobacter, Gluconacetobacter, Pseudomonas, and Serratia, among them, the species *Pseudomonas* sp. and *Bacillus* sp. are the most extensively studied [36]. The diversity of microorganisms associated with plant roots is enormous, in the order of tens of thousands of species. This complex microbial community associated with the rhizosphere is considered the second genome of the plant and is crucial for the health and nutrition of crops [37].

According to Finkel et al. [38], the study of plant microbiomes has drawn on both holistic ecological studies and mechanistic discoveries; both schools of thought are giving an increasingly close view of the ecological processes that govern the interactions between plants and microorganisms, as well as their molecular mechanisms.

Busby et al. [39] identified five priority themes for research in the study of the plant microbiome and its effect on agricultural sustainability: (1) development of microbiome-host model systems for cultivated and wild plants, with collections of associated microbial cultures and reference genomes, (2) definition of the main microbiomes and metagenomes in these model systems, (3) elucidate the rules of synthetic assembly in functional microbiomes, (4) determine the mechanisms of microorganisms in plant-microbiome interactions, and (5) characterize and refine the genotype-plant-environment-microbiome interaction. Achieving these goals could accelerate our ability to design and implement effective management of agricultural microbiomes and develop strategies that will in turn generate solutions for both consumers and producers for the global food supply.

It has been shown that there are bacterial strains capable of fixing N in non-legume species. Dent and Cocking [40] showed that strains of *gluconacetobacter diazotrophicus*, a non-nitrogen-fixing, non-nodule bacterium, isolated from the intercellular juice of sugar cane, were inoculated under specific conditions to intracellularly colonize the roots and shoots of cereals (wheat, corn, and rice) as well as in crops as diverse as potatoes, tea, oilseeds, and tomato significantly improved yields, both in the presence or absence of synthetic nitrogen fertilizers, possibly due to a combination of symbiotic intracellular nitrogen fixation, increased photosynthesis rate and the activity of additional plant growth factors. Van Deynze et al. [41] proposed a model for the association of nitrogen-fixing microbes with corn maize mucilage and identified the main functionalities for a productive diazotrophic association.

Pineda et al. [42] state that since insects can severely affect productivity in ecosystems, resistance to agricultural pests through the microbiome related to plants should be considered a key service to be included as a strategy in the management of a farm. Many of the ecosystem services of the soil microbiome are often not very effective in conventional production systems that use chemical pesticides and fertilizers and only become apparent when plants are exposed to abiotic stress conditions.

Berg et al. [43] suggest that combined strategies for genetic improvement and biocontrol should be developed to maintain the diversity and health of the ecosystem. The practices used in plant breeding, seed treatment, and agriculture, often caused by poor knowledge of the importance of endophytic fungi, are among the reasons for the loss of diversity of endophytic fungi in domesticated plants and also explain the efficacy reduced of some endophytic strains to confer benefits to plants.

According to Lugtenberg et al. [44], endophytic fungi play a key role in plant adaptation, resulting in higher yields and protection against biotic and abiotic stress, encoding a variety of secondary metabolites, including volatile organic compounds, especially in tolerant corn and rice. to a variety of stresses and for better postharvest control.

Horner et al. [45] indicate that modern agricultural practices have greatly increased crop production but have negatively affected soil health, suggesting that in the case of pea *(Pisum sativum)* the diversification of varieties can increase yield and promote interactions microbial, although the impacts on the associated microbial communities are unclear, despite the fundamental role in the functioning of the ecosystem.

5. CONCISE SUMMARY

To increase the level of sustainability, various agroecological management strategies of agricultural production systems have been reviewed, including the use of microorganisms in the process of plant domestication, genetic improvement, and the production of improved varieties. According to the references presented in the scientific literature, to increase agricultural production and the sustainability of production systems, future research should develop breeding methods that optimize symbiosis in crops [9], since the interactions with soil microorganisms could be important for the expression of heterosis in some species [20], proposing an alternative framework to produce new phenotypes by modifying genomic information and the microbiota associated with plants [3]; efficient interaction with certain beneficial microorganisms is expected to be an additional factor in plant production. These new agricultural practices can include the reproduction of the microbiome, the transplantation and engineering of specific microbiomes [30] and their transfer through seeds [3, 43] or *in vitro* culture [46]. This complex microbial community associated with the rhizosphere is considered the second genome of the plant and is crucial for the health and nutrition of crops [37].

AT A GLANCE

In this review, references to the use of microorganisms in the process of plant domestication, genetic improvement, and production of traditional and improved varieties have been identified. The domestication process may have had an adverse impact on the composition and functions of the associated microbiota and the microbiota associated with plants influences multiple regulatory processes of plants that together define their phenotype. According to scientific evidence, to increase agricultural production and the sustainability of production systems, future research should develop breeding methods that optimize the symbiosis between plants and microorganisms, to produce new plant phenotypes that result in the production of enough food to meet the needs of the human population.

REFERENCES

1. Velasco-Jiménez A, Castellanos-Hernández O, Acevedo-Hernández G, Aarland RC, Rodríguez-Sahagún A. Bacteriasrizosféricas con beneficiospotencialesen la agriculturaRhizospheric bacteria with potential benefits in agriculture. Terra Latinoamericana. 2020;38(2):333-345
2. Egamberdieva D. The Management of Soil Quality and Plant Productivity in stressed environment with Rhizobacteria. In: Maheshwari DK, editor. Bacteria in Agrobiology: Stress Management. Berlin Heidelberg: Springer-Verlag; 2012
3. Wei, Z., &Jousset, A. (2017). Plant breeding goes microbial. Trends in Plant Science, 22(7), 555-558 DOI: https://doi.org/10.1016/j.tplants.2017.05.009
4. Gopal M, Gupta A. Microbiome selection could spur next-generation plant breeding strategies. Frontiers in microbiology. 2016. DOI: 10.3389/fmicb.2016.01971
5. Walters DR. Induced resistance: Destined to remain on the sidelines of crop protection? Phytoparasitica. 2009;38(1):1-4. DOI: 10.1007/s12600-009-0067
6. Steensels J, Gallone B, Voordeckers K, Verstrepen KJ. Domestication of industrial microbes. Current Biology. 2019;29(10):381-393
7. Chen YH, Langellotto GA, Barrion AT, Cuong NL. Cultivation of domesticated rice alters arthropod biodiversity and community composition. Annals of the Entomological Society of America. 2013;106:100-110. DOI: 10.1603/AN12082
8. Chen YH, Gols R, Benrey B. Crop domestication and naturally selected species interactions. Annual Review of Entomology. 2015;60:35-58. DOI: 10.1146/annurev-ento-010814-020601
9. Porter SS, Sachs JL. Agriculture and the disruption of plant–microbial Symbiosis. Trends in Ecology & Evolution. 2020;35(5):426-439
10. Berg G, Raaijmakers JM. Saving seed microbiomes. The ISME Journal. 2018;12(5):1167-1170
11. Martínez-Romero E, Aguirre-Noyola JL, Taco-Taype N, Martínez-Romero J, Zuñiga-Dávila D. Plant microbiota modified by plant domestication. Systematic and Applied Microbiology. 2020;43(5):126106
12. Smýkal P, Nelson M, Berger J, Von Wettberg E. The impact of genetic changes during crop domestication. Agronomy. 2018. DOI: 10.3390/agronomy8070119
13. Pérez-Jaramillo JE, Carrión VJ, de Hollander M, Raaijmakers JM. The wild side of plant microbiomes. Microbiome. 2018;6(1):143
14. Martín-Robles N, García-Palacios P, Rodríguez M, Rico D, Vigo R, Sánchez-Moreno S, et al. Crops and their wild progenitors recruit beneficial and detrimental soil biota in opposing ways. Plant and Soil. 2020:1-15

15. Shi S, Chang J, Tian L, Nasir F, Ji L, Li X, et al. Comparative analysis of the rhizomicrobiome of the wild versus cultivated crop: Insights from rice and soybean. Archives of Microbiology. 2019;201(7):879-888
16. Leff JW, Lynch RC, Kane NC, Fierer N. Plant domestication and the assembly of bacterial and fungal communities associated with strains of the common sunflower, Helianthus annuus. New Phytologist. 2017;214(1):412-423
17. Pérez-Jaramillo JE, Carrión VJ, Bosse M, Ferrão LF, de Hollander M, Garcia AA, et al. Linking rhizosphere microbiome composition of wild and domesticated Phaseolus vulgaris to genotypic and root phenotypic traits. The ISME Journal. 2017;11(10):2244-2257
18. Peiffer, J. A., Spor, A., Koren, O., Jin, Z., Tringe, S. G., Dangl, J. L., Buckler, E .S. & Ley, R. E. (2013). Diversity and heritability of the maize rhizosphere microbiome under field conditions. Proceedings of the National Academy of Sciences, 110(16), 6548-6553
19. Brisson VL, Schmidt JE, Northen TR, Vogel JP, Gaudin AC. Impacts of maize domestication and breeding on rhizosphere microbial community recruitment from a nutrient-depleted agricultural soil. Scientific Reports. 2019;9(1):1-14
20. Wagner MR, Tang C, Salvato F, Clouse KM, Bartlett A, Sermons S, et al. Microbe-dependent heterosis in maize. bioRxiv. 2020:2020.05.05.078766
21. Zambonin G, Pacentchuk F, Lima FN, Huzar-Novakowiski J, Sandini IE. Response of maize crop hybrids, with different transgenic events, to inoculation with Azospirillumbrasilense. Applied Research & Agrotechnology. 2019;12(1):33-40
22. Walters WA, Jin Z, Youngblut N, Wallace JG, Sutter J, Zhang W, et al. Large-scale replicated field study of maize rhizosphere identifies heritable microbes. Proceedings of the National Academy of Sciences. 2018;115(28):7368-7373
23. Rangel-Lucio JA, Rodríguez-Mendoza M, Ferrera-Cerrato R, Castellanos-Ramos J, Ramírez-Gama RM, Alvarado-Bárcenas E. Afinidad y efecto de Azospirillum sp. enmaíz. AgronomíaMesoamericana. 2011;22(2):269-279
24. Iannucci, A., Fragasso, M., Beleggia, R., Nigro, F., & Papa, R. (2017). Evolution of the crop rhizosphere: Impact of domestication on root exudates in tetraploid wheat (Triticum turgidum L.). Frontiers inplant science, 8, 2124
25. Hassani, M. A., Özkurt, E., Franzenburg, S., &Stukenbrock, E. H. (2020). Ecological assembly processes of the bacterial and fungal microbiota of wild and domesticated wheat species. Phytobiomes Journal, PBIOMES-01
26. Carrillo J, Ingwell LL, Li X, Kaplan I. Domesticated tomatoes are more vulnerable to negative plant-soil feedbacks than their wild relatives. Journal of Ecology. 2019;107(4):1753-1766

27. Hardoim PR, van Overbeek LS, Berg G, Pirttilä AM, Compant S, Campisano A, et al. The hidden world within plants: Ecological and evolutionary considerations for defining functioning of microbial endophytes. Microbiology and Molecular Biology Reviews. 2015;79(3):294-320. DOI: 10.1128/MMBR.00050-14

28. Corbin KR, Bolt B, López CMR. Breeding for beneficial microbial communities using Epigenomics. Frontiers in Microbiology. 2020;11

29. Tosi M, Mitter EK, Gaiero J, Dunfield K. It takes three to tango: The importance of microbes, host plant, and soil management to elucidate manipulation strategies for the plant microbiome. Canadian Journal of Microbiology. 2020;66(7):413-433

30. Sessitsch A, Mitter B. 21st-century agriculture: Integration of plant microbiomes for improved crop production and food security. Microbial Biotechnology. 2015;8(1):32-33. DOI: 10.1111/1751-7915.12180

31. Arif I, Batool M, Schenk PM. Plant microbiome engineering: Expected benefits for improved crop growth and resilience. Trends in Biotechnology. 2020;S0167-7799(20):30121-30129 https://doi.org/10.1016/j.tibtech.2020.04.015

32. Bakker MG, Manter DK, Sheflin AM, Weir TL, Vivanco JM. Harnessing the rhizosphere microbiome through plant breeding and agricultural management. Plant and Soil. 2012;360(1-2):1-13

33. Prashar P, Kapoor N, Sachdeva S. Rhizosphere: Its structure, bacterial diversity and significance. Reviews in Environmental Science and Bio/Technology. 2014;13(1):63-77

34. Kennedy, A. C., & Smith, K. L. 1995. Soil microbial diversity and the sustainability of agricultural soils. In: The significance and regulation of soil biodiversity (pp. 75-86). Springer Netherlands

35. Brussaard L, De Ruiter PC, Brown GG. Soil biodiversity for agricultural sustainability. Agriculture, Ecosystems & Environment. 2007;121(3):233-244

36. Maheshwari, D.K. (2011). Bacteria in Agrobiology: Crop Ecosystems. DOI 10.1007/978-3-642-18357-7_2, Springer-Verlag Berlin, Heidelberg

37. Berendsen RL, Pieterse CM, Bakker PA. The rhizosphere microbiome and plant health. Trends in Plant Science. 2012;17(8):478-486

38. Finkel OM, Castrillo G, Paredes SH, González IS, Dangl JL. Understanding and exploiting plant beneficial microbes. Current Opinion in Plant Biology. 2017;38:155-163

39. Busby PE, Soman C, Wagner MR, Friesen ML, Kremer J, Bennett A, et al. Research priorities for harnessing plant microbiomes in sustainable agriculture. PLoS Biology. 2017;15(3):e2001793

40. Dent D, Cocking E. Establishing symbiotic nitrogen fixation in cereals and other non-legume crops: The greener nitrogen revolution. Agriculture & Food Security. 2017;6(1):7-9
41. Van Deynze A, Zamora P, Delaux P-M, Heitmann C, Jayaraman D, Rajasekar S, et al. Nitrogen fixation in a landrace of maize is supported by a mucilage-associated diazotrophic microbiota. PLoS Biology. 2018;16(8):e2006352
42. Pineda A, Kaplan I, Bezemer TM. Steering soil microbiomes to suppress aboveground insect pests. Trends in Plant Science. 2017;22(9):770-778
43. Berg G, Köberl M, Rybakova D, Müller H, Grosch R, Smalla K. Plant microbial diversity is suggested as the key to future biocontrol and health trends. FEMS Microbiology Ecology. 2017;93(5)
44. Lugtenberg BJ, Caradus JR, Johnson LJ. Fungal endophytes for sustainable crop production. FEMS Microbiology Ecology. 2016;92(12)
45. Horner A, Browett SS, Antwis RE. Mixed-cropping between field pea varieties alters root bacterial and fungal communities. Scientific Reports. 2019;9(1):1-10
46. Naqqash T, Hameed S, Imran A, Hanif MK, Majeed A, van Elsas JD. Differential response of potato toward inoculation with taxonomically diverse plant growth promoting Rhizobacteria. Frontiers in Plant Science. 2016;7:144. DOI: 10.3389/fpls.2016.00144

CHAPTER-7

APPROACHES TO ENHANCE BIOTIC STRESS RESISTANCE IN VEGETABLES THROUGH BREEDING

1. INTRODUCTION: AN OVERVIEW

The escalating challenges posed by biotic stresses in vegetable cultivation have spurred a concerted effort in the field of plant breeding to develop resilient varieties capable of withstanding such pressures. Biotic stresses, primarily induced by pathogens and pests, significantly impede crop productivity and quality, posing a constant threat to global food security. In response to this imperative, breeding approaches tailored to enhance biotic stress resistance in vegetables have gained prominence. These approaches involve the meticulous selection and development of cultivars with inherent or induced resistance mechanisms. By harnessing advancements in genetics, genomics, and molecular biology, plant breeders aim to fortify vegetable crops against the diverse array of biotic stressors, ensuring sustainable and robust agricultural practices. This quest for resilience not only addresses immediate concerns but also aligns with the broader goal of creating a more secure and resilient global food supply.

Globally the area under vegetable cultivation is growing annually at the rate of 4.12% and production by 6.48%. The mean productivity is 15.49mt/ha which is quite low. In vegetables infestation of biotic stresses reflect highly on production productivity and quality. Application of chemicals leaves chemical residues in vegetables above threshold levels. Resistance is a relative attribute and refers to the ability of the plant to withstand the pest or pathogen. The susceptible plant shows severe symptoms due to which yield loss occur. A completely resistant plant shows nil reaction and a moderately resistant or field tolerant plant develops less disease development. Plants have selective resistance to some pests or pathogens and susceptible to others. They are species-specific or strain-specific. The rate of spread depends on the pest load or population, spore count and multiplication rate of pest/pathogen.

Improvement of cultivated plants through tapping germplasm resources depends on introducing variability through traditional and molecular breeding techniques. Wild species provides a vast gene pool for resistance development. They have been used for decades to transfer genes of resistance or tolerance to the cultivated species. The use of wild species in breeding varieties particularly for increased vigor and resistance has been well recognized [1]. Introgression is the movement of genes or gene flow from one species into the gene pool. Inter specific hybridisation breaks the species barrier for gene transfer and makes it possible to transfer the resistant genes.

Complete exploitation of genetic variation enables the breeder to produce not only heterotic F1 hybrids but also recombinants with desirable attributes. Further, selection based on genetic nature will be highly useful to a great extent to screen out the parents and hybrids. Identification of resistance is also possible through quantifying the biochemical components present in the genotype. Further, in view of less marked host specificity, a plant breeding programme for insect resistance has to be handled separately from that of disease resistance.

In 3rd century B.C, Theophrastus observed that degree of resistance differ among varieties. It was later established in 1894, by Erikson that though pathogens are morphologically similar, they differ among each other in their ability to attack host plant. In 1911 Barrus narrated that various isolates of a pathogen differ in its ability to attack different varieties of the same plant species. This made the basis for the identification of physiological races and pathotypes. It was then called as pathogenecity *i.e* the infection of a host strain by a pathogen is genetically determined. In 1955, Flor formulated the of gene-for-gene hypothesis which denotes the relationship between host and pathogen. According to that disease resistance is determined by host and the genotype of the pathogen. The hosts differ in type of resistance while the pathogen differ in pathogenicity, but both are genetically controlled. The pathogen has the capacity to generate new variations in pathogenicity by reproduction methods and mutation. Therefore, the task of the breeder is to develop varieties resistant to the prevalent pathotypes of the pathogen and also for the new pathogen genotypes which will arise in future.

2. GENETIC BASIS OF RESISTANCE

According to the experimental results so far reported in vegetable crops, the genetic basis of insect resistance is monogenic. The resistance of muskmelon to melon aphid. The tolerance of muskmelon to western biotypes of *Aphis gossypii* in breeding line LJ 90234 was governed by a single dominant gene [2]. Inheritance studies of fruit fly resistance in pumpkin cultivar Arka Suryamukhi showed that the resistance was controlled by a dominant gene. Similar studies in water-melon also indicated that the resistance to fruit fly was governed by a single dominant gene. The work on *Cucurbita pepo* revealed that the resistance to squash bug, was controlled by at least 3 genes and gene action appeared to be additive in nature. In an interspecific cross between resistant *Cucumis callosus* and susceptible *Cucumis melo* it was revealed that the susceptibility to fruit fly was governed by two pairs of complimentary genes. While working with tomato for resistance to fruitworm Fery and Cuthbert [3] reported that the antibiotics factor present in *Lycopersicon hirsutum* appeared to be inherited recessively.

Since interest in resistant vegetable varieties started more than half a century ago work has been done on major insect pests. However, studies have shown that the Mendelian segregation has led to the identification of major genes and that the alleles for resistance were dominant over those for susceptibility in number of instances except in some where it was found to be additive or complementary gene action or recessive.

The genetics of disease resistance was first studied by Britten in 1905. Then Person and Sidhu [4] reviewed 1000 Published papers and concluded that regardless of species, resistance generally segregated in the mendelian ratios. Resistance was dominant over susceptibility. Resistance in vegetable crops have been reported to be governed by mono or oligo or polygenes and effect of genes may be additive or dominant or epistatic. The information on inheritance of various diseases of vegetables is very meager. However, some workers reported different kinds of nature of inheritance. Resistance to buck eye rot of tomato appeared to be dominant over susceptibility. Resistance to fusarium wilt of tomato was conditioned by a single dominant gene. Tomato leaf curl virus is transmitted by white fly and is most serious problem. According to Som and Chaudhary [5], resistance to TLCV was incompletely dominant and governed by polygenes.

Resistance to most of the diseases in watermelon is controlled by a single dominant gene. Walker [6] reported resistance to fusarium wilt in watermelon as recessive. Powdery mildew is a major limiting factor in the production of muskmelon in most of the parts of the world. Resistance to *Erysiphaecichoracearum* race- 1 and race-2 is monogenic dominant. A study on resistance to powdery mildew caused by *Sphaerotheca fuligina*, in two resistant varieties campo and PMR-6 indicated that they have the same locus/loci conferring resistance. Genetic studies of resistance to *E. cichoracearum* race-2 had indicated that resistance is partly dominant and controlled by Pm-2 [7]. Resistance to downy mildew (*Pseudoperonospora* cubensis) of muskmelon in PI 124111 is controlled by two independently dominant gene [8]. Whereas resistance in PI 124112 was controlled by two partially dominant genes [9].

Antonio *et al.* [10] studied the inheritance of resistance by antixenosis for tomato leaf miner and reported that the inheritance of antixenosis resistance of genotype BGH-1497 is ruled by a polygenes in epistatic interactions, with a phenotypic proportion of 13:3 between susceptible and resistant genotypes respectively. In another experiment Gabriele Vitelli et al. [11] reported three transgenic eggplant lines bearing a mutagenized *Bacillus thuringiensis* Berl. gene coding for the Cry3B toxin. The fruit production was almost twice in the highly resistant lines (3–2 and 9–8). The 6–1 transgenic line showed an intermediate level of resistance. Analysis by double antibody sandwich–enzyme linked immune sorbent assay (DAS–ELISA), performed on different tissues, revealed a lower amount of Cry3B protein in the 6–1 transgenic line.

The root knot nematode, *Meloidogyne* is one of the most economically damaging plant parasitic nematode and is widely distributed throughout world [12]. The genus *Meloidogyne* is composed of 100 species, with *M. arenaria, M. incognita, M. hapla and M. javanica* being considered as "major" species [13]. Natural resistance genes "R-genes" are responsible for inducing resistance against nematodes in tomato. The genes Mi-1, Mi-2, Mi-3, Mi-4, Mi-5, Mi-6, Mi-7, Mi-8, Mi-9, and Mi-HT confer resistance to the root Knot nematode [14].

3. BIOCHEMICAL BASIS OF RESISTANCE

The biochemical factors are more important than morphological and physiological factors in conferring resistance. Some biochemical constituents may act as feeding stimuli for insects. Occurrence at lower concentration or total absence of such biochemical leads to non preference, a form of insect resistance [15, 16]. The biochemical constituents like glycoalkaloid (solasodine), phenols, phenolic oxidase enzymes *viz.*, polyphenol oxidase and peroxidase are available in plants and these biochemical constituents possess insect resistant properties [17]. It was also recorded that the maximum polyphenol oxidase activity is available in fruit (0.388 in fruit as changes in OD min-1 g-1 of sample). Several workers have reported that the biochemical constituents act as stimulants of resistance mechanism. In brinjal, Praneetha [18] and Prabhu [19] have recorded that the biochemical constituents also contribute to confer resistance to shoot and fruit borer.

Studies on the biochemical basis of resistance to *Leucinodesorbonalis* and their correlation with shoot and fruit borer damage in five selected brinjal genotypes were done during June to December 2005 and it was shown that less susceptible genotypes for both shoot and fruit borer had higher amount of polyphenol oxidase (PPO), phenylalanine ammonium lyase (PAL), lignin and lower reducing sugar. Significant negative correlation was established with per cent infestation of shoot and fruit borer and PPO, PAL and lignin, whereas it was positively correlated with reducing sugar. Negative correlation was observed with the biochemical constituents, PPO, PAL, lignin and reducing sugar but PPO was positively correlated with PAL and lignin content and vice-versa.

4. BIOPHYSICAL BASIS OF RESISTANCE

In a study Silva et al. [20, 21] evaluated ninety-nine F3 families derived from an interspecific cross using *Solanum lycopersicum* and *Solanum pimpinellifolium* 'TO-937-15' (multiple pest resistance accession with type IV glandular trichomes and acylsugar accumulation) for their resistance against the whitefly. The higher resistance levels of BTR331 were associated with a positive combination of higher type IV trichome density and higher acylsugar levels. From the breeding stand point, the genetic similarity between *S. lycopersicum* and *S. pimpinellifolium* would allow a more efficient resistance introgression by facilitating recombination and minimizing the potentially undesirable linkage drag associated with this trait.

Niranjana et al. reported that the biophysical characteristics in brinjal genotypes *viz.*, shoot thickness at 2.5 cm below the tip, number of trichomes on lower surface of leaves, pedicel length, calyx length and diameter of fruit were correlated with the level of infestation by *L. Orbonalis*. Results revealed that the infestation in shoot was not significantly correlated with number of trichomes on leaves and positively correlated with shoot thickness. Fruit infestation was positively but not significantly correlated with length of pedicel and calyx whereas non-significant and negative correlation was recorded between fruit infestation and fruit characters *viz.*, length and diameter of fruit. The shape and color of fruit had no significant influence on the level of infestation.

5. CATEGORIZATION OF RESISTANCE

According to the response of the plant to the pathogen.

- **Susceptible:** In this the disease development is abundant and can not be checked by the plant.
- **Resistant:** It is lesser disease development than the susceptible and it is a relative attribute. In this the plants will be infected and establishment take place but the progression in the host plant will be limited. As a result these plants exhibit minor symptoms than the susceptible ones.
- **Tolerant:** Tolerance implies that the host is attacked by the pathogen but there is no less in biomass production or yield.

Based on number of genes governing the resistance trait as

- **Monogenic resistance**: Controlled by single gene. Easy to incorporate into plants by breeding. Easy to break also.

- **Oligogenic resistance**: Controlled by few genes
- **Polygenic resistance**: Controlled by many genes

Resistance is classified based on biotype reaction as.

- **Vertical resistance:** It is determined by one or few genes and is characterized by pathotype specificity. That is it is attacked by only one virulent pathotype. For all others, the host will be resistant. It is also called as specific resistance or race-specific resistance
- **Horizontal resistance:** It is determined by polygenes. Horizontal resistance does not prevent the development of symptoms of the disease, but it slows down the rate of spread of disease in the population. It is also known by race- nonspecific, partial and field resistance.

Resistance is classified based on population/Line concept

- **Pureline resistance**: Exhibited by lines which are phenotypically and genetically similar.
- **Multiline resistance**: Exhibited by lines which are phenotypically similar but genotypically dissimilar

Plants once infected by a specific pathogen become resistant to further infections by the same one. This was discovered in the beginning of 20th century. This concept is involved in viral cross-protection and induced systemic resistance. Induced systemic resistance in plants is of several types of which *Systemic Acquired Resistance (SAR)* is the most important one. It is long lasting and effective against viral, bacterial and fungal pathogens. It ranges from a oversensitive response to necrotic lesions. SAR is due to high level of salicylic acid which is essentially needed for the development of SAR. Salicylic acid acts as a phloem translocated signal that mediates SAR. It is also due to SAR genes which is different in monocotyledonous and dicotyledonous plants. In tobacco, SAR genes cover a set of non-allelic genes that can be classified on the basis of proteins they encode such as the pathogenesis related (PR) genes. These genes play an active role in the disease resistance as their expression in transgenic plants impart significant disease resistance. SAR genes in various species differ in considerable extent.

6. SOURCES OF RESISTANCE

Plants that may be less desirable in other ways, but carry a useful disease resistance trait. Ancient known plant varieties and wild species, cultivated varieties and land races are very important to preserve because they are the most common sources of enhanced plant disease resistance. Source of resistance are available for melon aphid, striped and spotted cucumber beetles, squash bug, squash borer, pickleworm, red pumpkin beetle, fruit fly in different cucurbits, cabbage maggot and aphid in cabbage and spinach, fruit and shoot borer in brinjal, jassids in okra, potato leaf hopper, melon fly in different beans, pea aphid and weevil. Genetic basis of insect resistance has been reported to be monogenically dominant in muskmelon aphid and in pumpkin and watermelon for fruit fly, whereas, additive gene action have been reported for resistance to both striped and spotted cucumber beetles and squash bug in squash. In interspecific crosses of muskmelon with wild melon two pairs of complimentary genes are reported to be involved for resistance to fruit fly. Maternal influence has also been indicated in inter-varietal crosses of squash for resistance to spotted cucumber beetle.

In India work on resistance of cucurbits to red pumpkin beetle and fruit fly was initiated at IARI, New Delhi as early during 1962 and at Indian Institute of Horticultural Research, Bangalore during 1969. A highly resistant source to fruit fly was obtained in pumpkin, which was utilized, in breeding a resistant variety Arka Suryamukhi. Some of the pumpkin lines were fairly resistant to red pumpkin beetle. It was observed that among the different species *Citrullus colocynthis* was highly damaged by fruit fly. In *Lycopersicon* genus *L.hirsutum* and *L. hirsutum f. glabratum* is resistant to fruitworm, and also indicated that since these were cross-compatible with *L. esculentum* and it was possible to transfer the resistance factor in the cultivated varieties. It was further indicated that *L. hirsutum f. glabratum* was also resistant to tobacco flea beetle and carmine spider mite. It was possible to incorporate resistance to more than one insect species in one genotype.

In Onion sources of resistance were identified, cause and mechanism studied and suitable varieties were developed. In Okra difference in varietal response to jassid attack was observed and fruit of the resistant line was found to have strong prickly hairs and was highly susceptible yellow vein mosaic virus. In spinach variety Manchuria was reported as resistant to aphid as early as in 1920. In Carrot resistance to fruit fly was reported and one of the amaranthus lines was observed to have high field resistance to grasshopper in Nigeria. With monogenic inheritance available for melon aphid in cantaloupe, for fruit fly in pumpkin and in watermelon it would be possible to utilize backcross method for incorporating resistant gene in commercial varieties.

7. BREEDING METHODS

Introduction: Collections of related materials from other countries, particularly from areas where the pathogen and host species may have co-evolved, sometimes provide rich pools of resistance genes [22]. In vegetable peas, early introductions from Europe and USA were found quite successful and popular in India. These included Arkel (early maturing, dwarf type, introduction from England in 1970s) and Bonneville (main season, late maturing, tall type, introduction from USA in 1970s). These introductions were obtained at IARI, New Delhi and were released for commercial cultivation after preliminary evaluation. Early Badger a dwarf, wrinkled seeded variety introduced from USA has resistance to *Fusarium* wilt.

8. UTILIZATION OF WILD SPECIES

The major bottleneck in the resistance breeding programme is the lack of resistant source in the cultivated germplasm. This has necessitated breeders to search resistance for genes in wild species that are taxonomically related and compatible. The use of wild forms in breeding crop plants, particularly to obtain vigor and resistance hasbeen well recognized [1]. In vegetables, several experiments involving wild species have been carried out. Selection of a genotype with high yield and resistance reduces the yield loss on one hand and increases the availability of the produce to market which is fairly free from residue on the other hand. Generally the source of genes for resistance are (i) provided by the variability within the crop species, (i) varieties from original home of insect, (ii) varieties from centres of great insect occurrence, (iii) varieties from areas of greatest morphological diversity.

In muskmelon gene for resistance is not available in the cultivated species, but Cucumis africanus was found to be fairly resistant to fruit fly. Similarly, Cucumis heptadactylis was resistant to red pumpkin beetle. An attempt to incorporate the resistance to shoot and fruit borer in the cultivated egg plant genotypes was made in interspecific hybridprogenies of the cross S. melongena x Solanum viarum and evaluating the direct segregating progenies of such interspecific crosses so as to identify recombinant inbred plants with high yield and shoot and fruit borer resistance. The data recorded from direct F9 generation derivatives of EP 65 x Solanumviarum were utilized to study thoroughly and three progenies each out of thirty in F9 generations were selected for further studies. All these selected progenies have performed very well with respect to shoot and fruit borer resistance. From this evaluation studytwo hybrid derivatives were selected and designated as HD 1 and HD 2. The progenies of the culture HD1 recorded minimum shoot (7.69%) and fruit borer infestation (6.67%). The HD2 progenies recorded the minimum shoot (9.09%) and fruit borer infestation as 6.85%. The selected progenies viz., HD 1 and HD 2 showed profuse flowering and fruiting and also cluster bearing habit. The color of the fruit was bright purple while the fruit surface was smooth, glossy along with tightly packed seeds in its flesh which again act as physical barrier for mandibles of fruit borer to chew and bore into the flesh of fruits [18].

Tomato is a self pollinated crop, which is a high demand vegetable crop. The wild species are reservoir of important genes in tomato. Solanum pimpinellifolium is the only red-fruited wild species of tomato. Because of the close phylogenetic relationship between the two species, there is little or no difficulty in initial crosses in subsequent generations of pre-breeding and breeding activities. Nineteen accessions from seven Lycopersicon species were bio assayed for their resistance to

Heliothisarmigera by [23]. It was found that among the various Lycopersicon spp. bioassayed, accessions of L. hirsutum f. glabratum is most potential for breeding H. armigera resistant cultivars.

Wilt, little leaf and phomopsis blight are the serious diseases of brinjal. Solanum incanum is resistant to Fusarium wilt. In humid tropical areas brinjal is highly infected with bacterial wilt. Wild species of Solanum viz., S. torvum, S. xanthocarpum, S. nigrum, and S. sisymbriifolium are resistant [24]. The wild species Solanum viarum showed no infection and was immune, whereas the species S. incanum and S. sisymbrifolium were resistant to little leaf disease.

Pinheiro et al. [25] conducted two assays, to evaluate the resistance to root knot nematode, M. incognita race 1 in Citrullus lanatus cv. Citroides, Lagenaria vulgaris, Sicanaodorifera, Cucurbita facifolia, Cucurbita moschata, Cucurbita moschata x Cucurbita maxima, Luffa sp., Cucumis melo and Cucumis metuliferus accessions. The results revealed that three accessions of Cucumis metuliferus ('Kino') were resistant to M. incognita race 1 in the first experiment. In the second experiment conducted to evaluate the reaction to nematode M. incognita race 1, M. javanica and M. enterolobii all the seedlings in pots were inoculated with 2nd stage (J2) juveniles and 5000 eggs of each Meloidogyne species. The observations on egg mass index (IMO), gall index (IG), number of eggs per gram of root (NEGR) and reproduction factor (RF) was observed on 53 and 84 days after inoculation, respectively. The melon Cucumis metuliferus was resistant to root-knot nematode.

9. SCREENING VARIETIES

Varietal difference with regard to resistance for fruit and shoot borer was observed in brinjal. Choudhary *et al.* [26] screened eight varieties against *L. Orbonalis.* The order of susceptibility ofbrinjal varieties was recorded as Pant Samrat< Pant Rituraj<Manjarigota<Pusa Purple Long < Pant Brinjal-5 < Kavach< MHB-80 < BR-112 during 2014–2015 and 2015–2016, respectively. In another study, Amit et al. [27] screened twenty five brinjal varieties against brinjal shoot and fruit borer and red spider mite and IBH-3, IBL-116, Rajindra brinjal, KS-356, JB-24, JBH-8, IBH-02 andCHBR-1 were found tolerant. The research carried out at RARS, Jamalpur, Bangladesh showed that, the brinjal varieties Jumki-1 and Jumki-2 were highly resistant (HR), Islampuri-3, BL-34 and Muktakeshi were fairly resistant (FR), Singnath long and Singnath-4 were tolerant to brinjal shoot and fruit borer [28].

Similarly, in okra Rehman *et al.* [29] screened four varieties (SabzPari, SadaBahar, PusaSawani, Arka Anamika) and those varieties showed some degree of resistance against sucking insect pests. Okra variety Sada Bahar was less infested with jassid (1.30/leaf) and whitefly (5.36/leaf) compared to other tested varieties and resulted in maximum yield (1529.62 kg/ha). Number of fruits pods per plant was found non significantly different on all the tested okra varieties.

In another experiment, Jackson and Bohac [30] evaluated sweet potato accessions by using bio assay techniques using the adults of banded cucumber beetle and spotted cucumber beetle. A single beetle was placed on a piece of sweetpotato peel that was embedded periderm-side up in plaster in a petridish. Feeding and longevity of insects on sweet potato genotypes were evaluated. Durability of feeding with respect to banded cucumber beetles on sweet potato peels ranged from 12 d for the most-resistant genotype to 123 d for a susceptible control cultivar (SC1149–19). The feeding longevity of spotted cucumber beetles was slightly shorter than banded cucumber beetles. For the highly resistant genotypes, both the species exhibited a significant delay in feeding initiation, and most beetles died before they had fed the sweet potato. Thus it was evident that both antibiosis and non preference (antixenosis) are important mechanisms of resistance in sweetpotato genotypes.

Seventy seven eggplant genotypes were tested for resistance to *root*-knot nematode by classical testing. As a result it was determined that P29 and P52 genotypes were resistant.*Ditylenchus destructor* and *Ditylenchus dipsaci* are economically important plant-parasitic nematodes, affecting potato production mostly in temperate climates. Mwaura and Vidal [31] screened 25 potato varieties for resistance to and tolerance for *D. destructor* and *D. dipsaci* infections. Reproduction factor (RF) and relative susceptibility (RS) were used to evaluate resistance, Based on Reproduction factor, sixteen varieties were assessed as susceptible (S) and five were identified as resistant (R) to *D.* destructor. The varieties Innovator, Aveka and Spunta were identified as resistant to *Ditylenchus dipsaci*. The highly susceptible one for *D. destructor* and *D. dipsaci* in both experiments was Desiree and was used as the standard susceptible control variety for the calculation of RS. An 1–9 scale was used to assess and classify the potato varieties based on level of resistance to *D. destructor* and *D. dipsaci*, where 9 indicated the highest level of resistance. Among the varieties screened six had significantly lower relative susceptibility (RS) to *D. dipsaci* than the standard susceptible control. Few varieties were also observed to be tolerant to both the nematodes. The suitable indices for resistance and tolerance determination were relative susceptibility (RS) and external potato tuber damage.

Nayak and Pandey [32], screened one hundred fifty brinjal varieties/cultivars against root-knot nematode, only twenty varieties have shown resistant reaction with least gall index (1.1 to 2.0) viz., Gachhabaigan, Azadkranti, Kantabaigen, Athagara Local, Kamaghara local, Solanum indicum, PBR 129–5, ARU-1, BB1–3, BB 45-C, BB-49, KS-224, Utkal madhuri, BR-112, LB-13, LB-25, LB-28, LB-30, LB-44, LB-55.

Akhter and Khan [33], screened thirty brinjal varieties for their resistance/susceptibility to root-knot nematode (*Meloidogyne incognita* race-1) infestation. Out of 30 brinjal varieties, eighteen varieties viz., Black Beauty, Brinjal 1 hybrid, Brinjal No.38, Chamak, Govinda, Green round, Nagina, Nav Kiran, Neel Kamal, Nishant, P.K-123, Prabha Kiran, Prasad, Sukhda, Surya Kiran, i9Utkal, VNR-51 and VNR-60 were highly susceptible, seven varieties (Brinjal Advance, Brinjal BSS1013, Green long, Harshit, Prapti, Shamli and Ujjwal) were susceptible, two varieties (Mahy 112 and Mahy Ruby) were tolerant, two varieties (Hybrid green and JK Kajal) were moderately resistant and only one variety Mahy 80 was resistant against *Meloidogyne incognita* race-1. Mahy 80 variety was ported to be resistant against root-knot nematode, *M. incognita* race-1 for the first time.

10. SELECTION

Selection is an important method for breeding of varieties resistant to biotic stresses. It is an important means of isolating or identifying sources of disease resistance. Normally the sources of resistance are available in natural populations, wild species, introductions and spontaneous mutants. In the earlier periods selection was accomplished by sequestering the resistant survivors of natural epiphytotics. Now a days advanced artificial epiphytotics are being created and selection of resistant types are being done rather than escaped suscepts [34]. Selection of resistant plants from a commercial variety is the cheapest and quickest method of developing a resistant variety. IIHR Bangalore developed tomato varieties through pureline selection viz., Arka Alok and Arka Ahuti which showing resistant against bacterial wilt.

The production of garden pea is seriously limited by major diseases namely, wilt, powdery mildew and rust. *Fusarium* species cause root rot (*F. solani* f.sp. *pisi* and *F. avenaceum*) or wilt (*F. oxysporum* f.sp. *pisi*). Coyne *et al.* [35] developed three breeding lines (W6 26,740, W6 26,743, W6 26,745) having high level of resistance to Fusarium root rot caused by *F. solani* f.sp. *pisi* with acceptable agronomic traits. In melons the genotypes A19, A32, A30, JAB-11, JAB-20,JAB-3, JAB-7, C384, C67, JAB-9, JAB-18 were identified for powdery mildew resistance.

Pinheiro et al. [36] studied thirty seven pepper genotypes, *Capsicum chinense*, *C. annuum* and *C. frutescens*, were characterized for resistance to three root-knot nematode species (*Meloidogyne javanica, M. incognita* race 1 and *M. enterolobii*). Three experiments were carried out, in 2013, 2014 and 2016, in a greenhouse. Among the genotypes of *Capsicum frutescens* evaluated all were resistant or immune to *M. javanica* and *M. incognita* race 1. In *C. chinense* six accessions were susceptible. In the second experiment all genotypes of *C. chinense* and *C. annuum,* evaluated were resistant to *M. incognita.* In the third experiment, with *C. annuum* genotypes, most were susceptible to *M. incognita* race 1 while CNPH 30118 and CNPH 6144 were resistant to *M. enterolobii*, the most aggressive species. A greater degree of resistance was observed in few accessions of *C. chinense* and *C. frutescens.*

11. HYBRIDIZATION

The common method used for resistance breeding is hybridization. In this the resistance is transferred by two means. In the first, by backcross method the resistance is transferred from a wild species or a variety with undesirable horticultural attributes to a susceptible but otherwise a desirable variety. In pedigree method the resistance is combined with some desirable characters of one variety and superior characteristics of another variety.

Heterosis a complex biological phenomenon manifested in the superiority of hybrids over parental forms due to the rate of development of one or more complex traits. Heterosis values should be negative for getting tolerant hybrids through heterosis breeding. As far as pest infestation is concerned a hybrid with least incidence might be due to complementation of genes. With the genetic knowledge available on insect resistance in vegetable crops it would be necessary to employ hybridization to transfer single, multiple or additive genes into commercial varieties in the advanced filial generations. The major achievement is expected through inter specific hybridization. In order that fruit quality, yield and resistance to insect pest may preferrably be incorporated in one genotype, it may possibly involve number of parents to achieve desired results.

Inter-varietal and inter-specific crosses, followed by selection, have accounted for the development of resistant hybrids. The production of inter-specific hybrids is useful for the transfer of desirable genes from wild to cultivated species. The Source, mechanism, biochemical and genetical basis of resistance in squash, muskmelon, cucumber and watermelon were studied by several authors.

By utilizing interspecific breeding technique and recombination, borer-free brinjal can be developed to protect the high yield and satisfy the preference of consumers. The parental lines EP 65 and Pusa Uttam had recorded higher yield together with lower fruit and shoot borer incidence. Both of them were found to be good combiners for the above traits. So these parents can be involved in multiple crossing programmes to transfer the resistant genes and to isolate desirable hybrids with high yield and low fruit and shoot borer incidence.

Introgression breeding has been extensively used in the genetic improvement of potato and tomato. In potato twelve traits have been introgressed from wild species viz., *S.demissum, S. spegazzinii, S. stoloniferum, S. chacoense, S. acaule, S. vernei* and in tomato from the wild species *S. peruvianum, S. cheesmanii, S. pennellii* and *S. chilense.* In watermelon, the F_1 (between resistant and susceptible) showed pronounced resistance to fruit fly.

Kishaba *et al.* [37] studied resistance by using several melon aphid-susceptible (MAS) recurrent parents from an initial cross of 'PMR 45' with P1 414,723. Fifteen advanced melon breeding lines with different levels of melon aphid resistance (MAR), their recurrent parents and P1 414,723 were compared in a naturally infested field test for susceptibility to feeding damage by CB. None of the MAR entries were more susceptible than their recurrent parents for fruit damage by CB. P1 414,723 was found to have a low level of resistance to seedling damage, and a high level of resistance to fruit damage from feeding by Cucumber beetle.

Hybridization was undertaken in brinjal with *Solanum viarum* to combine the resistance trait with high yield. The derivatives of the inter specific cross of *Solanum melangena* and *Solanum viarum* EP 65 (accession of *Solanum melangena*) x *Solanum viarum* were assessed till F9 generation. As a result two recombinant progenies *viz.*, 7 and 9 were chosen and carried to the next generation as they had recorded high marketable yield and least infestation of shoot and fruit borer. Molecular confirmation with RAPD primers was done which depicted the introgression of the genes from donor parent *Solanum viarum* to brinjal and thus hybridity was confirmed.

Backcross method: This method is widely used for incorporation of resistant gene from a wild species or any variety with undesirable traits to a susceptible variety which is good in other agronomic attributes. The parents from which the resistant gene is to transferred is called as donor parent or non-recurrent parent. The susceptible parent in which the resistance gene is transmitted is known as the recurrent parent. The generation up to which the selection process has to be effected in a backcross program would differ according to the allelic relationship of the resistance genes *i.e.*, whether it is resistance or dominant to the allele. Normally the backcross method help in the recovery of recurrent parent phenotype along with the transfer of resistant genes.

[38] investigated hybrid progenies of late blight resistant potato somatic hybrids developed through hybridization with common potato varieties, and also linked ISSR markers with resistant parent/progenies. Potato somatic hybrids (*Solanum tuberosum* di haploid 'C-13' + *S. pinnatisectum*) were back-crossed with potato varieties (*S. tuberosum*) and true potato seed (TPS) were produced. TPS-raised seedlings were advanced to back-cross progenies clones (BC1-C1, BC1-C2, BC1- C3 and BC1-C4) during the five years based on tuber traits in field trials and field resistance to late blight. The BC1-C2 progenies were profiled by ISSR markers and alleles linked to late blight resistant somatic hybrid parent P8 and their progenies (P8 × Kufri Jyoti) were identified. Eight promising advanced hybrids for late blight resistance were identified. This is the first ever report in India towards widening the genetic base of potato by exploitation of interspecific somatic hybrids. The methodology followed was as follows (Figure 1).

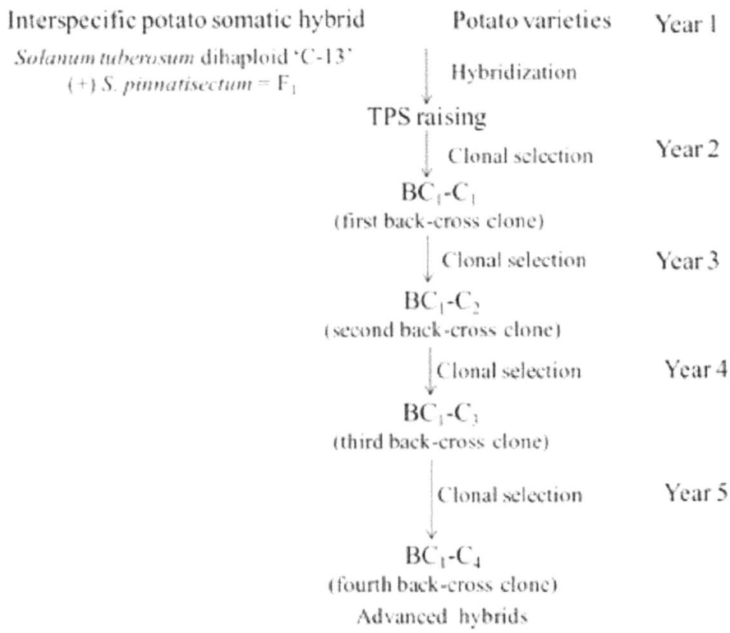

Figure 1. A schematic presentation of development of interspecific potato somatic hybrids (S. tuberosum + S. pinnatisectum).

Powdery mildew is a serious disease of pea and for which new good resistant donors are available, a typical backcross breeding approach as applicable to a character governed by a single recessive gene was outlined by Gritton [39].

Pedigree method: It is quite suited when the resistance is governed by horizontal or polygenic. In breeding for disease resistance, artificial disease epidemics are generally produced to help in selection for disease resistance. Pedigree breeding is mainly used when the resistance is governed by major genes and have higher heritability. If it is used for low heritability traits then the selection process will be time consuming as it takes several generations usually F5 or F6 to attain homozygosity [40, 41], implemented a quantitative trait locus (QTL) mapping approach to study the inheritance of anthracnose resistance in an F2 population derived from the pedigree of C. annuum × C. chinense.

The major crops in which break through research was done and the resistant varieties developed through introgression breeding methods is tabulated below.

Crops	Diseases	Sources of resistance	Remark	References
Bottle gourd	Gummy stem blight	IIHRBGH-10 (BG 114–3 x BG 95)	Hybridization Method	IIHR, Annual report, 2019
Tomato	ToLCD+BW + EB	Arka Apeksha (ITHR 2834 X ITHR 2918	Hybridization Method	IIHR, Annual report, 2019
		Arka Vishesh (ITHR 2834 X ITHR 2917)		
Cauliflower	Downy mildew and Alternaria leaf spot	Arka Spoorthi	Mass pedigree method	IIHR:https://www.iihr.res.in/varieties
		Arka Vimal		
Amaranthus	White rust	Arka Arunima	Pure line selection	IIHR:https://www.iihr.res.in/varieties
Garden Pea	Powdery mildew and rust	Arka Pramodh	Pedigree method of selection	IIHR:https://www.iihr.res.in/varieties
Okra	YVMV	Arka Abhay	Interspecific hybridization	IIHR:https://www.iihr.res.in/varieties
Watermelon	Powdery mildew, Downey mildew and anthracnose	Arka Manik	Mass pedigree method	IIHR:https://www.iihr.res.in/varieties
Brinjal	Bacterial wilt	Arka Anand	Hybridization Method	IIHR:https://www.iihr.res.in/varieties

12. MUTATION BREEDING

Literature on mutation breeding of vegetable crops for resistance to insect attack is meager. In IIHR, Bangalore varietal differences, have been indicated for resistance to jassids in the M1and M 2 populations of okra (dry seed irradiated with 55 to 60 KR gamma rays), to aphid in the M population of muskmelon (30–40 KR gamma rays); and to aphid in the MQ population of watermelon (dry seed irradiated with 50 KR gamma rays). This was indicative that there was a great potential in this approach for resistance genes. It is particularly applicable in case of muskmelon to fruit fly, brinjal fruit and shoot borer, tomato fruit borer, melon fruit fly and others where useful source of resistance has not been obtained. Further in cases where one or two sources have been located in nature, it would be desirable to obtain more sources of resistance through mutation breeding. It is evident that very little experience has been gained in the use of induced mutations in resistance breeding against insect pests but the prospects of this approach are great.

Selection of spontaneous and induced mutant plants with resistance through the use of mutagenesis. In germplasm repositories sometimes natural spontaneous mutations may occur. During that occurrence some resistant types will be developed naturally. If such instances does not occur then mutations can be induced by artificial methods. The short comings in this method is that mostly duplication of naturally occurring genes will be the outcome and induced mutants will have mono factorial behavior and the inheritance is short-lived [4]. The genetic stability has to be tested by repeated screening of mutants for the confirmation of resistance.

Potato is a major vegetable crop and is infected by many diseases which induces losses in yield and quality. Artificial mutations through irradiations were tried for developing resistant varieties against stem canker and black scurf of potato caused by *Rhizoctonia solani*. The mutants developed were resistant to these diseases. Aslı Kara and Şerife Evrim Arici [42] investigated the effect of gamma radiation on the susceptibility of the potato plant to *Rhizoctonia solani* at 22, 33, 54, 57 and 109 Gy. The best results were found with a dose of 22 Gy. The application of gamma irradition in this study may offer a new approach for potato breeders for developing plants resistant to *R. solani*.

13. BIOTECHNOLOGICAL APPROACHES

The Bt-brinjal has been developed by inserting cry1Ac gene from a soil bacterium called Bacillus thuringiensis through Agrobacterium tumefaciens mediated method. Bt brinjal contains three genes, cry1Ac gene, which encodes an insecticidal protein Cry1Ac, is derived from a common soil bacterium and is driven by CaMV 35S promoter (cauliflower mosaic virus 35S). It also has nptII gene (neomycin phosphor transferase-II) which contains an antibiotic resistance marker and another marker gene "aad "for amino glycoside adenyl transferase. The cry1AC protein formed in Bt brinjal is analogous in structure and function to that found in nature. The resistance against fruit and shoot borer of brinjal is provided by cry1Ac genes and it minimizes damages and facilitate for the reduction in pesticide sprays and thus it is eco-friendly. Bacillus thuringiensis and B. t var. Kenyae (B.t.k) microbial formulations have been found to be highly specific to target insect pests, and do not have deleterious effects on non-target organisms such as beneficial insects, birds, fish, and mammals including human beings. The confirmation by ELISA revealed the presence of the Cry1Ac protein. The quantitative estimates established significant levels of Cry1Ac protein (2.46–4.33 ng ml-1) in the leaf extract of the transformed plants. The expression of this insecticidal protein in high levels resulted in significant amount of mortality of larvae and also stunted the growth of any surviving larva on transformed plant tissue.

Incongruity occurs in inter-specific crosses as a result of a lack of genetic information in one partner to complete pre- and post-pollination processes in the other. After fertilization the growth of the embryo is restricted due to some post-fertilization barriers. For the developmental process an equilibrium has to be established between embryo and endosperm for sharing the nutrients. When this equilibrium in the development of the zygote is disturbed, first division is delayed and abortion of the young embryo or disintegration of endosperm happens. This abortion may occur in any stages of development of the young seed. Based on the stage of embryo abortion in vitro methods have been developed to overcome post-fertilization barriers in a number of plant species. Embryo rescue techniques are the oldest and most successful in vitro procedures.

In many instances, progeny from wild crosses of inter-varietal and inter-specific is difficult to produce owing to several barriers like pre zygotic and post-zygotic barriers. The development of young zygote may be arrested by hybrid breakdown, hybrid sterility and hybrid non viability.Post-zygotic barriers such as endosperm abortion and, at later stages, embryo degeneration are of common occurrence, leading to low fertility but these have been overcome through the use of embryo rescue. Embryo culture is one of the earliest forms of in vitro culture applied to practical problems that has proven of greatest value to breeders. Among the very important strategies hybrid embryo rescue, and related applications like ovule/ovary/placental cultures through sequential embryo cultureis especially useful in vegetable crops and culture of embryos has also been demonstrated in tomato, brinjal, capsicum, hot pepper, onion, potato, tomato including cucurbits for rescuing useful hybrids.

Ovule and ovary culture are more suitable than embryo culture for small seeded species or very young embryos. When abortion occurs in a very young stage and maternal tissue has no negative influence on the development of seeds, ovary culture can be applied. Ovary culture has been applied in many Brassica species. The ovule culture is applied in Lycopersicon. Depending on the genotypic combination of the inter-specific crossing, the percentage of seedlings obtained from ovule culture varied from 0.5–22.5 per cent, whereas in the in vivo situation on the plant no seeds could be harvested.

Embryo rescue technique was attempted in an interspecific hybridisation of tomato variety MT-3 and Kashi Amrit with wild relative *S. peruvianum* (WIR-3957). The optimum time for rescuing the embryos was standardized as twenty five days after pollination. The most effective media for germination of the immature putative hybrid embryos was Murashige and Skoog's medium supplemented with 1 mg/L GA3, 0.1 mg/L NAA and 0.5 mg/L BAP. The confirmation of hybridity of this inter-specific crosses was done using RAPD markers. Verba et al. [43] made inter-specific reciprocal crossings between *Solanum melongena, S. aethiopicum.*

Confirmation of resistance: The confirmation of resistance in the developed varieties/hybrids are usually undertaken by artificial screening studies.

Artificial Screening: This is artificial epidemics created by inoculation of pathogen onto the plant population. The most common methods used to inoculate *Colletotrichum* on chili plants either involve using the fruit puncture method or the spraying method in the laboratory or in the field [44]. The injection of very small amount of conidia suspension into the fruit pericarp is known as microinjection or fruit puncture method. The resistance can be known by the lesions shown at the injection or inoculation area [20, 21, 45]. Secondly, spraying method in which the conidia suspension is sprayed on plants at flowering and fruiting. But this method is not safe and risky. So the puncture of detached fruits in the laboratory has paid an outstanding development of anthracnose resistance evaluation in chili.

Different artificial screening methods are employed based on the mode of spread of the diseases as given below

- **Soil borne diseases**- Sick plots are created for testing resistance to such diseases.
- **Air borne diseases**- Spraying a suspension of spores.
- **Seed borne diseases**-dry spores are dusted on seeds or seeds may be soaked in a suspension of pathogen spores

14. CONCISE SUMMARY

In vegetables the biotic stresses are pests, diseases and nematodes. The damages induced by these factors affect the production, productivity and quality. Breeding of resistant varieties offer the cheapest means of pest/disease/nematode management. Resistant varieties obviate the use of chemicals, thus reduce environmental pollution. Effectiveness of resistant varieties depend upon the stability of their performance in various environmental conditions and changing climatic scenario. However like any other concept, resistant breeding also has its own limitations. That is breakdown of resistance. However, horizontal resistance being durable but difficulty relates to an accurate & reliable assessment of the level of resistance. A variety resistant to a stress may be susceptible for other. So future planning should require far greater effort for introgression of genes from different resistant sources and develop multiple resistant varieties against several biotic stresses than that required for single.

AT A GLANCE

In vegetables the factors for biotic stress are pests, diseases and nematodes. The damages induced by these factors reflect highly on production, productivity and quality. Although application of pesticides/fungicides and nematicides has managed these stresses, excessive use of unsafe chemicals results in environmental pollution and leave residues in vegetables which are above threshold levels and also promote the development of new races/biotypes of pests and pathogens. Therefore vegetable improvement works concentrate on high yielding varieties with multiple resistance to these biotic stresses. For such studies, the knowledge on the genetic basis of resistance and plant-pest/pathogen interactions is necessary which will in turn improve the efficiency of the breeding programmes by introducing resistant genes and result in high-yielding genetically resistant cultivars. For the development of resistant varieties and pre-breed lines, information on sources of resistance is prerequisite and serve as a backbone in the breeding programme. Further, gene action responsible for the inheritance of characters helps in the choice of suitable breeding methods for the improvement of the crop. Work has been done by using the various breeding methods and resistant varieties have been bred and they offer the cheapest means of pest/disease/nematode control. Resistant varieties obviate the use of chemicals, thus reduce environmental pollution and facilitate safe food for human consumption.

REFERENCES

1. Sarvayya V. 1936. The first generation of an interspecific cross in Solanums between Solanum melongena and S.xanthocarpum. Madras Agric. J., 24(4): 139-142
2. Bonn GW, Kishaba AIT, Principe JA, Toba HH. Tolerance to melon aphid in Cucumismejlo L. J. Amer. Soc. Hort. Sci. 1973;98(l):37-40
3. Fery RL, Cuthbert. Antibiosis in Lycopersicon to the tomato fruit worm (Heliothiszea). J. Am. Soc. Sci. 1975;100:276-278

4. Person, C., Sidhu, G. 1971. Genetics of host-parasite interrelationships. Mutation breeding for disease resistance. 3 1-38 IAEA, Vienna 249 pp
5. Som MG, Chaudhary B. Veg. Sci. 1975;2:93-99
6. Walker, 1941. Fusarium wilt of watermelons. Effect of soil temperature on the wilt disease and the growth of watermelon seedlings. Fla. Agr. Exp. Sta., Bull.363: 1-29. 1941
7. Bohn GW, Whitaker TW. Genetics of resistance to powdery mildew race 2 in muskmelon. Phytopathology. 1964;54:587-591
8. Cohen, Y. Cohen S. and C.E. Thomas (1985) Inheritance of resistance to downy mildew in Cucumis melo PI 124111 CGC 8:36-38
9. Kenlgsbuch and Cohen, Y. (1992). Plant Disease76: 615-617
10. Antonio, A.C., D. J. H. Silva, M. C. Picanço, N. T. Santos and M. E. Fernandes. 2011. Tomato plant inheritance of antixenotic resistance to tomato leafminer. Pesq. agropec. bras., Brasília,46(1):74-80
11. Gabriele Vitelli NA, Arpaia S, Mennella G, Sunseri F, Rotino GL. Transgenic resistance to the Colorado potato beetle in Bt-expressing eggplant fields. HortScience. 2000;35(4):722-725
12. Jones JT, Haegeman A, Danchin EGJ, Gaur HS, Helder J, Jones MGK, et al. Top 10 plant-parasitic nematodes in molecular plant pathology. Molecular Plant Pathology. 2013;14:946e961
13. Elling AA. Major emerging problems with minor Meloidogyne species. Phytopathology. 2013;103:1092-1102
14. Ahmed H. El-Sappah, Islam M. M., Hamada H. El-awady, Shi Yan, Shiming Qi, Jingyi Liu, Guo-ting Cheng and Yan Liang, 2019. Tomato natural resistance genes in controlling the root-knot nematode. Genes 2019, 10, 925 p
15. Singh BD. Breeding for Resistance to Biotic Stresses II. Insect Resistance In: Plant Breeding Principles and Methods. Ludhiana, India: Kalyani Publishers; 1983a. p. 494
16. Singh B.D. 1983b, Plant breeding, Breeding for biotic stress-Disease resistance, Kalyani publishers, Ludhiana, India, pp-359-368
17. Kalloo G. Biochemical Basis of Insect Resistance in Vegetables. In: Vegetable Breeding. Florida: Vol. II. C.R.C. Press. Inc. Boca Raton; 1988. p. 125
18. Praneetha, S. 2002. Breeding for shoot and fruit borer (Leucinodesorbonalis G.) resistance in brinjal (Solanum melongena L.). Ph.D., (Hort.) Thesis, Tamil Nadu agricultural university, Coimbatore
19. Prabu 2014. Breeding for high yield with shoot and fruit borer (Leucinodesorbonalis guen.) resistance in brinjal (Solanum melongena Ph.D

thesis submitted to Tamil Nadu Agricultural University, Coimbatore, India pp 217

20. Silva, K.F.A.S., M. Michereff-Filho. M. E.N. Fonseca, J. G. Silva-Filho, C.A. Texeira, A.W. Moita, J. B. Torres, R. Fernández and L. S. Boiteux. 2014a. Resistance to Bemisiatabaci biotype B of Solanum pimpinellifolium is associated with higher densities of type IV glandular trichomes and acylsugar accumulation. EntomologiaExperimentalis Et Applicata. 218-230
21. Silva SAM, Rodrigues R, Goncalves LSA, Sudre CP, Bento CS, Carmo MGF, et al. Resistance in Capsicum spp. to anthracnose affected by different stages of fruit development during pre- and postharvest. Trop. Plant Pathology. 2014b;39:335-341
22. Leppik EE. Gene centers of plants as sources of disease resistance. Annual Review of Phytopathology. 1970;8:323-344
23. Kashyap, R. K., M. K. Banerjee, Kalloo and A. N. Verma. 1990. Survival and development of fruit borer, HeliothisArmigera (Lepidoptera: Noctuidae) on Lycopersicon Spp. (Hübner), Intl J. of Tropical Insect Science, 11 (6) : 877-881
24. Sugha SK, Pathania NK, Kumar S. Evaluation of Brinjal germplasm against Phomopsis disease. Capsicum & Eggplant Newsl. 2002;21:106-109
25. Pinheiro JB, Silva GO, Oliveira VR, Amaro GB, Morais AA. 2019. Prospection of genetic resistance resources to root-knot nematodes in cucurbit genotypes. Horticultura Brasileira 37: 343-347. DOI - doi:10.1590/S0102-053620190314
26. Choudhary RS, Rana BS, Kumar A, Murdia A, Singh B. Screening of brinjal varieties against, Leucinodesorbonalis (Guenee) infestation. Journal of Entomology and Zoology Studies. 2018;6(2):194-198
27. Amit K, Kaul V, Shankar U. Response of Brinjal (Solanum melongena Guen.) varieties for the resistant reaction against Brinjal shoot and fruit borer (BSFB) and red spider mites (RSM). Journal of Entomology and Zoology Studies. 2018;6(2):1369-1373
28. MannanMA,BegumA,HossainMM.Screeningoflocalandexoticbrinjalvarieties/cultivars for resistance to brinjal shoot and fruit borer (Leucinodesorbonalis). Bangladesh Journal of Agricultural Research. 2009;34:705-712
29. Rehman H, Ayyaz M, Nadeem M, Begum HA. Screening of okra varieties resistance against insect pests under agro climatic conditions of Deraismail khan, Pakistan. Russian Agricultural Sciences. 2017;43(2):149-152
30. Jackson DM, Bohac JR. Resistance of sweetpotato genotypes to adult Diabroticabeetles. Journal of Economic Entomology. 2007;100(2):566-572

31. Mwaura, B. Niere & S. Vidal 2014. Resistance and tolerance of potato varieties to potato rot nematode (Ditylenchus destructor) and stem nematode (Ditylenchus dipsaci). Annals of Applied Biology ISSN 0003-4746. doi:10.1111/aab.12180
32. Nayak DK, Pandey R. Screening and evaluation of brinjal varieties/cultivars against root-knot nematode, Meloidogyne incognita. International Journal of Advanced Research. 2015;3(10):476-479
33. Akhter G, Khan TA. Response of brinjal (Solanum melongena L.) varieties for resistance against root-knot nematode, Meloidogyne incognita race-1. The Journal of Phytopharmacology. 2018;7(3):222-224
34. Hayes, H. K., Immer, F. R., Smith, D. C. 1955. Methods of plant breeding. 2nd ed. McGrawHill: New York 5 5 1 pp
35. Coyne CJ, Porter LD, Inglis DA, et al. Registration of W6 26740, W6 26743 and W6 26745 green pea germplasm resistant to fusarium root rot. J Plant Registration. 2008;2:137-139p
36. Pinheiro, JB; Silva, GO; Macêdo, AG; Biscaia, D; Ragassi, CF; Ribeiro, CSC; Carvalho, SIC; Reifschneider, FJB. 2020. New resistance sources to root-knot nematode in Capsicum pepper. Horticultura Brasileira 38: 33-40. DOI - doi:10.1590/S0102-053620200105
37. Kishaba AN, Castle SJ, Coudriet DL, McCreight JD, Bohn GW. Resistance to western spotted and striped cucumber beetle in melon. Cucirbitacea. pp. 1998;101-105
38. Jagesh Kumar Tiwari, SK Luthra, Sapna Devi, Vinod Kumar, Nilofer Ali, Rasna Zinta and SK Chakrabarti (2018). Development of advanced back-cross Progenies of potato somatic hybrids and linked ISSR markers for late blight resistance with diverse genetic base- first ever produced in indian potato breeding. Potato J. 45 (1): January – June
39. Gritton ET. Pea breeding. In: Basset MJ, editor. Breeding Vegetable Crops. Westport, Connecticut: AVI Publishing Company, Inc.; 1986. pp. 283-319
40. Collard BC, Mackill DJ. Marker-assisted selection: An approach for precision plant breeding in the twenty-first century. Philos. Trans. R. Soc. Lond. B Biol. Sci. 2008;363:557-572
41. Voorrips RE, Finkers R, Sanjaya L, Groenwold R. QTL mapping of anthracnose (Colletotrichum spp.) resistance in a cross between Capsicum annum and C. chinense. Theor. Appl. Genet. 2004;109:1275-1282
42. Aslı Kara and Şerife Evrim Arici (2019). Determination of Gamma Rays Efficiency against Rhizoctonia solani in Potatoes. Open Chem.,17: 254-259

43. Verba VM, Mamedov MI, Pyshnaya ON, Suprunova TN, Shmykova NA. Isolation of eggplant interspecific hybrids by the method of embryo culture. Agricultural Biology. 2010;5:66-71

44. Susheela K. Evaluation of screening method for anthracnose disease in chili. Pest Manag. Hortic. Ecosyst. 2012;18:188-193

45. Mahmodi F, Kadir J, Puteh A. Genetic diversity and pathogenic variability of Colletotrichum truncatum causing anthracnose of pepper in Malaysia. Journal of Phytopathology. 2014;162:456-465

CHAPTER-8

BREEDING STRATEGIES FOR HIGH TEMPERATURE TOLERANCE IN CROP PLANTS

1. INTRODUCTION: AN OVERVIEW

The escalating challenges posed by climate change, particularly the intensification of high temperatures, have underscored the critical need for developing crop varieties with enhanced heat tolerance. Rising temperatures can significantly impact crop productivity, making it imperative to explore innovative breeding mechanisms that can confer resilience to heat stress. This necessitates a comprehensive understanding of the physiological and genetic factors influencing high-temperature responses in crop plants. In this context, breeding mechanisms aimed at augmenting heat tolerance have emerged as a focal point for researchers and agricultural practitioners alike.Increase in global temperature had major impact on crop productivity especially in tropical and sub tropical regimes. Based on climate model predictions, around 1.8–4.0°C rise in air temperature was expected in 21st century [1]. The increase in temperature beyond a certain threshold level tends to induce detrimental effects in plant growth and development. In general, the elevation in temperature of 10–15°C above ambient triggers heat shock in crop plants. The extent of induced heat stress depends on the duration, intensity and rate of increase in global air temperature [2]. Indian lowlands share 15 per cent of global wheat production. The change in global climate would shift these fertile lowlands into heat stressed unproductive environment [3]. Similarly, the cultivation of cereals in Southern Africa and South East Asia was predicted to be heat stressed zone in near future [4]. Around 4–14% yield decline in rice was encountered due to elevated temperature of 1°C in South-East Asia [5]. The declined productivity due to elevated temperature imposes the urgent need for development of climate resilience genotypes. Evolving heat tolerant cultivars would highly benefit the livelihood of developing countries as around 70–80% of population relies on agriculture. Understanding the effect of heat stress on crop plants and its adaptation mechanisms would help in framing out the breeding strategies for high temperature tolerance.

Heat tolerance in crop plants is a complex mechanism involving adaptations through altered physiological process, morpho-anatomical features and induction of several biochemical pathways. On exposure to high temperature, several signal transduction pathways were triggered leading to changes in gene expression. As a result, varied stress related proteins were synthesized contributing heat tolerance in plants [6]. The tolerance mechanism to high temperature stress varies within genotypes of a plant species. The existing variation between and within species provide scope for evolving heat tolerant lines through conventional breeding approaches [7]. Dissecting out genetic information through molecular tools would hasten the development of climate resilient cultivars contributing to food security in near future. A brief review on plant response, adaptation mechanisms and genetic approaches to combat heat stress were presented in this chapter.

2. EFFECT OF HEAT STRESS ON CROP PLANTS

Heat stress had varying impact on different phenological stages *viz.*, germination, seedling, vegetative, flowering and reproductive of crop plants [8]. The plant response to heat stress depends on the duration, degree of rise in temperature and plant type. Under tropical regimes, high temperature with intense solar radiation poses a major limiting factor for yield by inducing leaf abscission, leaf senescence, scorching of leaves, branches and stems, growth inhibition, pollen infertility and poor seed formation [9, 10]. A significant decline in relative growth rate, shoot dry weight and net assimilation rate was recorded in sugarcane, maize and pearl millet on exposure to high temperature stress [11]. High reduction in grain quality was recorded in most of the cereal crops grown under heat stress environments [12]. Several physiological processes such as partitioning of assimilates, plant-water relations and shoot growth was affected due to heat stress in common bean [13]. In general, the susceptibility to heat stress was found higher at reproductive stage of plant development. An excessive yield loss is recorded in legumes on exposure to high temperature (30–35°C) during anthesis stage [14]. Drastic reduction in grain number and weight was observed in wheat at high temperature regimes [15]. Heat stress affects several metabolic pathways leading to accumulation of reactive oxygen species (ROS) which is a major component for oxidative stress in crop plants [16]. The photosystem centres (PS I and PS II) of chloroplast, mitochondria and peroxisomes are the major sites for generation of ROS in plants [17]. High temperature stress disrupts the stability of cell membrane through protein denaturation [18]. The induction of ROS due to high temperature stress was correlated with premature leaf senescence in *Gossypium* sp. [19]. Accumulation of ROS in root cells was evidenced in wheat on exposure to high temperature for two days [20].

3. ADAPTATION MECHANISMS

Plants tend to adapt several complex mechanisms through phenological and morphological changes to combat high temperature stress (Figure 1). On heat stress regimes, plants exhibit varied short term escape/avoidance mechanisms *viz.*, altered leaf orientation, transpirational cooling, altered membrane lipid properties, early maturation and so on for its survival. Plants show varied degree of leaf rolling upon intensity of solar radiation. A significant tolerance to high temperature was observed in wheat by maintenance of water potential in flag leaf through adoption of leaf rolling under heat shock conditions [21]. Increase in trichomatous and stomatal densities, waxy layer on leaves, and larger xylem vessels are the common features induced during heat stress [22]. On contrary, plants also evolve long term tolerance mechanisms for its effective survival and productivity under high temperature. Induction of osmoprotectants, antioxidants, late embryogenesis abundant proteins, dehydrins, and heat shock proteins are the major factors involved in counteracting the heat shocks. Accumulation of osmolytes such as proline, trehalose, and glycine betaine plays a vital role in imparting tolerance *via* cellular osmotic adjustment, detoxification of ROS, stabilization of enzymes and membrane proteins [23]. Several enzymatic and non-enzymatic antioxidant defense components are also involved in protection against oxidative stress induced by free radicals [24]. The activities of ROS scavenging enzymes are temperature specific. In general, most of the antioxidant enzymes show increased activity with elevation in temperatures. It is also influenced by genotype, growing season and phenological stages of plant [25]. Under high temperature conditions, several signaling molecules such as nitrous oxide, Ca-dependent protein kinases, Mitogen mediated protein kinase, sugars, and phytohormones play a role in stimulation of stress responsive genes *via* transduction pathways [26]. Evolving adaptation mechanisms (either tolerance or avoidance) to high temperature and drought would be more rewarding at arid conditions as it is often correlated.

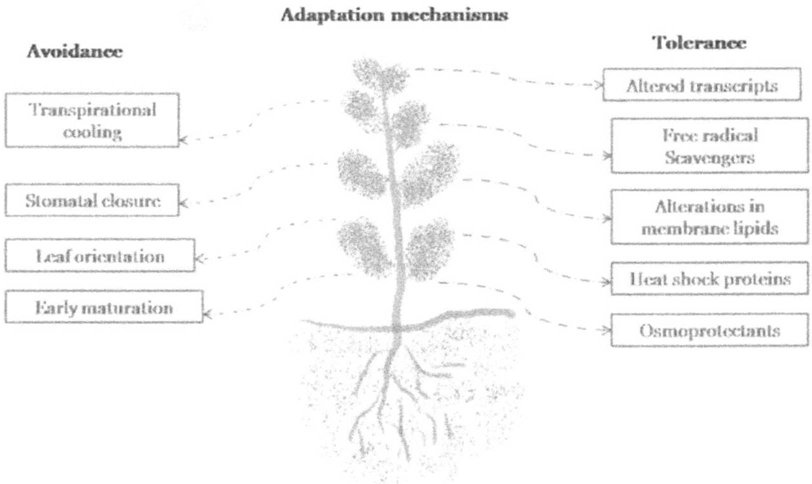

Figure 1. Adaptation mechanisms for high temperature tolerance in crop plants.

4. THERMO-TOLERANCE THROUGH BREEDING STRATEGIES

4.1 Screening criteria

Breeding for high temperature tolerance requires an essential knowledge on plant adaptation response to heat shocks. In general, the genotypes exhibiting less detrimental effect on photosynthesis and reproductive development tend to survive well under heat prone areas [27]. Involvement of these two components in selection criteria would be beneficial in evolving thermo tolerant cultivars. Tolerant genotypes evolve several morphological, physiological and biochemical alterations in response to heat shocks. Knowledge on sensitivity of several phenological stages to high temperature will pave way for trait specific improvement. High temperature is often correlated with other environmental factors which poses a major limitation for selection under field conditions. At present, varied selection criteria has been developed by scientists, which favors delineation of superior variety at prevailing environment [28]. Heat tolerant index has been evolved for sorghum which depicts the proportion of growth recovery after exposure to high temperature stress. It is the ratio of increase in coleoptile growth in a heat stress environment [50°C] to the enhancement in coleoptile length under normal environment (non-stress) [29]. It proves cost effective and rapid method to screen a large population size within shorter period. A proper validation of such technique would facilitate the development of tolerant lines in other crop species. Pollen viability and fruit set was considered as major selection criteria to predict yield under high temperature stress in tomato [30]. Physiological based trait selection such as harvest index, photosynthetic efficiency, respiration rate, delayed senescence and canopy architecture will also contribute towards increased tolerance to heat stress [31, 32].

4.2 Genetic resources for thermo tolerance

Inter-mating among closely related individuals for improvement of economic traits resulted in decline of genetic variability in a crop species [33]. Characterization of gene pool including land races and wild relatives would offer several tolerant genes for abiotic tolerance. Extensive efforts were made in screening of heat tolerant genotypes which can be directly introduced as a cultivar or utilized to introgress gene into new genetic background [34]. Thermo-tolerant lines were successfully isolated from wild gene pool in wheat [35]. High magnitude of variation was observed in wild progenitor "*Aegilops tauschii*" of wheat for cell viability and membrane stability [36]. Similarly, a heat tolerant source for reproductive stage was identified in *A. geniculata* and *A. speltoides* Tausch which would pave way in development of thermo-tolerant hexaploid wheat cultivars in near future [37]. A higher growth rate and improved photosynthetic efficiency was observed in wild relative "*Oryza meridionalis*" of rice at high temperature [38]. Indirect selection on pollen viability led to identification of thermo-tolerant accessions in soybean (DG 5630RR) [39], chickpea (ICC15614 & ICC1205) [40], maize (AZ100) [41], and several other crop species. Direct selection based on yield under target environment (heat stress) resulted in development of tolerant lines in many tropical grain legumes. Four tolerant genotypes/accessions *viz.*, SRC-1-12-1-48, SRC-1-12-1-182, 98012-3-1-2-1 and 98020-3-1-7-2 were isolated in common bean by employing stress tolerant indices [42]. Nine thermo-tolerant wild accessions were delineated in USDA upland cotton germplasm by employing chlorophyll fluorescence technique [43].

4.3 Conventional breeding approaches

Evolving thermo-tolerance through conventional breeding approach proves promising in many crop species. Breeding for early maturing genotype in broccoli had improved head quality by avoiding heat stress at flowering stage [44]. In general, breeding programmes are carried out in hotter regions which promote selection of thermo-tolerant traits. Physiological based trait breeding was practiced at International Maize and Wheat Improvement Center (CIMMYT) for development of heat tolerant cultivars in wheat. The parental genotypes were characterized through various crossing schemes and appropriate breeding programme was framed for improvement of thermo related traits [45]. A wild ancestor "*T. tauschii*" was utilized as a gene donor for achieving increased grain size and filling percent under high temperature through recurrent selection [46]. Similarly, three cycles of recurrent selection had led to improved yield under heat stress regimes in potato [47]. Thermo tolerant alleles were introgressed into heat sensitive cultivar "Paymaster 404" from a donor accession "7456" of *G. barbadense* through backcross breeding [48]. A significant improvement in yield was realized under heat stress environment by adoption of gametic selection in maize [41]. A deep rooted cultivar "Nagina 22 (N22)" of aus rice exhibited high pollen viability and spikelet fertility (64–86%) under heat stress [49]. The thermo-tolerance of N22 was successfully introgressed into Xieqingzao B line through backcross method [50]. Dissecting out the genetic and physiological basis of thermo-tolerance will hasten up the development of resilient cultivars suited to hotter regions.

4.4 Advanced breeding approaches for thermo tolerance

The genetic basis of thermo-tolerance is not clearly understood because of complex trait inheritance. Advances in molecular approaches such as DNA marker identification and genotyping assay had paved way in determination of several QTL's associated with high temperature tolerance [51]. In wheat, QTL's were identified for canopy temperature, and chlorophyll fluorescence imparting tolerance to heat stress [52]. A major QTL "Htg 6.1" in lettuce was involved in enhancement of seed germination capacity at high temperature [53]. A recessive QTL for increased spikelet fertility under high temperature was dissected out in rice at chromosome 4. The identified QTL were found in several populations of heat tolerant rice cultivars [54]. Six QTL's were involved to enhance fruit set at high temperature in tomato [55]. Five thermo tolerant QTL's were identified in *Brassica campestris* by employing random amplified polymorphic DNA (RAPD) and amplified fragment length polymorphism (AFLP) markers [56]. In maize, eleven major QTL's for increased pollen germination and pollen tube growth under high temperature was mapped using restriction fragment length polymorphism (RFLP) markers [57]. Identification of candidate QTL's would pave way in precise introgression of heat tolerant genes into superior cultivars through marker assisted breeding approach.

The closely associated markers with targeted QTL will hasten the recovery of superior genotypes with heat tolerant traits in a population. A marker assisted breeding approach was employed in rice to derive heat tolerant line with superior grain quality. Two flanking markers *viz.*, ktIndel001 and RFT1 enclosing 1.5 Mb chromosomal region was transferred from tolerant cultivar "Kokoromachi" to Tohoku 168. Significant improvement in grain quality under high temperature was observed in the derived NIL's compared to susceptible cultivar "Tohoku 168" [58]. Fourteen SSR markers linked to heat susceptibility index of grain filling per cent and single kernel weight was identified in bread wheat which was employed in marker assisted selection (MAS) to screen genotypes for thermo tolerance [59]. Utilization of MAS approach for heat tolerance remains less efficient because of high gene x environment and epistatic interactions. The low breeding efficiency can be resolved by genomic selection (GS) approach which involves wide number of molecular markers exhibiting high genome coverage. High genetic gain is realized in GS approach due to close association between predicted and true breeding value over generations [60].

At present, transgenic approach also proves to be desirable tool for designing thermo tolerant lines *via* introgression of genes from diverse gene pools [61]. The genetic transformation was focused primarily on transcription factors, induction of heat shock proteins, molecular chaperones, osmolytes, antioxidant components and growth regulators [62]. Heat shock proteins play a primary role in imparting thermo tolerance in crop species. It is functionally associated with diverse group of molecular chaperones that is involved in restoration of degraded proteins to their native structure under high temperature. Induction of heat shock proteins through genetic manipulation was achieved in *arabidopsis* [63], maize [64], rice [65], soybean [66], and pepper [67]. The DREB gene family was also reported to impart heat tolerant response in many crop species. Over expression of ZmDREB2A in maize [68] and GmDREB2A in soybean [69] was associated with increased survival and adaptation under high temperature. Transgenic techniques were employed to alter membrane lipid properties for thermo-tolerance in crop species. High proportion of saturated fatty acid in membrane had increased tolerance under heat stress. Suppression of omega-3 fatty acid desaturase gene in chloroplast had reduced the accumulation of trieonic fatty acid in transgenic tobacco [70] and tomato [71] leading to thermo-tolerance. A significant accumulation of glycine betaine (osmolyte) was achieved in *arabidopsis* through transfer of "cod gene" from *Arthrobacter globiformis* [72]. High proportion of glycine betaine protects the PSII component by inhibiting the ROS activities under heat stress. Implementation of transgenic approaches in other crop species will accelerate the development of resilient genotypes suited to high temperature regimes.

5. CONCISE SUMMARY

Development of thermo-tolerant lines has to be prioritized to meet out the future climatic change coupled with food demands. Knowledge on plant response and adaptation mechanisms to heat stress is required for framing out breeding strategies. It remains a challenging task in evolving resilient genotypes suited to high temperature because of less efficient screening protocols at field conditions. The existence of low genetic variation for heat response related traits limited the progress of conventional breeding approach in many crop species. Use of molecular breeding strategies had opened up several heat tolerant related QTL's in crop species. However, still precise research work involving huge marker data is needed for attaining high breeding efficiency for thermo tolerance. Recently, the involvement of transgenic approach paved way for utilization of tolerant source from diverse gene pools. Study on induction of heat shock proteins led to increased thermo tolerance in many crop species. Similarly, other heat response related traits such as induction of antioxidant components, osmolytes, and chaperones were also included in transgenic approach for inducing heat stress tolerance. Thus, high economic yield could be realized at elevated temperature regimes with the involvement of combined breeding approaches.

AT A GLANCE

Increase in global warming poses a severe threat on agricultural production thereby affecting food security. A drastic reduction in yield at elevated temperature is a resultant of several agro-morphological, physiological and biochemical modifications in plants. Heat tolerance is a complex mechanism under polygenic inheritance. Development of tolerant genotypes suited to heat extremes will be more advantageous to tropical and sub tropical regimes. A clear understanding on heat tolerance mechanism is needed for bringing trait based improvement in a crop species. Heat tolerance is often correlated with undesirable traits which limits the economic yield. In addition, high environmental interactions coupled with poor phenotyping techniques limit the progress of breeding programme. Recent advances in molecular technique led to precise introgression of thermo-tolerant genes into elite genetic background which has been reviewed briefly in this chapter.

REFERENCES

1. Intergovernmental Panel on Climate Change (IPCC). The Physical Science Basis. Contribution of Working Group I to the Fourth Assessment Report of the Intergovernmental Panel on Climate Change. 2007. Available from: http:// www.cambridge.org /features/earth_environmental/ climatechange/wg1. htm"/t"blank"
2. Wahid A, Gelani S, Ashraf M, Foolad MR. Heat tolerance in plants: an overview. Environmental and Experimental Botany. 2007;61:199-223
3. Bita C, Gerats T. Plant tolerance to high temperature in a changing environment: scientific fundamentals and production of heat stress tolerant crops. Frontiers in Plant Science. 2013;4:273
4. Fischer R, Edmeades GO. Breeding and cereal yield progress. Crop Science. 2010;50:85-98. DOI: 10.2135/cropsci2009.10.0564
5. Lobell DB, Burke MB, Tebaldi C, Mastrandrea MD, Falcon WP, Naylor RL. Prioritizing climate change adaptation needs for food security in 2030. Science. 2008;319:607-610
6. Iba K. Acclimative response to temperature stress in higher plants: approaches of gene engineering for temperature tolerance. Annual Review of Plant Biology. 2002;53:225-245
7. Camejo D, Rodríguez P, Morales MA, Dell'amico JM, Torrecillas A, Alarcon JJ. High temperature effects on photosynthetic activity of two tomato cultivars with different heat susceptibility. Journal of Plant Physiology. 2005;162:281-289

8. Hasanuzzaman M, Nahar K, Fujita M. Extreme temperatures, oxidative stress and antioxidant defense in plants. In: Vahdati K, Leslie C. Abiotic Stress - Plant Responses and Applications in Agriculture. ed. InTech: Rijeka, Croatia; 2013. p. 169-205
9. Vollenweider P, Günthardt-Goerg MS. Diagnosis of abiotic and biotic stress factors using the visible symptoms in foliage. Environmental Pollution. 2005;137:455-465
10. Ismail AM, Hall AE. Reproductive stage heat tolerance, leaf membrane thermostability and plant morphology in cowpea. Crop Science. 1999;39:1762-1768. DOI: 10.2135/cropsci 1999.3961762x
11. Wahid A. Physiological implications of metabolites biosynthesis in net assimilation and heat stress tolerance of sugarcane sprouts. Journal of Plant Research. 2007;120:219-228
12. Maestri E, Klueva N, Perrotta C, Gulli M, Nguyen HT, Marmiroli N. Molecular genetics of heat tolerance and heat shock proteins in cereals. Plant Molecular Biology. 2002;48:667-681
13. Koini MA, Alvey L, Allen T, Tilley CA, Harberd NP, Whitelam GC, et al. High temperature-mediated adaptations in plant architecture require the bHLH transcription factor PIF4. Current Biology. 2009;19:408-413
14. Siddique KHM, Loss SP, Regan KL, Jettner RL. Adaptation and seed yield of cool season grain legumes in Mediterranean environments of South-Western Australia. Australian Journal of Agricultural Research. 1999;50:375-387
15. Ferris R, Ellis RH, Wheeeler TR, Hadley P. Effect of high temperature stress at anthesis on grain yield and biomass of field grown crops of wheat. Annals of Botany. 1998;34:67-78
16. Asada K. Production and scavenging of reactive oxygen species in chloroplasts and their functions. Plant Physiology. 2006;141:391-396
17. Soliman WS, Fujimori M, Tase K, Sugiyama SI. Oxidative stress and physiological damage under prolonged heat stress in C3 grass Lolium perenne. Grassland Science. 2011;57:101-106
18. Rodríguez M, Canales E, Borrás-Hidalgo O. Molecular aspects of abiotic stress in plants. BiotechnologiaApplicada. 2005;22:1-10
19. Young LW, Wilen RW, Bonham-Smith PC. High temperature stress of Brassica napus during flowering reduces micro and megagametophyte fertility, induces fruit abortion, and disrupts seed production. Journal of Experimental Botany. 2004;55:485-495
20. Miller G, Schlauch K, Tam R, Cortes D, Torres MA, Shulaev V, et al. The plant NADPH oxidase RbohD mediates rapid, systemic signaling in response to diverse stimuli. Science. Signal. 2009;2:ra45

21. Sarieva GE, Kenzhebaeva SS, Lichtenthaler HK. Adaptation potential of photosynthesis in wheat cultivars with a capability of leaf rolling under high temperature conditions. Russian Journal of Plant Physiology. 2010;57:28-36

22. Srivastava S, Pathak AD, Gupta PS, Shrivastava AK, Srivastava AK. Hydrogen peroxide-scavenging enzymes impart tolerance to high temperature induced oxidative stress in sugarcane. Journal of Environmental Biology. 2012;33:657-661

23. Verbruggen N, Hermans C. Proline accumulation in plants: a review. Amino Acids. 2008;35:753-759

24. Apel K, Hirt H. Reactive oxygen species: metabolism, oxidative stress and signal transduction. Annual Review of Plant Biology. 2004;55:373-399

25. Almeselmani M, Deshmukh PS, Sairam RK, Kushwaha SR, Singh TP. Protective role of antioxidant enzymes under high temperature stress. Plant Science. 2006;171:382-388

26. Ahmad P, Bhardwaj R, Tuteja N. Plant signaling under abiotic stress environment. In: Ahmad P, Prasad MNV. Environmental Adaptations and Stress Tolerance of Plants in the Era of Climate Change. ed. Springer: New York,USA; 2012.p.297-324

27. Prasad P, Pisipati S, Mutava R, Tuinstra M. Sensitivity of grain sorghum to high temperature stress during reproductive development. Crop Science. 2008;48:1911-1917. DOI: 10.2135/cropsci2008.01.0036

28. Hall AE. Breeding Cowpea for Future Climates. In: Yadav SS, Redden R, Hatfield JL, LotzeCampen H, Hall AJW. Crop Adaptation to Climate Change. ed. Hoboken: John Wiley & Sons; 2011

29. Young TE, Ling J, Geisler-Lee CJ, Tanguay RL, Caldwell C, Gallie DR. Developmental and thermal regulation of maize heat shock protein, HSP101. Plant Physiology. 2001;127:777-791

30. Berry SZ, Rafique-Uddin M. Effect of high temperature on fruit set in tomato cultivars and selected germplasm. HortScience. 1988;23:606-608

31. Cossani CM, Reynolds MP. Physiological traits for improving heat tolerance in wheat. Plant Physiology. 2012;160:1710-1718

32. Gupta PK, Balyan HS, Gahlaut V, Kulwal PL. Phenotyping, genetic dissection, and breeding for drought and heat tolerance in common wheat: status and prospects. Plant Breeding Reviews. 2012;36:85

33. Gur A, Zamir D. Unused natural variation can lift yield barriers in plant breeding. PLoS Biol. 2004;2:1610-1615

34. Jagadish SVK, Craufurd PQ, Wheeler TR. Phenotyping parents of mapping populations of rice for heat tolerance during anthesis. Crop Science. 2008;48:1140-1146

35. Reynolds MP, Balota M, Delgado MIB, Amani I, Fischer RA. Physiological and morphological traits associated with spring wheat yield under hot, irrigated conditions. Functional Plant Biology. 1994;21:717-730

36. Gupta S, Kaur S, Sehgal S, Sharma A, Chhuneja P, Bains NS. Genotypic variation for cellular thermo tolerance in Aegilops tauschii Coss., the D genome progenitor of wheat. Euphytica. 2010;175:373-381

37. Pradhan GP, Prasad PVV, Fritz AK, Kirkham MB, Gill BS. High temperature tolerance in aegilops species and its potential transfer to wheat. Crop Science. 2012;52:292-304

38. Scafaro AP, Haynes PA, Atwell BJ. Physiological and molecular changes in Oryza meridionalis Ng, a heat-tolerant species of wild rice. Journal of Experimental Botany. 2010;61:191-202

39. Salem MA, Kakani VG, Koti S, Reddy KR. Pollen based screening of soybean genotypes for high temperatures. Crop Science. 2007;47:219-231

40. Devasirvatham V, Gaur P, Mallikarjuna N, Raju TN, Trethowan RM, Tan DKY. Effect of high temperature on the reproductive development of chickpea genotypes under controlled environments. Functional Plant Biology. 2012;39:1009-1018

41. Petolino JF, Cowen NM, Thompson SA, Mitchell JC. Gamete selection for heat stress tolerance in maize. In: Ottaviano E. Angiosperm Pollen and Ovules. ed. Springer: Verlag, New York; 1992. p. 355-358

42. Porch TG. Application of stress indices for heat tolerance screening of common bean. Journal of Agronomy and Crop Science. 2006;192:390-394

43. Wu T, Weaver DB, Locy RD, McElroy S, Van Santen E. Identification of vegetative heat-tolerant upland cotton (Gossypium hirsutum L.) germplasm utilizing chlorophyll fluorescence measurement during heat stress. Plant Breeding. 2013;133:250-255

44. Farnham MW, Bjorkman T. Breeding vegetables adapted to high temperatures: a case study with broccoli. HortScience. 2011;46:1093-1097

45. Reynolds M, Manes Y, Izanloo A, Langridge P. Phenotyping approaches for physiological breeding and gene discovery in wheat. Annals of Applied Biology. 2009;155:309-320

46. Gororo NN, Eagles HA, Eastwood RF, Nicolas ME, Flood RG. Use of Triticum tauschii to improve yield of wheat in low yielding environments. Euphytica. 2002;123:241-254

47. Benites FRG, Pinto CABP. Genetic gains for heat tolerance in potato in three cycles of recurrent selection. Crop Breeding and Applied Biotechnology. 2011;11:133-140

48. Rodriguez-Garay B, Barrow JR. Pollen selection for heat tolerance in cotton. Crop Science. 1988;28:857-859

49. Poli Y, Basava RK, Panigrahy M, Vinukonda VP, Dokula NR, Voleti SR, et al. Characterization of a Nagina22 rice mutant for heat tolerance and mapping of yield traits. Rice. 2013;6:36

50. Jiang-lin L, Hong-yu Z, Xue-lian S, Ping-an Z, Ying-jin H. Identification for heat tolerance in backcross recombinant lines and screening of backcross introgression lines with heat tolerance at milky stage in rice. Rice Science. 2011;18:279-286

51. Pinto RS, Reynolds MP, Mathews KL, McIntyre CL, Olivares-Villegas JJ, Chapman SC. Heat and drought adaptive QTL in a wheat population designed to minimize confounding agronomic effects. Theoretical Applied Genetics. 2010;121:1001-1021

52. Vijayalakshmi K, Fritz AK, Paulsen GM, Bai G, Pandravada S, Gill BS. Modeling and mapping QTL for senescence-related traits in winter wheat under high temperature. Molecular Breeding. 2010;26:163-175

53. Argyris J, Dahal P, Hayashi E, Still DW, Bradford KJ. Genetic variation for lettuce seed thermo inhibition is associated with temperature-sensitive expression of abscisic acid, gibberellin, and ethylene biosynthesis, metabolism, and response genes. Plant Physiology. 2008;148:926-947

54. Ye C, Argayoso MA, Redon ED, Sierra SN, Laza MA, Dilla CJ, et al. Mapping QTL for heat tolerance at flowering stage in rice using SNP markers. Plant Breeding. 2012;131:33-41

55. Ventura G, Grilli G, Braz LT, Gertrudes E, Lemos M. QTL identification for tolerance to fruit set in tomato by fAFLP markers. Crop Breeding and Applied Biotechnology. 2007;7:234-241

56. Shuancang Y, Yongiian W, Xiaoying Z. Mapping and analysis QTL controlling heat tolerance in Brassica campestris L. ssp. Pekinensis. Acta HorticulturaeSinica. 2003;30:417-420

57. Frova C, Sari-Gorla M. Quantitative trait loci (QTLs) for pollen thermotolerance detected in maize. Molecular and General Genetics. 1994;245:424-430

58. Shirasawa K, Sekii T, Ogihara Y, Yamada T, Shirasawa S, Kishitani S, et al. Identification of the chromosomal region responsible for high-temperature stress tolerance during the grain filling period in rice. Molecular Breeding. 2013;32:223-232

59. Sadat S, Saeid KA, Bihamta MR, Torabi S, Salekdeh SGH, Ayeneh GAL. Marker assisted selection for heat tolerance in bread wheat. World applied sciences journal. 2013;21:1181-1189

60. Habier D, Fernando RL, Dekkers JCM. Genomic selection using low-density marker panels. Genetics. 2009;182:343-353. DOI: 10.1534 / genetics. 108. 100289

61. Ashraf M. Inducing drought tolerance in plants: recent advances. Biotechnology Advances. 2010;28:169-183

62. Grover A, Mittal D, Negi M, Lavania D. Generating high temperature tolerant transgenic plants: achievements and challenges. Plant Science. 2013;205:38-47

63. Prandl R, Hinderhofer K, Eggers-Schumacher G, Schoffl F. HSF3, a new heat shock factor from Arabidopsis thaliana, derepresses the heat shock response and confers thermotolerance when over expressed in transgenic plants. Molecular and General Genetics. 1998;258:269-278

64. Nieto-Sotelo J, Martınez LM, Ponce G, Cassab GI, Alagon A, Meeley RB, et al. Maize HSP101 plays important roles in both induced and basal thermotolerance and primary root growth. The Plant Cell. 2002;14:1621-1633

65. Katiyar-Agarwal S, Agarwal M, Grover A. Heat-tolerant basmati rice engineered by over-expression of hsp101. Plant Molecular Biology. 2003;51:677-686

66. Chen XJ, Ye CJ, Lu XY, Xu MX, Li W, Zhang LM, et al. Cloning of GmHSFA1 gene and its overexpression leading to enhancement of heat tolerance in transgenic soybean. Yi Chuan=. Hereditas. 2006;28:1411-1420

67. Dang FF, Wang YN, Yu L, Eulgem T, Lai Y, Liu ZQ, et al. CaWRKY40, a WRKY protein of pepper, plays an important role in the regulation of tolerance to heat stress and resistance to Ralstonia solanacearum infection. Plant, Cell and Environment. 2013;36:757-774

68. Qin F, Kakimoto M, Sakuma Y, Maruyama K, Osakabe Y, Tran LS, et al. Regulation and functional analysis of ZmDREB2A in response to drought and heat stresses in Zea mays L. The Plant Journal. 2007;50:54-69

69. Mizoi J, Ohori T, Moriwaki T, Kidokoro S, Todaka D, Maruyama K, et al. GmDREB2A; 2, a canonical DEHYDRATION-RESPONSIVE ELEMENT-BINDING PROTEIN 2-type transcription factor in soybean, is post translationally regulated and mediates dehydration-responsive element-dependent gene expression. Plant Physiology. 2013;161:346-361

70. Murakami Y, Tsuyama M, Kobayashi Y, Kodama H, Iba K. Trienoic fatty acids and plant tolerance of high temperature. Science. 2000;287:476-479

71. Wang HS, Yu C, Tang XF, Wang LY, Dong XC, Meng QW. Antisense-mediated depletion of tomato endoplasmic reticulum omega-3 fatty acid desaturase enhances thermal tolerance. Journal of Integrative Plant Biology. 2010;52:568-577

72. Alia HH, Sakamoto A, Murata N. Enhancement of the tolerance of Arabidopsis to high temperatures by genetic engineering of the synthesis of glycine betaine. The Plant Journal. 1998;16:155-161

CHAPTER-9

PLANT REACTIONS TO SALT STRESS

1. INTRODUCTION: AN OVERVIEW

As global agricultural landscapes contend with the repercussions of climate change and environmental degradation, the phenomenon of soil salinity has become a significant challenge affecting crop productivity. High salinity levels in soils can exert detrimental effects on plants, triggering a cascade of physiological and biochemical responses that collectively define the plant's strategy to cope with salt stress. Understanding these intricate mechanisms is pivotal for developing resilient crops capable of thriving in saline environments. The most important factor in the survival of humanity is food. A person, who cannot access to enough food for survive, cannot continue its development and eventually lose its life. For meeting the food needs of growing population, agricultural production has to be increased by 87% by 2050 [1]. For fulfilling the food needs of world besides increasing the cultivated lands, it is necessary to take maximum efficiency and benefit from the yield. Food, which can be called the main source of our lives, is produced by plants. When plants contain essential external and internal factors, they go under phase of production the food that enables the living population of the world to survive. However, the needed food sometimes cannot be obtained from the plant due to external factors. In plants, the external factor reducing growth of plant, decreasing yield of plant, inhibition plant development, is called stress. Stress causes reactions in altered gene expression in plants, cell metabolism, growth rates, yield and many other areas. In addition to these factors, stress causes death of plant and the loss of quality and quantity in plants. Generally, stress can be biotic stress caused by living factors such as microorganisms, wild plants, pathogens or can be abiotic stress caused by non-living factors such as temperature, mineral toxicity, various gasses [2]. For surviving against these negative factors, plants develop some response mechanisms. Mostly they are in two tendencies. Plants can prevent the activities of stress factors with developing mechanisms or they may try to continue their lives by protecting themselves against external factors with tolerance mechanisms. Soil salinity is one of the biggest problems that are considered among abiotic stress factors and decrease the usability of our agricultural lands today.

Soil salinity is considered as one of the most important problems of agriculture throughout history [3]. It limits agricultural production by especially harming the crop yields [4]. 1125 million ha of cultivated lands in world are coping with salinity, 76 million ha of agriculture lands are in the effect of human-induced salinity and sodicity [5]. One out of five irrigated lands are affected by salinity and every year 1.5 million ha agriculture lands lose their suitability for agricultural production. And if the conditions continue this way, 50% of the cultivated lands will be at the edge of loss by 2050 [6, 7]. Salinity is a condition of the reason of high concentration soluble salts and when the ECe value is 4 dS/m^{-1} and more, the soil is considered as salty. Soil salinity creates stress in two ways [4]. The salts with high concentrations in the soil complicate to get water to cells for roots and salts with high concentrations in the plant causes toxicity. The salt outside of plant root affects cell expansion and cell growth directly. Toxic concentrations of salt spend time for accumulation before affecting the plant [4].

The effect of salt on cell growth and expansion, plant membrane irregularity, ion toxicity, changing metabolic process, the mechanism of germination, photosynthetic activity, shoot and root lengths, leaf development is incontroversible [8, 9]. And plants develop some mechanisms to get rid of from these negative effects. Since NaCl is the most soluble and common salt, all plants develop mechanisms for regulating the accumulation of NaCl [10]. Halophytes, plant species of high salinity soils, maintain better this extracting from plant than glycophytes, which do not have any tolerance to high salinity soils [11]. Because the salinity is common in arid and semi-arid regions, adaptation mechanisms of plants occur according to this low water potentially areas [4]. For coping with harmful effects of salinity, plants create a lot of different morphological, physiological and biochemical adaptations [12].

2. ABIOTIC STRESS AND SALT STRESS

2.1 Abiotic stress

Abiotic stress is a type of stress caused by environmental factors that affects plant growth, development, yield and seed quality in a negative way. Abiotic stress usually affects plant with factors like drought or floods, excessive light, excessive high or low temperature, lack of minerals, excessive pH in soil. Abiotic stress cannot directly occur, they are caused by multiple factor interactions. For example, "acid stress usually occurs because of the interaction aluminium toxicity" [13]. Plants create responses in 4 stages against abiotic stress effect. 1-Beginning alarm phase,

2-acclimation phase, 3-repair phase, 4-exhaustion phase [14]. The effect of stress is linked to plant sensitivity. When some species are so sensitive to external factors, some can tolerate it. From abiotic stresses, after drought stress the most effective factor on our world is mineral stress. In the occurrence of mineral stress, salinity is the most effective factor [2]. Big part of our agricultural lands suffers from salinity and each day this situation is getting impassable.

2.2 Salinity and salt stress

Salinity occurs due to problems like wrong usage of agricultural lands, lack of rain, excess evaporation, lack of drainage. Soils that have salt concentrations that prevent plant growth (Ece > 4 dS/m-1) and soils that do not have salt concentrations that disturb the structure of soil (ESP < 15) is called saline soil. Due to the excessive accumulations of these salts, affecting features like plant growth, development, yield, seed quality is called salt stress. Today, irrigation is being made in 17% of arid and semi-arid lands in the world and due to wrong treatments approximately 20% of these irrigated lands are being unproductive, plant presence is under negative cycle with it [2]. There is a little bit different situation in our country. Irrigation needed lands covers 2% of surface areas and 74% of these lands (approximately 12 thousand hectares) are exposed to negative effects of salt [15]. Soil salinity puts plant into stress with complicating ground-water flow from roots due to high concentrations of salts and with causing toxicity due to accumulation of high concentrations of salt in the plant [4]. The plants that struggle with salinity try to continue their life cycles by giving morphological, biochemical, physiological responses. More than 800 million ha agricultural lands are under negative effects of salinity in the world. This ratio composes more than 6% of world total agricultural lands [16]. Salinity separates as primary and secondary salinity. While primary salinity occurs by natural factors like oceans, corrosion of rocks, human induced secondary salinity occurs by excessive irrigation in agricultural lands, deterioration of agricultural land structure [4]. This situation shows its effect more day by day. Being estimated that 50% of cultivated agricultural lands will be under salt stress by 2050 [17]. NaCl and Na2SO4 salts are the main reasons affecting the salinity of agricultural lands [18]. It is easier to understand whether the plant is salt tolerant at non-lethal salt concentrations during germination and early seedling stages, the most damages occurs in these stages [19]. Some factors should be considered in controlling salinity [20]. "In addition to well known principles such as drainage and management of irrigation resource, cultural practices and agricultural land development works are also important too. When it comes to cultural practices, fertilization, planting method, irrigation treatment, land leveling factors come to mind. Agricultural land developments are development of drainage, land leveling, breeding irrigations." [20]. The most important factor affecting salinity is lack of drainage. It causes million of fertile agricultural lands to be destroyed.

3. EFFECT OF SALINITY ON PLANTS

Salinity composes stress by damaging ionic and osmotic balances in plants. Osmotic stress caused by increasing the amount of salt in soil, decreases the amount of water that plant use and as a result physiological drought occurs. After these conditions, ionic stress occurs in the plant with deterioration of plant ion balance. Na and Cl ions which increases in medium with ionic stress, get in competition with essential nutrients such as K^+, Ca^{2+}, Mg^{2+} lead to nutrient deficiency in plant. While the direct effect of salinity is osmotic and ionic stresses, deteriorations in structure and synthesis of toxic components composes secondary effect [21].

3.1 Secondary effect of salt stress

The main secondary factors caused by NaCl are, complication in taking of K^+ into cells (it is efficient in closure of stomatas), decreasing photosynthetic activity, generation of reactive oxygen species (ROS) and programmed cell deaths [22]. K element is one of the most vital elements for plants for this reason Na^+ ions compete with K^+ ions for getting into the cell. Na^+ composes stress with blocking K^+ influx into the cell. Ion and hyperosmotic stress causes secondary metabolic effects in plant and the plant has to decrease these stresses for maintain its development [8].

4. THE CHANCES CAUSED BY SALT STRESS IN PLANT

Salt stress occurs as a result of excessive salt accumulation in the soil and the plant cannot take water it needs with roots. Salt stress affects plants by toxicity caused by osmotic stresses and ions [23]. As a result of these effects, some negative changes occur in plants.

4.1 The changes in growth characteristics

Plants are the most sensitive at germination and seedling growth stages. In these stages, many activities expected to occur in natural course of the plant slow down or stop. Along with these limitations, many limitations such as physiological drought, sterility, stunted growth, reduction in leaf area, slow or lack blooming, irregularity of membrane, generation of reactive oxygen species (ROS), reduction in photosynthetic activity can happen. Also, high concentration of salinity causes reduction in leaf fresh and dry weights, with low humidity it causes reduction in shoot and root growth on plants. With affecting stomata, salt stress stops permanence of stomatal reactions. Formation of root nodules, plant sprouts and leaves are affected from salt stress. According to the studies, these negative results can be counted as the effects of salt stress on plants but the main responses of plants are still not fully known [24].

4.2 Plant root and shoot length

Root length, root length density and thick roots which are the features of the root, are very vital in development of subsoil parts of plant by taking the existing water. A fertile root system during early seedling stage which is the most sensitive stage, provides advantage in accelerating growth. Since the water which is taken by plant in danger of being lost easily as a result it has to be taken from non-deep layers of soil [24].

Roots are one of the vulnerable parts of the plant. While under salt stress reduction in root length is being seen, in addition to salt stress not being able to use the existing water negatively affects root and shoot growth of plant. Besides all of these negative effects, roots are also known as surprisingly strong when they directly exposed to salt stress [24].

4.3 Cell level effect

4.3.1 Organelle level effect

Without doubt, the most affected organelle in plant from salt stress is chloroplast. Stress mostly affects thylakoids and stroma in chloroplast [2]. Chloroplasts tend to generate reactive oxygen species such as H_2O_2, O_2^-, OH^-. Reactive oxygen species seriously affects plant metabolic activities by causing oxidative damage to lipids that result in protein breakdown and membrane lipid peroxidation, causing the thylakoids to swell and turn into a wavy shape [2, 25]. The stress caused by salt, causes another negative situation starch accumulation which is not known how it occurred fully in chloroplast. Another negative situation caused by salt stress is deterioration of grana lamellae. Salt stress disrupts electrical charges that composes grana lamellae by changing ionic composition.

As a result of stress mitochondria, the another organelle affected from salt stress, is exposed to negative effects such as structural fragmentations, accelerations in vacuole forming, swelling and decrease in crystal [26].

4.3.2 Effect of ion toxicity

Accumulation of Na^+ and Cl^- ions in plant makes negative effect by developing competitive system with limiting intake of the other ions into the plant. Soluble salts with high concentrations in soil can cause physiological drought [23]. With accumulation of soluble salts in plant root area, decrease in plant water intake occurs. Accumulated salts in root area creates ion imbalance in cells by getting into plant cells and this imbalance causes growth problems in plant tissues such as leaves, seeds etc. Excessive Na^+ accumulation in plants causes necrosis in old leaves and high concentrations of Na^+ ions in shoots cause metabolic and osmotic problems [27]. Accumulation of soluble salts in high concentrations in soil causes negative effects.

4.4 Effect to photosynthesis

It is known that the effect occurring in photosynthesis which is negatively affected by high and low salt concentrations, may be from stomata, non-stomata, or both limitations [2]. Several factors can be counted as a reason of decrease in photosynthetic activity. First reason is reduction in cell permeabilization of CO_2 as a result of dehydration of membranes. High concentration of salt in soil creates high osmotic potential in plant by limiting water reaching of plant but with decrease in water potential osmotic stress in plant occurs. Photosynthetic electron transport is affected negatively. With ion toxicity caused by Na^+ and Cl^- ions, essential nutrients cannot be taken and this situation leads to limitation of photosynthesis and generation of reactive oxygen species (ROS). The changes in enzyme activities also cause decrease in photosynthesis ratio [24].

4.4.1 Closure of stomata

High concentration salinity in soil causes osmotic stress formation by limiting water availability of plant with roots whereas closure of stomatas is a first response of plants. This response of plant limits transpiration and as a result stomata conductivity decreases [4]. Closure of stomata happens two ways as hydroactive closure and hydropassive closure [28]. Plants synthesis chemical signal molecules in occurring of hydroactive closure. ABA is one of the important synthesized chemical signal molecules and it is effective in plant growth and creating water balance. Under low water potential ABA molecules are transported into stomata by roots and old leaves via xylem. Low water potential is sensed by root tip and ABA molecules are synthesized at the root and transported to shoots with the help of xylem. The reason of synthesis of ABA molecules is ensuring regulation of stomata conductivity in low water potential conditions and as a result of this situation, it causes decrease in leaf water content [2].

4.5 Oxidative stress

With decrease in photosynthesis formation, reactive oxygen species increases and increasing of reactive oxygen species increases production of enzymes that enable detoxify of these reactive oxygen species. While plants adapt to changing environment, they undergo the activation of biochemical processes that prevent oxidative damages against photosystem, leaf morphology, chloroplast pigment composition and many other changes. Reactive oxygen species can divide the activities of plant. For example, DNA, proteins and lipids make negative effects by mixing into plant metabolism [29].

5. EFFECT OF SALINITY ON SOME CROPS

Until 2050, for fulfilling the food needs of increasing population and developing world, the yield of the crops obtained from crops, which have undeniable effects in human life, has to increase by 50% [30]. The most effective factor in limiting product yield is undoubtedly salt stress. Plants try to tackle with stress for removing the pressure on them. But the method of tackle and the effect of each species differ. The process of field crops under salinity and their responses vary from species to species.

5.1 Rice (*Oryza sativa* L.)

Rice is a monocotyl warm-season cereal crop. It is grown in many parts of the world, especially Asia, it is an indispensable food of approximately 50% of the world population [23]. Rice shows more different responses to salt stress compared to other field crops. Rice crop is severely affected by salinity stress. Especially in the early growth stage, reduction is observed in the plant growth. In high concentration of salt, reduction is observed in plant seed growth and plant wet weight [31]. The first organ of the plant affected by salinity is root. As a result of excessive accumulation of Na^+ in the root, reduction is observed in plant root and shoot growth [32]. Salinity has negative effects on cell division and cell wall. As a result of high concentration of salt, the salt effect is severely observed in leaf length, root and yield [9]. Rice gives very rapid response under salinity stress. For example, in early response to stress, production of ethylene phytohormone contributes to plant survival [23, 33]. Salt stress causes poor development of inner and outer spikelets and sterility of cluster, this situation results in decreased grain yield [23].

5.2 Wheat (*Triticum aestivum* L.)

Wheat is a cereal genus belonging to the *Poaceae* family. In today, approximately 36% of humans use wheat as main product [34]. Salt stress affects wheat too. Yield in wheat starts to decrease in 6 dS/m^{-1} salinity ratio [35]. The presence of highly concentrated salts in the soil prevents the plant from bringing water to cells through its roots, causing osmotic stress and causing ion stress as a result of ion irregularity in plant. As a result of accumulation of Na^+ ion in plants in high concentrations, causes plant ionic stress more toxic by becoming dominant against other ions [34]. Osmotic stress caused by the absence of water to the plant causes a decrease in germination rates of the wheat varieties against the salt stress they are exposed to [12]. In addition to the negativities occurring in germination, there are some reductions in the shoot and roots, leaves and cells of the plant. Most sensitive stages against salt stress in wheat plants are early growth stages [23]. Excessive accumulation of Na^+ and Cl^- ions in chloroplast, which is the organelle most affected by salt stress, leads to decrease in photosynthetic ratio [36]. Decreases and stops in photosynthetic activity occurs when the salinity falls below 150 mM NaCl concentration [37].

5.3 Maize (*Zea mays* L.)

Maize plant which is not very sensitive to salinity, is a C4 plant belonging to *Poaceae* family [38]. The ratio of salt that allows maize to live without harming its growth and development, without causing any harm, is 0.25 mM NaCl or 1.8 dS/m^{-1} [39]. When there is salt accumulation above this ratio, disruptions in plant life cycle occurs. Like the other field crops, maize plant is most susceptible against salt stress during early growth stages including germination and seedling stages [40]. Salinity in seed germination may lead to fatal results by a) causing toxicity of Na^+ or Cl^- or both ions, b) preventing the plant from retaining water [39]. Early stages of stress, the typical reaction of plant before Na^+ ions reaches toxic degrees is decrease in growth as a result of the osmotic reaction [41]. Shoot growth in maize is also disrupted by osmotic stress [23]. The only reason for maize that the stress is caused by ion imbalance and that it increases and turns into a toxic effect is that Na^+ ions make it difficult to absorb ions such as K^+ and Ca^{2+} and as a result of this, it causes water loss and causes necrosis [41].

5.4 Sorghum (*Sorghum bicolor* L.)

Sorghum which is mostly grown in Australia in the world, is a monocotyl C4 plant that belongs to the *Poaceae* family. Even though sorghum has a high tolerance against salt concentrations, salt stress has a big role in plant growth deceleration [42]. In soils with high salinity, ion toxicity and toxicity-induced mineral deficiency are observed [43]. Tolerable salt concentration for grain sorghum is 6.8 dS/m^{-1} and when salinity reaches 7 dS/m^{-1} 25% decrease is observed in yield and when salinity reaches 10 dS/m^{-1} a 50% decrease is observed [44]. Under salinity, while germination rate decreases, germination time increases [45]. Plant stem yield and soluble carbohydrate decrease with increasing salinity [46].

5.5 Cotton (*Gossypium hirsutum* L.)

Production of cotton in the world, one of the most important fiber and oil plants, has increased by 2.5% compared to previous season with 121.6 million bales [47]. 35% of world fiber usage provided by cotton. Cotton cultivation and production is mostly done in arid and semi-arid regions [48]. Salinity causes negative effects on leaf area affected by osmotic stress, plant growth, root and shoot growth, in addition to these effects salinity causes decreases in photosynthetic activity, fiber quality, metabolic activities [49]. Salinity after flowering stage causes decrease in fiber quality of plant [23]. Undoubtedly, these occurring negative effects tend to change depending on the type of cotton, magnitude and time of stress exposed to.

6. RESPONSES OF PLANT TO SALT STRESS

In order to determine the responses of plants to salt stress, firstly we have to know the factors cause this stress. It is a priority to know whether the toxic effect caused by excessive salt accumulation in the plant or the osmotic stress caused by soluble salts in the soil in which the plant is restricting growth. While plants give rapid responses to external induced osmotic stress, they give slower responses to accumulation of Na^+ ions in the leaves [4, 50].

There are 2 types of struggle with salinity, human-help responses and the plant's natural adaptation responses. Natural struggle strategies among plants leans on 3 strategies. 1) extracting Na^+ ions from cytoplasm due to low intake, 2) the desire of Na^+ ion to enter the vacuole, 3) accumulation in leaves due to preference. Genotypes with high concentrations of Na^+ ions in leaves have proven to be highly susceptible to salinity, generally those who tolerate high concentrations are those that transmit Na^+ ions to the vacuoles of leaf cells. Salt tolerant plants get rid of harmful effects of NaCl [23].

6.1 Morphological adaptation responses against salinity in plants

The factors such as the type of plant (halophyte) that can adapt to the environment in which the plant is located or the type of plant (glycophyte) that is poorly affected by environment, the time it is exposed to salinity, salt concentrations in irrigation water have effects on growth and these factors cause plants to develop different mechanisms against negative effects [4].

6.1.1 Germination

Seed germination is one of the vital stages for the plant but it is prevented by salinity. Salt stress causes negative effects on plant imbibition and root growth [51]. Salinity induced decrease in germination and reduction in plant root growth are connected to ion toxicity and osmotic stress [52]. Decrease in germination is observed in plants growing under salinity. Especially wheat among field crops, is severely affected by salinity and decrease in germination is observed. Salinity also delays germination time [53].

6.1.2 Seedling development

In order for the plant to continue its vital events, seedling development under salinity stress play an important role. Plant biomass accumulation and stunted growth of plant are among the results of salinity, the most impact is its role in leaf area expansion [51]. Although some salt tolerant plants appear to increase biomass combination under high salinity, there is an inverse relationship between seedling development and salinity. Salinity also negatively affects seedling fresh and dry weights, plant length, and root surface area in plant [51].

6.1.3 Photosynthesis

Plant produce their foods with photosynthesis [51]. Photosynthesis is affected by salinity in long-term or short-term. While it can get rid of the effects in short-term with stomata restrictions that cause a decrease in carbon accumulation, it can get rid of the effects caused by salt accumulation in the leaves in long term [4, 54]. Closure of stomata prevents plant from losing water through transpiration. Deteriorations in thylakoid membranes and decrease in activities of Calvin cycle enzymes are the most important factors caused by salinity [51]. As a result of salinity, deterioration in PSII receptors cause yield loss in PSII [51]. Reduced chlorophyll content due to salinity may be depend on increased pigment degradation or impaired biosynthesis. Generally, the plant shows its response to salinity by decreasing photosynthetic activity and restricting the production of the factors that make up this activity [51].

6.1.4 Water relations

Excessive amounts of Na^+ and Cl^- ions prevents water intake of plant by increasing osmotic potential of soil, this situation causes negativities in plant growth by decreasing water consent in plant cells [9]. Under the salinity increase, inverse correlation against stress between osmotic potential and water potential occurs, which means that while salinity increases, decrease in both of these factors is observed [51]. Water based osmotic stress causes closure of stomata and by going further causes disruption in photosynthesis by preventing CO_2 flow. Regulation of water flow is the key solution in eliminating these negative factors [55]. In addition, in tackling salinity especially in field crops using canal waters instead of salty groundwater can be preferred. Gypsum usage is also among the options where canal water is not accessible [56].

6.1.5 Ion toxicity

Salt stress occurs due to accumulation of high amounts of Cl- or Na+ ions in soil and causes ion toxicity in plant. Although plant responses against salinity varies from species to species, generally excessive intake of Na+ causes nutrient imbalance [51, 52, 53, 54, 55, 56, 57]. Na+ ion has toxic concentration earlier than Cl- ion [4]. While high concentrations of Na+ ions cause negative effects in photosynthetic activity, they may lead to worse results in salt sensitive plants [58]. While the effect of Na+ ions may be negative in some plants, Cl- ion may be more negative in plants such as soybean and citrus [59]. Cl- ions can also deteriorate photosynthetic activity and may cause ion toxicity in plants like Na+ ions. In response to ion toxicity ABA hormone, which is produced by plant itself, is important. The amount of synthesis increases during stress, it is used to prevent disruption in the growth and development mechanisms of cell [60, 61].

6.1.6 Osmotic stress

Root water conductivity is impaired as the plant's high salinity restricts water intake from the soil, resulting in osmotic stress [4, 5, 6, 7, 8, 9, 10]. Salinity related osmotic stress causes closure of stomata [50]. Cell division and elongation are negatively affected in decreasing turgor pressure [62]. Cell permeability decreases, water intake to plant deteriorates and this decreases water intake and transpiration rate in leaf water potential [63]. Most of these effects are observed in maize plants, which are sensitive to salt stress, from field crops. But when the responses to salt stress are examined, responses of each plant are different.

6.1.7 Oxidative stress

Salt stress constitutes reactive oxygen species (ROS). Reactive oxygen species such as H_2O_2, O^{2-}, OH^- are produced in plant cells, in chloroplast and mitochondria [64]. Decrease in photosynthesis rate increases formation of reactive oxygen species [65]. Reactive oxygen species increase under stress and balance between antioxidant defense systems deteriorates and as a result oxidative stress occurs [66]. Under salt stress the negative effects such as disruption in membrane integrity, oxidation of carbohydrates and nucleic acids are observed as most harmful effects of reactive oxygen species [51]. Disruption of a gene related to oxidative stress tolerance may block the plant's tolerance to any abiotic stress [4]. Malondialdehydes are considered as important symptoms of oxidative stress, accumulation of high concentrated malondialdehyde in cells also cause oxidation of complementary structural components of cells [9].

6.2 Physiological and Biochemical Adaptation Responses Against Salinity in Plants

6.2.1 Ion Homeostasis

In plant growth, some ions must be inside of cell. Although ions such as nitrogen, potassium, calcium are present in soil, they cannot enter the plant cells as a result of competition with other ions with high concentrations. Presence of salt with high concentration in soil complicates the intake of these ions which are effective in plant development. Plants make some responses to ensure that this negative effect is eliminated and for the continuity of low ion concentration. Dividing the ions to be added to plants instead of intaking them inside of cells as one time is considered as important action for plant growth and development [51]. In maintaining the low concentration of ions, in transportation to plants the cell membrane is important [67]. Thanks to proteins, channel proteins and semptomers, ions can be moved to the plant [68]. Antiporters are also used in transportation. These transporters are located in vacuolar membranes. V-ATPase are known as channels needed for the continuity of plant under salt stress [69]. Due to the excessive accumulation in the soil, when Na^+ ion enters the cytoplasm, it wants to move to vacuoles and this transport is carried out by Na^+/H^+ antiporters. In the cell metabolism of plant, the other role is cytoplasmic K^+ homeostasis. Under salinity, K^+ concentration undergoes a severe decrease [64]. K^+ ions, which can be transmittable to cells by K^+ transporters and membrane channels, have low concentration under salinity stress. There is an important factor in cell recruitment. When the extracellular K^+ concentration is low, K^+ transporters mediating high affinity of K^+ uptake mechanisms allow affinity if extracellular K^+ concentration is high. As a result of this, concentration of Na^+ ion increases under salinity and with this increase Na^+ competes with K^+ and reduces K^+ uptake into the cell [4, 51]. More K^+ retention of roots in plants such as wheat, maize, beans have been observed as one of the mechanisms applied by plants to withstand salt stress [51, 52, 53, 54, 55, 56, 57, 58, 59, 60, 61, 62, 63, 64]. Accumulation of K^+ ions in cell increases under salinity stress [51, 70].

6.2.2 Biosynthesis of osmoprotectants

Osmoprotectants are high rated soluble compounds [71]. Certain organic compounds, which are sugars, amino acids, proline and osmoprotectants that are interchangeable compounds, are synthesized by plants under stress conditions according to their stress levels. Osmolytes are in charge of providing adaptation of plant to salt stress [71]. Quaternary ammonium compounds as betalain betanin which is synthesized only by a few members of *Plumbaginacease* family [72]. Amino acids like proline present in plant types are present in different type of plants from ammonium compounds like betalain betanin present in plant types. These osmoprotectant compounds can replace each other [73]. While accumulation of compounds varies with salt stress, most of the osmolytes try to make it easier to maintain osmotic balance and structure of plant cells with an uninterrupted water flow [8].

6.2.2.1 Amino acids

Free amino acids take part in reducing osmotic stress caused by high concentrations of salt [74]. Amino acids such as arginine, glycine, alanine, proline, leucine, valine, serine take part in regulation of cell [75]. Accumulation of these amino acids is done for solving the problems that the plant creates against salt stress. But amino acids such as methionine, arginine, which make up the majority of amino acids, decreases under stress condition unlike proline, which increases under salinity [51].

6.2.2.2 Proline

Proline, which has wide usage, is one of the most common osmolytes [76]. In high plants, which are found to be abundant, even under salt stress this content does not decrease, conversely it tends to increase [57]. Accumulation of proline is accepted as an important precaution to avoid salinity stress [76, 77]. The effect of proline accumulation under salinity conditions varies from species to species. Increasing the proline content in plant cells is a positive factor in preventing negative effects occurring at cell level [76]. Even in negative conditions, the accumulation of proline helps the plant to grow [76].

6.2.2.3 Glycine betain

Glycine betain, the variety of which is known to be found in organisms as well as plants, is accepted as the most common quaternary ammonium compound [78]. Under salt stress, it takes part in protecting membrane structure of plant, regulation of plant cells [76, 77, 78, 79]. According to the studies, glycine betain accumulated in chloroplast is more effective than its accumulation in cytosol [79]. In conducted study, while in rice seedlings, which are under 150 mM salt stress, fragmentation of grana, swelling of thylakoids, deterioration in mitochondria may be observed, it is observed that these effects disappear with the pretreatment of glycine betaine [80]. As a result of accumulation of exogenously applied glycine betain, the plant has a stronger system against stress [76].

7. CONCISE SUMMARY

With increasing population, satisfying the food need of humanity is important. Responding to this need gets harder day by day. Most of the agricultural lands in the world are exposed to salt stress. Other than the natural factors causing salt stress, unfortunately the wrong practices made by people also invite salt stress. Plants develop some response mechanisms for surviving against the negative effects occurring on them. These mechanisms which we can group as physiological, morphological, biochemical, depend on the magnitude and effect of stress and vary from species to species. As a result of these responses, the tolerance of plants to the environment they live in increases, they may be affected less by external factors and they may continue their life cycle.

AT A GLANCE

Salt stress is one of the harmful abiotic stress factors. It makes agricultural lands especially in arid and semi-arid regions useless despite the efforts. More than six percent of total world agricultural lands are on the edge of vanishing due to salt stress. Salinity in soil occurs as a result of the factors such as lack of drainage, improper irrigation, excessive accumulation of soluble salts. Salinity limits the growth of plants. Despite the main results, some results of plants due to these limitations vary from species to species. The negative effects get morphological, biochemical and physiological reactions from plants. Slowed or stopped growth of roots and shoots, closuring of stomata, germination slowing, decreased or stopped development of seedling, deterioration of photosynthetic activity are the main reactions of plants to stress. On the other hand, plants also develop tolerance mechanisms as a result of some auxiliaries for surviving under adverse conditions.

Plants have tendency to protect themselves from salinity with osmotic protectants synthesized by them such as sugars, proline, amino acids, glycine betaine. In this review, the responses of plants to salt stress were investigated and gathered.

REFERENCES

1. Kromdjik J, Long SP. One crop breeding cycle from starvation? How engineering crop photosynthesis for rising CO_2 and temperature could be one important route to alleviation. Proceedings of the Royal Society B: Biological Sciences. 2016;283(1826)20152578. DOI: 10.1098/rspb.2015.2578
2. Çulha Ş, Çakırlar H. Tuzluluğun bitkiler üzerine etkileri ve tuz tolerans mekanizmaları. Afyon Kocatepe University Journal of Sciences. 2011;11:11-34.
3. Nandal M, Hooda, R. Salt tolerance and physiological response of plants to salinity: A review. International Journal of Scientific & Engineering Research. 2013;4:44-67
4. Munns R, Tester M. Mechanisms of salinity tolerance. The Annual of Plant Biology. 2008;651-681
5. Wicke B, Smeets E, Dornburg V, Vashev B, Gaiser T, Turkenburg WC, Faaij APC. The global technical and economic potential of bioenergy from salt-affected soils. Energy & Environmental Science. 2011;4(8):2669-2681. DOI: 10.1039/C1EE01029H
6. Hasanuzzaman M, Nahar K, Alam M, Bhowmik PC, Hossain A, Rahman MM, Majeti P, Ozturk M, Fujita M. Potential use of Halophytes to remediate saline soils. Journal of Biomedicine and Biotechnology. 2014;2014(8):1-12. DOI: 10.1155/2014/589341
7. Hossain S. Present scenario of global salt affected soils,its management and importance of salinity research. International Research Journal of Biological Sciences. 2019;1(1):1-3
8. Hasegawa PM, Zhu JK, Bressan RA, Bohnert JH. Plant cellular and molecular responses to high salinity. Annual Review of Plant Physiology and Plant Molecular. 2000;51:463-499. DOI: 10.1146/annurev.arplant.51.1.463
9. Munns R. Comparative physiology of salt and water stress. Plant Cell and Environment. 2002; 25(2): 239-250. DOI: 10.1046/j.0016-8025.2001.00808.x
10. Munns R. Genes and salt tolerance: Bringing them together. New Phytologist.2005;167(3):645-663. DOI: 10.1111/j.1469-8137.2005.01487.x
11. Garthwaite AJ, Colmer TD, Bothmer RV. Salt tolerance in Wild Hordeum species is associated with restricted entry of Na^+ and Cl^- into the shoots. Journal of Experimental Botany. 2005;56(419):2365-2378. DOI: 10.1093/jxb/eri229

12. Motos JA, Ortuño MF, Vivancos PD, Vicente AB, Hernandez JA, Sánchez BM. Plant responses to salt stress: Adaptive mechanisms. Agronomy.2017;7(1):18. DOI: 10.3390/agronomy7010018

13. Rhodes D, Nadolska-Orczyk A. Plant Stress Physiology. Encyclopedia of life sciences. Nature Publishing Group. 2001;10. DOI: 10.1038/npg.els.0001297

14. Kosová K, Renaut J, Prasil IT, VítámvásP.Plantprotome changes under abiotic stress. Journal of Proteomics. 2011;74(8):1301-1322. DOI: 10.1016/j.jprot.2011.02.006

15. Kendirli B, Çakmak B, Uçar Y. Salinity in the Southeastern Anatolia Project (GAP), Turkey: issues and options. Irrigation and Drainage. 2005;54(1):115-122. DOI: 10.1002/ird.157

16. Roy RN, Finck A, Blair GJ, Tandon HLS. Plant nutrition for feed security: A guide for integrated nutrient management.FAO; 2006. 348 pages

17. FAO. How to feed the World in 2050. In: Proceedings of the Expert Meeting on How to Feed the World in 2050;24-26 June 2009; Rome; 2009

18. Pessarakli M, Szabolcs I. Soil Salinity and Sodicity as Particular Plant/Crop Stress Factors. In: PessarakliM,editor. Handbook of Plant and Crop Stress . 4th ed. Routledge Handbooks Online; 2010. p .1-19. DOI: 10.1201/B10329-3

19. Munns R, Gilliham M. Salinity tolerance of crops- what is the cost ? New Phytologist. 2015;208(3):668-673. DOI: 10.1111/nph.13519

20. Ayers RS, Westcot DW. Water quality for agriculture. 2nd ed. FAO Irrigation and Drainage; 1985. 174 pages

21. Botella MA, Rosado A, Bressan R, Hasegawa PM. Plant adaptive responses to salinity stress. In: Jenks MA, Hasegawa PM editors. Plant Abiotic Stress. USA, Iowa: Blackwell Publishing; 2007. p. 37-70. DOI: 10.1002/9780470988503.ch3

22. Serrano R, Rios G, Ros R, Marquez JA, Proft M, Mulet JM, Mendrizabal I, de Larrinoa I.F, Ahuir-Pascual A, Montesinos C. A glimpse of the mechanisms of ion homeostasis during salt stress. Journal of Experimental Botany. 1999;50:1023-1036. DOI: 10.1093/jxb/50.Special_Issue.1023

23. Hussain S, Bai Z, Huang T, Cao X, Zhu L, Zhu C, Khaskheli MA, Zhang C, Jin Q, Zhang J. 1-Methylcyclopropene modulates physiological,biochemical and antioxidant responses of Rice to different salt stress levels. Journal of Integrative Agriculture. 2019;10(124):1-18. DOI: 10.3389/fpls.2019.00124

24. Shah S, director. Morphological, Anatomical and Physiological Responses of Plants to Salinity [Internet]. 2014. Available from: https://www.youtube.com/watch?v=rihGVDoeq8A&t=244s

25. Suo J, Zhao Q , Chen S, Dai S, David L. Salinity response in chloroplasts: insights from gene characterization. International Journal of Molecular Sciences. 2017;18(5) :1011. DOI: 10.3390/ijms18051011

26. Koyro HW. Ultrastructural effects of salinity in higher plants. In: A. Lauchli, U. Lüttge,editors. Salinity: Environment-Plants-Molecules. Springer: Dordrecht;2002. p. 139-157. DOI: 10.1007/0-306-48155-3_7

27. Tester M, Davenport R. Na^+ tolerance and Na^+ transport in higher plants. Annals of Botany. 2003;91(5):503-527. DOI: 10.1093/aob/mcg058

28. Mahajan S, Pandley GK, Tuteja N. Calcium and salt-stress signaling in plants: shedding light on SOS pathway. Archives of Biochemistry and Biophysics. 2008;471(2):146-158. DOI: 10.1016/j.abb.2008.01.010

29. Demiral T, Türkan İ. Comparative lipid peroxidation,antioxidantdefense systems and proline content in roots of two rice cultivars differing in salt tolerance. Environmental and Experimental Botany. 2005;53(3):247-257. DOI: 10.1016/j.envexpbot.2004.03.017

30. Godfray C, Beddington JR, Crute IR, Haddad L, Muir JF, Lawrence D, Pretty NJ, Robinson S, Thomas SM, Toulmin C. Food security: The challenge of feeding 9 billion people . Science. 2010;327(5967):812-818. DOI: 10.1126/science.1185383

31. Kazemi K , Eskandari H. Effects of salt stress on germination and early seedling growth of Rice(Oryza sativa L.) cultivars in Iran. African Journal of Biotechnology.2011;10(77):17789-17792. DOI: 10.5897/AJB11.2219

32. Hussain S, Zhong C, Bohr JA, Hu JJ, Jin Q , Cao XC, Zhang JH, Zhu LF. Effects of salt stress on Rice growth and development characteristics and the regulating ways: A review. Journal of Integrative Agriculture. 2017;16(11):2357-2374. DOI: 10.1016/S2095-3119(16)61608-8

33. Sheeren A, Mumtaz S, Raza S, Khan A, Solangi S. Salinity effects on seedling growth and yield components of different inbred Rice lines. Pakistan Journal of Botany. 2014;37(1):131-139

34. Hasanuzzaman M, Nahar K, Rahman A, Anee TI, Alam MU, Oku H, Fujita M, Bhuiyan TF. Approaches to enhance salt stress on Wheat. In: Wanyera R, OwuocheJ,editors. Wheat Improvement,Management and Utilization. IntechOpen: 2017. p. 152-187. DOI: 10.5772/67247

35. Royo A, Abió D. Salt tolerance in Durum Wheat cultivars. Spanish Journal of Agricultural Research. 2003;1(3):27-35. DOI: 10.5424/sjar/2003013-32

36. Wahid A, Gelani S, Basra MA, Perveen M. Pretreatment of seed with H_2O_2 improves salt tolerance of Wheat seedlings by alleviation of oxidative damage and expression of stress proteins. Journal of Plant Physiology. 2006;164(3):283-294. DOI: 10.1016/j.jplph.2006.01.005

37. Zhang X, Li K, Liu S, Zou P, Chen X, Qin Y, Li P, Yu H, Xing R. Relationship between the degree of polymerization of Chitooligomers and their activity affecting the growth of Wheat seedlings under salt stress. Journal of Agricultural and Food Chemistry. 2017;65(2):501-509. DOI: 10.1021/acs.jafc.6b03665

38. Chinnusamy V, Jagendorf A, Zhu J. Understanding and improving salt tolerance in plants. Crop Science. 2005;45(2):437-448. DOI: 10.2135/cropsci2005.0437

39. Menezes-Benavente L, Kernodle SP, Scandalios JG, Margis- Pinheiro M. Salt-induced antioxidant metabolism defenses in maize (Zea mays L.) seedlings. Redox Report. 2004;9(1):29-36. DOI: 10.1179/135100004225003888

40. Khodarahmpour Z, Ifar M, Motamedi M. Effects of NaCl salinity on Maize (Zea mays L.) at germination and early seedling stage. African Journal of Biotechnology. 2012;11(2):298-304. DOI: 10.5897/AJB11.2624

41. Farooq M, Hussain M, Wakeel A, Siddique KHM. Salt stress in Maize: effects, resistance mechanisms, and management.A review. Agronomy for Sustainable Development. 2015;35(2):461-481. DOI: 10.1007/s13593-015-0287-0

42. Almodares A, Hadi MR, Ahmadpour H. sorghum stem yield and soluble carbohydrates under different salinity levels. African Journal of Biotechnology. 2008;7(22):4051-4055

43. Netondo GW, Onyango JC, Beck E. Sorghum and salinity:I.response of growth,waterrelations,and ion accumulation to NaCl salinity. Crop Science. 2004;44(3):797-805. DOI: 10.2135/cropsci2004.0797

44. Maas EV, Hoffman GJ. Crop salt tolerance- current assessment. Journal of the Irrigation and Drainage Division,American Society of Civil Engineers. 1977;103(2):115-134

45. Almodares A, Hadi MR, Dosti B. Effects of salt stress on germination percentage and seedling growth in sweet sorghum cultivars. Journal of Biological Sciences. 2007;7(8):1492-1495. DOI: 10.3923/jbs.2007.1492.1495

46. Almodares A, Hadi MR, Kholdebarin B, Samedani B, Kharazian ZA. The response of sweet sorghum cultivars to salt stress and accumulation of Na^+, Cl^-, K^+ ions in relation to salinity. Journal of Environmental Biology. 2014;35(4):733-739

47. Meyer LA. Cotton and Wool Outlook. United States Department of Agriculture. 2020. 6 pages

48. Viglioni Penna JC, Verhalen LM, Kirkham MB, McNew RW. Screening Cotton genotypes for seedling drought tolerance. Genetics and Molecular Biology. 1998;21(4):545-549. DOI: 10.1590/S1415-47571998000400023

49. Muhammad S, Ditta A, Iqbal MS, Hussain SB, Khan MI, Ramzan M, Malik TH. Effect of salinity stress on Cotton growth and role of marker assisted breeding and agronomic practices (chemical,biological and physical) for salinity tolerance. Scholars Report. 2018;4(1):1-13

50. Munns R. Physiological processes limiting plant growth in saline soils: Some dogmas and hypoteses. Plant Cell and Environment. 1993;16:15-24. DOI: 10.1111/j.1365-3040.1993.tb00840.x

51. Shahzad B, Fahad S, Tanveer M, Khan BA, Saud S. Plant responses and tolerance to salt stress: profiling and counteraction. In: Hasanuzzaman M, Nahar K, Fujita M, Oku H, Islam T, editors. Approaches for Enhancing Abiotic Stress Tolerance in Plants. 1st ed. CRC Press; 2019. p. 61-78. DOI: 10.1201/9781351104722-3

52. Katembe WJ, Ungar IA, Mitchell JP. Effect of salinity on germination and seedling growth of two Atriplex species (Chenopodiaceae). Annals of Botany. 1998;82(2):167-175. DOI: 10.1006/anbo.1998.0663

53. Lin J, Li X, Zhang Z, Yu X, Gao Z, Wang Y, Li Z, Mu C, Wang J. Salinity- alkalinity tolerance in Wheat, seed germination, early seedling growth,ion relations and solute accumulation. African Journal of Agricultural Research. 2012;7(3):467-474. DOI: 10.5897/AJAR11.1417

54. Parida AK, Das AB. Salt tolerance and salinity effects on plants : A review. Ecotoxicology and Environmental Safety. 2005;60(3):324-349. DOI: 10.1016/j.ecoenv.2004.06.010

55. Flexas J, Bota J, Galmés J, Medrano H, Ribas- Carbó M. Keeping a positive carbon balance under adverse conditions : Responses of photosynthesis and respiration to water stress . Physiologia Plantarum. 2006;127(3):343-352. DOI: 10.1111/j.1399-3054.2006.00621.x

56. Zaka AM, Schmeisky H, Hussain N, Rafa H. Utilization of brackish and canal water for reclamation and crop production. In: The International Conference on Water Conservation in Arid Regions; 12-14 October 2009; Jeddah, Saudi Arabia: ICWCAR; 2009. p. 1-16

57. Doğan M, Tıpırdamaz R, Demir Y. Salt resistance of Tomato species grown in sand culture. Plant,Soil and Environment. 2010;56(11):499-507. DOI: 10.17221/24/2010-PSE

58. Jamil M, Jung KY, Lee DB, Ashraf M, Lee SC, Rha ES. Effect of salt (NaCl) stress on germination and early seedling growth of four vegetable species. Journal of Central European Agriculture. 2006;7(2):273-282

59. Grattan SR, Grieve CM. Salinity-mineral nutrient relations in Horticultural crops. Scientia Horticulturae. 1998;78(1-4):127-157. DOI: 10.1016/S0304-4238(98)00192-7

60. Carillo P, Pontecorvo G, Woodrow P, Fuggi A, Annunziata MG. Salinity stress and salt tolerance. In: Shanker A, Venkateswarlu B, editors. Abiotic Stress in Plants-Mechanisms and Adaptations. 1st ed. Intech; 2011. p. 21-38. DOI: 10.5772/22331

61. Algül BE, Tekintaş FE, Günver Dalkılıç G. Bitki büyüme düzenleyicilerinin kullanımı ve içsel hormonların biyosentezini arttırıcı uygulamalar. Journal of Adnan Menderes University Agricultural Faculty. 2016;13(2):87-95. DOI: 10.25308/aduziraat.294100

62. Zeng L, Shannon MC, Grieve CM. Evaluation of salt tolerance in Rice genotypes by multiple agronomic parameters. Euphytica. 2002;127:235-245. DOI: 10.1023/A:1020262932277

63. Chaudhuri K, Choudhuri MA. Effects of short-term NaCl stress on water relations and gas exchange of two Jute species. Biologia Plantarum. 1997;39:373-380. DOI: 10.1023/A:1001013913773

64. Abbasi GH, Javaid A, Ul-haq MA, Ali S, Chen ZH, Malik W. Exogenous potassium differentially mitigates salt stress in tolerant and sensitive maize hybrids. Pakistan Journal of Botany. 2014;46(1):135-146

65. Apel K, Hirt H. Reactive oxygen species : metabolism, oxidative stress, and signal transduction. Annual Review of Plant Biology. 2004;55:373-399. DOI: 10.1146/annurev. arplant.55.031903.141701

66. Büyük İ, Aydın Soydam S, Aras S. Bitkilerinstreskoşullarınaverdiğimoleküllercevaplar. Türk Hijyenve Deneysel Biyoloji Dergisi. 2012;69(2):97-110. DOI: 10.5505/TurkHijyen.2012.40316

67. Sairam RK, Tyagi A. Physiology and molecular biology of salinity stress tolerance in plants. Current Science. 2004;86(3):407-421. DOI: 10.1007/1-4020-4225-6

68. Blumwald E. Sodium transport and salt tolerance in plants. Current Opinion in Cell Biology. 2000;12(4):431-434. DOI: 10.1016/S0955-0674(00)00112-5

69. Dietz KJ, Tavakoli N, Kluge C, Mimura T, Sharma SS, Harris GC, Chardonnens AN, Golldack D. Significance of the V-Type ATPase for the adaptation to stressful growth conditions and its regulation on the molecular and biochemical level. Journal of Experimental Botany. 2001;52(363):1969-1980. DOI: 10.1093/jexbot/52.363.1969

70. Chakraborty K, Sairam RK, Bhaduri D. Effects of different levels of soil salinity on yield attributes, accumulation of nitrogen,and micronutrients in Brassica spp. Journal of Plant Nutrition. 2016;39(7):1026-1037. DOI: 10.1080/01904167.2015.1109105

71. McNeil SD, Nuccio ML, Hanson AD. Betaines and related osmoprotectants. Targets for metabolic engineering of stress resistance. Plant Physiology. 1999;120(4):945-949. DOI: 10.1104/pp.120.4.945

72. Hanson AD, Rathinasabapathi B, Rivoal J, Burnet M, Dillon MO, Gage DA. Osmoprotective compounds in the Plumbaginaceae: A natural experiment in metabolic engineering of stress tolerance. Plant Biology. 1994;91(1):306-310. DOI: 10.1073/pnas.91.1.306

73. Jantaro S, Mulo P, Incharoensakdi A, Mäenpää P. Content and biosynthesis of polyamines in salt and osmotically stressed cells of Synechocystis sp. PCC 6803. FEMS Microbiology Letters. 2003;228(1):129-135. DOI: 10.1016/S0378-1097(03)00747-X

74. Ashrafijou M, Izadi A, Sadat-Noori SA, Saghafi S. Effect of salinity and radiation on proline accumulation in seeds of Canola (Brassica napus L.). Plant,Soil and Environment. 2010;56(7):312-317. DOI: 10.17221/2/2010-PSE

75. Mansour MMF. Nitrogen containing compounds and adaptation of plants to salinity stress. Biologia Plantarum. 2000;43:491-500. DOI: 10.1023/A:1002873531707

76. Saxena SC, Kaur H, Verma P, Prakash B, Rao V, Majee M. Osmoprotectants: potential for crop improvement under adverse conditions. In: Tuteja N, Gill SS,editors. Plant Accimilation to Environmental Stress. 1st ed. Springer; 2013. p . 197-232. DOI: 10.1007/978-1-4614-5001-6_9

77. Ahmed CB, Rouina BB, Sensoy S, Boukhriss M, Abdullah FB. Exogenous proline effects on photosynthetic performance and antioxidant defense system of young olive tree. Journal of Agricultural and Food Chemistry. 2010;58(7):4216-4222. DOI: 10.1021/jf9041479

78. Rhodes D, Hanson AD. Quaternary ammonium and tertiary sulfonium compounds in higher plants. Annual Reviews of Plant Physiology and Plant Molecular Biology. 1993;44:357-384. DOI: 10.1146/annurev.pp.44.060193.002041

79. Sakamoto A, Murata AN. Metabolic engineering of Rice leading to biosynthesis of glycinebetaine and tolerance to salt and cold. Plant Molecular Biology. 1998;38:1011-1019. DOI: 10.1023/A:1006095015717

80. Rahman, S, Miyake H,Takeoka Y. Effects of exogenous glycinebetaine on growth and ultrastructure of salt-stressed Rice seedlings (Oryza sativa L.). Plant Production Science. 2002;5(1):33-44. DOI: 10.1626/pps.5.33

CHAPTER-10

DEVELOPING CROPS WITH RESISTANCE TO WATER SCARCITY

1. INTRODUCTION: AN OVERVIEW

In response to the escalating challenges posed by climate change and the increasing frequency of drought events, there is a growing imperative to enhance the resilience of agricultural crops to water scarcity. Breeding for drought resistance stands at the forefront of innovative strategies aimed at securing global food production in the face of unpredictable and arid environmental conditions. This endeavor involves the deliberate selection and cultivation of plant varieties with inherent adaptive traits, ultimately contributing to the development of more robust and water-efficient crops. In this pursuit, scientific advancements and genetic insights converge to propel agricultural breeding programs towards sustainable solutions for ensuring food security in an era of climatic uncertainty.

Many crops grown in diverse environmental conditions and are subjected to different stress conditions among them drought is the major yield limiting factor in major crops. Food for future generations is challenged by increasing demand and threatened by deteriorating water availability, nearly 23 million hectares of rainfed rice in South and Southeast Asia are drought prone areas. In some states of India, due to severe drought conditions nearly 40% yield loss, amounting to $800 million is affected. As the Southwest monsoon are irregular it results in moderate to severe drought in rainfed regions, particularly in eastern India, as a result many morphological, physiological and phenological traits have been reported to improve the performance of many crops to challenge the drought condition Drought Tolerance is defined as the capacity of plants to uphold a certain level of physiological activity over the regulation and well fine-tuning of thousands of genes and numerous metabolic pathways to decrease the resulting damage [1, 2].

Various biotic and abiotic environmental factors interfere with the complete genetic potential of crop plant are called stress. Moisture stress occurs when plants are unable to meet evapotranspiration demand. Drought is induced by absence of water due to irregular rainfalls or insufficient irrigation but it can be impaired by other factors like soil salinity and physical properties and high air or soil temperature. Drought is insufficiency of water availability, including precipitation and soil moisture storage capacity, in quantity and supply the life cycle of a crop to restrict the maximum genetic grain yield possibility of the crop.

2. EFFECTS OF DROUGHT ON PLANT GROWTH AND DEVELOPMENT

Water stress has significant effect on plant cellular processes, growth, development and economic yield. At cellular level, it effects structure of membranes and organelles, hydration and structure of proteins and nucleic acids, pressure differential across the membrane cell wall complex. Water stress is usually measured as leaf water potential since leaves are directly involved with production of assimilates for growth and yield. Osmotic adjustment or Osmoregulation is active accumulation of solutes in the cells during stress period. Different solutes like sugars, fructans and most importantly inorganic ion K^+. Osmotic adjustment is finite and sufficient time is required for solute accumulation.

The development of water stress in the field and its effects on the crop is as follows. Initially, a moderate level of stress develops at which leaf expansion ceases but photosynthesis continues a part of photosynthate is used for osmotic adjustment. Thus plant growth is inhibited by moderate level of stress that reduce cell enlargement, increased root/shoot ratio, leaf area prevention. Osmotic adjustment would occur which will protect cell from extreme desiccation and allow a continued gas exchange. As water stress increases, older leaves senesce to various degrees, reduced leaf area, decline in water use, increased hydraulic resistant, stomata remain open and photosynthesis may continue as long as bulk leaf turgor is maintained in leaves due to osmotic adjustment. The level of water stress in reproductive meristems is lower than that in the transpiring leaves of a plant at any given time. During increased water stress, turgor is lost, stomata close fully, growth ceases, leave roll up, gas exchange drops to zero, carbon is lost by respiration, tissue water continues to decrease slowly, plant enters the pre lethal, non-reproductive stage of survival, leaves temperature increases to lethal levels, and the meristem dies then plant is considered as died. Time scale of plant processes which may influence their drought tolerance includes turnover of some proteins, stomatal movement within minutes of stress. In hours, production of heat shock proteins (hsp) or dehydrins, leaf movement, wilting, osmotic adjustment response to ABA. Cellular hardening, induction of housekeeping genes, floral initiation in 1–2 days of stress.

3. DROUGHT RESISTANCE

Drought resistance is mechanisms causing minimum loss of yield in a drought environment. Different mechanisms through which a crop is capable of minimize the loss in yield due to drought stress are grouped in to three categories: drought escape, dehydration avoidance and dehydration tolerance.

3.1 Drought escape

Drought escape describes the situation where drought susceptible variety performs well in a drought environment simply by avoiding the period of drought. Early maturity is an important attribute of drought escape, suitable for environments subjected to late season drought stress. Early varieties usually contain lesser leaf area index, minor total evapotranspiration and inferior yield potential.

3.2 Dehydration avoidance

It is the ability of a plant to retain a relatively higher level of hydration under conditions of soil or atmospheric water stress. Dehydration avoidance can be achieved either by reducing transpiration (water savers) or increased water uptake (water spenders). The following plant characteristics are responsible for dehydration avoidance. i) Reduced transpiration by closure of their stomata in water saving xerophytic species in response to water deficit well before wilting. In less irrigate plants, stomata possibly will remain open during morning hours and close as solar radiation increases. In CAM plants stomata closes during the night, when CO_2 is fixed by them. ii) Osmotic adjustment is an important mechanism positively affects growth and yield under stress. iii) Increased concentrations of Abscisic acid a stress hormone plays an important role in water stress avoidance by effecting stomata closure, reduction in leaf expansion and promotion of root growth.

Many of stress responsive genes encodes such proteins e.g., dehydrin, osmotin, Lea proteins etc. also produced in response to ABA; genes encoding such proteins are called ABA responsive (ABAR) genes. iv) Deposition of wax within and over the cuticle reduces transpiration during water stress. v) Leaf characteristics like leaf pubescence, erect leaf angle, leaf rolling etc. vi) increased water uptake by deep root system.

3.3 Dehydration tolerance

Dehydration tolerance of a genotype means that a significantly lower level of changes in it than those in another genotype when both of them are subjected to the same level of dehydration. Various measurements of dehydration tolerance are as follows. i) maintenance of membrane integrity by low levels of solute leakage. ii) seedling growth parameters like seedling survival or recovery, seedling growth, seed germination under osmotic stress created by Polyethylene glycol (PEG), stem reserve mobilization, presence of awns in cereals, high levels of proline accumulation. Plant drought response mechanisms and main related traits (Figure 1).

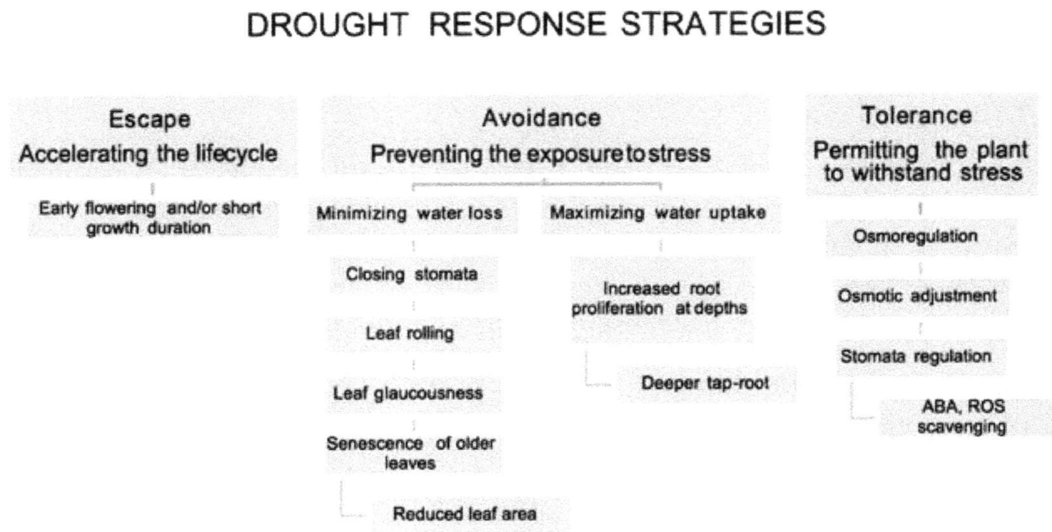

Figure 1. Plant drought response mechanisms and main related traits.

4. CHARACTERIZATION OF DROUGHT RESISTANCE TRAITS

4.1 Traits associated with drought tolerance

Sl.no	Category	Traits
1.	Morphological &Anatomical:	Yield; More Root length, Root Volume, Root Dry Weight, Root Thickness; Root surface area, More Plant Biomass; Harvest index; Leaf drying; Leaf tip firing; Delay in flowering.
2.	Phenological:	Early to maturity, Late Flowering; Anthesis, Silking Interval; Seedling vigor; Weed competitiveness; Photosensitivity; perennially.
3.	Physiological & Biochemical:	Osmotic Adjustment; Carbon Isotope Discrimination; Stomatal conductance; Remobilization of stem reserves; Specific leaf weight; ABA; Electrolyte leakage; leaf rolling, tip firing, Stay-green; Epicuticular wax; Feed forward response to stress; Heat shock proteins; Cell wall proteins; Leaf water potential; Water use efficiency; Aquaporins; Nitrogen use efficiency; Dehydrins.

Despite many decades of research on drought tolerance in several crops, little progress has been reported in terms of genetic enhancement of crop productivity under water-deficits environments. Breeders and crop physiologists need to work closely in testing the viability/validity of the trait-based approaches for drought tolerance. This has not happened to any great extent previously, but a few success stories have been recently reviewed [3].

Identification of simple to observe morphological and phenological traits, reflective of mechanisms and processes that confer drought tolerance is a priority activity in drought research. An appropriate screening trait for drought stress tolerance should fill the following criteria: (i) a strong link with higher or more stable grain yield in the target stress environment, (ii) a high level of heritability, and (iii) the expression of tolerance must be easily measurable, with adequate replication (Table 1).

Table 1. Screening procedure for drought tolerance.

S. No.	Instruments/techniques used Screening for the purpose of	References
1.	Infrared thermometry used to measure the moisture uptake efficiency	[4]
2.	Banding herbicide metribuzin by deep in the soil, and usage of iodine-131 and hydroponic culture in stress at 15 bar towards growth of the roots	[5, 6]
3.	Adaptation of psychometric process Evaluation of osmotic	[7, 8]
4.	Diffusion porometry method Leaf water conductance	[9]
5.	Mini-rhizotron method Root penetration, supply and thickness in the field	[10]
6.	Infrared aerial photography Dehydration postponement	[11]
7.	Carbon isotope discrimination Improved water-use efficiency	[12]
8.	Drought index measurement Total yield and number of fruits	[13, 14]
9.	Visual scoring or extent Maturity, leaf molding, leaf length, angle, orientation, root morphology and other morphological traits	[2]

5. BREEDING APPROACHES FOR DROUGHT RESISTANCE

To improve crop at genetic level, plant breeders are usually more concerned in the use of intra specific variation, easily exploitable without any genetic barriers. Intra-specific crosses follow normal Mendelian segregation and selection of F2 at future generations to identify the suitable pant progenies including pure lines. Many breeding methods like mass selection, pure line recurrent selection, methods can be applied, in stress or non-stress conditions besides some other methods like in-vitro selection and usage of somaclonal variations can also be employed.

- **Mass selection:** Mass selection, the easy method of selection and is used to develop the large population by combining positive or negative mass selection method. Mass selection is employed within populations for characters with high narrow sense heritability. Environment has major influence in mass selection it is a main disadvantage towards development, phenotypical expression of single plants. It also helps in selecting the varieties appropriate for local performance.

- **Pure Line Selection:** In this method a parent with drought tolerance is crossed with high yielding parent to develop segregating individuals. New combination of alleles obtains due to recombination and segregation of traits. In F2 generation best performing plants are selected in both drought stress and irrigated situations. Seeds from the chosen plants are collected to raise new plant progenies. New plant types are grown to screen best progenies and superior plants among the population to develop pure lines. Disadvantages of this methods very few recombinants are formed F_1 to F_3 generation there is a very less probability to change the genotypes in further generations because homozygosity is achieved nearly 87.5% in F4 generation.
- **Recurrent Selection:** In this method, the individual plants are selected from original population and are subjected to Progeny testing further the individuals are to cross among each individual in all possible combinations to produce seeds to generate the new base population. The selfing of F_2 plants results heterozygosity, and creates novel genetic variation in each population and generates new recombinants and help to involve promising alleles into a single genotype. Among these individual's gene linkage is slowly removed and new recombination are introduced. The process is repeated for many times and novel genetic recombination's are formed in each cycle of selection. Recurrent selection is widely used to enhance yield ability and resistance to drought. After each cycle precise selection is subjected to increase the genetic variability. Many studies have proven that using recurrent selection in drought increase in gain yield is observed [15].
- **Manipulation of Somaclonal Variation:** Variations are phenotypic variations occur due epigenetically changes occur in cells derivative from somatic or gametic explants. These variations are usually not desirable because they decrease the homogeneity nature of the restored plants. However, these variations are spontaneous provides the plant breeders to identify the new variations among the individuals among the emerged plants. These variations can be permanent or temporary. The variations which caused due to physiological effects within the plants which are not heritable are temporary in nature. Permanent variations may arise due to mutations, polyploidy, endopolyploidy, chromosomal aberrations.

It is significant that drought resistance is combined in to materials with more yield potential. Drought tolerance ability can be seen in improved cultivated, land races, related wild species or varieties developed using molecular breeding techniques. Breeding for drought resistant varieties is difficult to creation drought stress situations and selection of individuals is a difficult task which has to be carried out in rainout shelter conditions.

To identify drought resistance varieties and with high yielding ability both yield and yield related characters are evaluated in both irrigated and water stress conditions. Stability in yield for selected individuals are evaluated over different environments in both irrigated and stress conditions. The breeding approach should be able to combine high yielding with wide adaptability. Selection of individual progenies is selected by considering the average performance over wide range of environmental conditions under stress. The selected individuals are grown under best environments and in progeny selections are supported out for yield and yield characters. The breeding material evolved from breeding environments are lastly assessed in a wide range of environments and those showing high mean performance with high stability may be released for cultivation. A breeding approach to develop drought resistant variety for different environmental conditions (Figure 2).

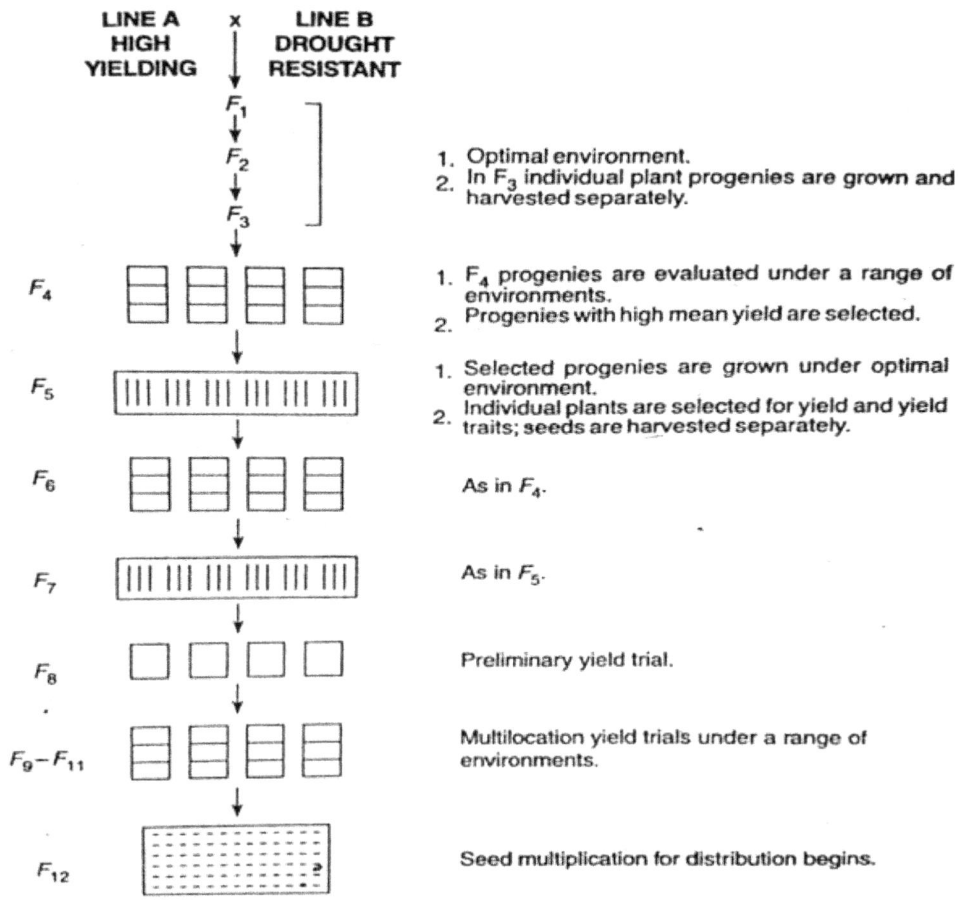

Figure 2. A breeding approach to develop drought resistant variety for different environmental conditions.

6. DIFFICULTIES IN DROUGHT RESISTANCE BREEDING

- Identifying drought prone areas in which the variety should be developed is a difficult task because it differs from place to place and time to time.
- It is difficult to maintain the plant population in controlled conditions like moisture stress and temperature in both field and green house conditions.
- The plant breeder should use combined selection devices with other useful resources
- Estimating of all drought resistant characters is a difficult task and not simple.
- There is an inverse relationship between drought resistant characters and yield potentiality of the crop so breeder has to make additional breeding efforts to increase the yielding ability of the plant.
- Combined breeding scheme is to be developed by plant breeder to enhance yield and drought resistance characters.
- Usage of primitive or wild cultivars will be a source of drought resistance will be difficult.

6.1 Drought hardening

Drought hardening refers to an agronomic practice to improve the resistance of genotype to drought as a consequence of seed/seedling treatment. Pre-sowing and post sowing treatments improve drought hardening. In pre-sowing treatments, seed is soaked in water for 24 hr. and dried in sun before they are sown in the main field. In post sowing treatment, a slight moisture stress applied to young seedlings to recover their drought resistance during later stages of growth.

7. PROLINE CONTENT

All plants are capable of detecting and responding to [16, 17]. To overcome the effect of stress, plants have advanced using adaptive mechanisms which may be classified into four categories. Three of these adaptations are developmental traits (e.g., time of flowering), structural traits (e.g. leaf waxiness) and physiological mechanisms (e.g. ability to exclude salt while maintaining the absorption of water and the capacity to sort ions with in vacuoles) involve complex interaction. The fourth one is the metabolic responses such as alteration in photosynthetic metabolism and accumulation of organic osmolytes, most commonly proline. One mechanisms utilized by the plants for overcome the water stress effects might be via accumulation of compatible osmolytes, such as proline and soluble sugars. Production and accumulation of free amino acids, especially proline by plant tissue during drought, salt and water stress is an adaptive response. Proline has been proposed to act as a well-suited solute that regulates the osmotic potential in the cytoplasm. Thus, proline can be used as a metabolic marker in relation to stress. Moreover, under drought stress, the accumulation of total soluble sugars in different plant parts would be increased. However, the rate of additional production or accumulation of proline and soluble sugar is different in different plant parts. Proline content increases in a large variety of plant under stress up to 100 times the normal level, which makes up to 80% of the total amino acid pool. Proline accumulation is maximum during the flowering stage and minimum at vegetative stage. Proline source can be either from the synthesis from glutamate or hydrolysis of proteins. The proline accumulated in response to drought stress or salinity stress in plants is primarily restricted in the cytosol (Figure 3) [18, 19, 20].

GLUTAMATE

↓

γ - GLUTAMATEPHOSPHATE

↓

GLUTAMIC-γ- SEMIALDEHYDE

↓

Δ^1 – PYRORPLINE-5-CARBOXYLATE

↓

PROLINE

Figure 3. Flow chart showing proffered rout of proline biosynthesis in plant in stress full condition [21, 22]. Proline biosynthesis path way undergoes by plant in drought stress condition.

7.1 Estimation of proline

Proline content was assessed for the leaves exposed to control and water-stressed conditions. Leaves (100 mg) from control and water-stressed plants were separately standardized in 10 mL of 3% sulphosalicylic acid using mortor and pestle and centrifuged at 5000 rpm for 10 min and the supernatant was collected to estimate the Proline. Ninhydrin (1.25 g) was liquified in 30 mL of glacial acetic acid and then 20 mL of 6 M phosphoric acid was added and kept for 24 h at 40°C. To 2 mL of plant extract, 2 mL of acid Ninhydrin and 2 mL of glacial acetic acid were added and the mixture was boiled at 100°C for 1 h in a water bath. At that time, the solution continued to cool and the reaction was completed. About 4 mL of toluene was added to the contents and mixed vigorously for few sec and OD values for the colored component was restrained at 520 nm using toluene as the blank. From the OD values, proline content (μmoles/g fresh wt.) was calculated individually.

8. CONCISE SUMMARY

Finally, studying all traits related to shoot and root its morphological, physiological, biochemical, phenological, anatomical or responses to environment shows additional opportunities to increase drought resistance in crops. The characters of the plant have to be synchronized with the suitable agronomical practices or better expression of diverse characters. This is a best approach to plant breeding for drought resistance.

AT A GLANCE

Drought is the most severe abiotic stresses in many parts of the world and is one of the major problems in present-day climatic scenario. Drought tolerant varieties are with high demand which seems to be a great challenging task to plant breeders however difficulties are combined by the difficulty of crop yield on the genetic and physiological bases. Drought resistance may be defined as the mechanism(s) causing minimum loss of the yield in a drought environment relative to the maximum yield in a constant-free of optimal environment for the crop. Several researchers explained the plant reaction to drought through drought escape, dehydration avoidance, and/or dehydration tolerance mechanisms. Drought stress decreases size of the leaves, stem extension and root proliferation inside the soil, it also disturbs plant water relations and reduces water-use efficiency ultimately reduces the yielding ability of the plant so, breeding for Drought resistance is a good approach, following different breeding strategies and approaches to develop a drought resistant variety combining both conventional and molecular approaches. Considering the parameters like root morphology studies, proline estimation, leaf rolling etc., Selection based on a comprehensive approach of testing might be more effective in breeding better drought-tolerant cultivars.

REFERENCES

1. Mitra J (2001) Genetics and genetic improvement of drought resistance in crop plants. Curr. Sci. 80(6): 758-763
2. Chapman SC, Edmeades GO (1999) Selection improves drought tolerance in tropical maize populations: II. Direct and correlated responses among secondary traits. Crop Sci 39:1315– 1324
3. ichards, R. 2006. Physiological traits used in breeding of new cultivars for water scarce environments. Agric. Water Manage. 80:197-211
4. Blum A, Mayer J and Golan G (1982) Infrared thermal sensing of plant canopies as a screening technique for dehydration avoidance in wheat. Field Crops Res. 57: 137-146
5. Robertson BM, Hall AE and Foster KW (1985) A field technique for screening for genotypic differences in root growth. Crop Sci. 25: 1084-1090
6. Ugherughe PO (1986) Drought and tropical pasture management. Z. Acker-u. Pflanzenbau. J. Agron. Crop Sci., 157: 13-23
7. Morgan JM (1980) Osmotic adjustment in the spikelet and leaves of wheat J. Exp. Bot. 31: 655-665
8. Morgan JM (1983) Osmoregulatiom as selection criterion for drought tolerance in wheat. Aust. J. Agric. Res. 34: 607614
9. Gay AP (1986) Variation in selection for leaf water conductance in relation to growth and stomatal dimensions in Lolium perenne L. Ann. Bot. 57: 361-369
10. Bohm W (1974) Mini-rhizotrons for root observations under field conditions. Z. Acker-u. Pflanzenbau. J. Agron. Crop Sci. 140: 282-287
11. Blum A, Schertz KF, Toler RW, Welch RI, Rosenow DT, Johnson JW and Clark LE (1978) Selection for drought avoidance in sorghum using aerial infrared photography. Agron. J. 70: 472-477
12. Farquhar GD and Richards RA (1984) Isotopic composition of plant carbon correlates with water-use efficiency of wheat genotypes. Aust. J. Plant Physiol. 11: 539-552
13. Clarke JM, Townley-Smith TF, McCaig TN and Green DG (1984) Growth analysis of spring wheat cultivars of varying drought resistance. Crop Sci. 24: 537-541
14. Ndunguru BJ, Ntare BR, Williams JH and Greenberg DC (1995) Assessment of groundnut cultivars for end of season drought tolerance in a Sahelian environment. J. Agric. Sci. Cambridge 125: 79-85
15. Giménez C, de Garcia E, Xena de Enrech N, Blanca I (2001) Somaclonal variation in banana: cytogenetic and molecular characterization of the somaclonal variant CIEN BTA-03. In vitro Cell Dev Biol Plant 37:217– 222

16. Bartels, D. and Sunkar, R. 2005. Drought and salt tolerance in plants. Crit. Rev. Plant Sci. 24(1): 23-58

17. Leigh, R.A., Ahmad, N. and Wyn Jones, R.G. 1981. Assessment of glycine betaine and proline compartmentation by analysis of isolated beet vacuoles. Planta. 153: 34-41

18. Ketchum, R.E.B., Warren, R.C., Klima, L.J., Lopez-Gutierrez, F. and Nabors, M.W. 1991. The mechanism and regulation of proline accumulation in suspension cultures of the halophytic grass Distichlis spicata L. J. Plant Physiol. 137: 368-374

19. Pahlich, E., Kerres, R. and Jager, H.J. 1983. Influence of water stress on the vacuole/extravacuole distribution of proline in protoplasts of Nicotiana rustica. Plant Physiol. 72(2): 590-591

20. Al-Khayri JM, Al-Bahrany AM (2004) Growth, water content, and proline accumulation in drought-stressed callus of date palm. Biol Plant 48(1):105-108

21. Bohnert, H.J., Nelson, D.E. and Jensen, R.G. 1995. Adaptations to environmental stresses. Plant Cell. 7(7): 1099-1111

22. Bates, L.S., R.P. Waklren, 8nd I.D. Te8re. 1973. Rapid determination of free proline water stress studies. Plant Soil. 39: 205-207

SECTION-3

MOLECULAR BREEDING OF CROPS

CHAPTER-11

EXPLORING MARKER-ASSISTED SELECTION AND ITS UTILIZATION IN THE FIELD OF PLANT BREEDING

1. INTRODUCTION: AN OVERVIEW

Delving into the realm of plant breeding, the application of Marker-Assisted Selection (MAS) stands out as a revolutionary approach that has significantly enhanced the precision and efficiency of breeding programs. As we embark on this exploration, we unravel the intricate web of molecular markers and their pivotal role in identifying desirable traits with unparalleled accuracy. This cutting-edge technology has not only streamlined the breeding process but has also opened new avenues for addressing challenges in agricultural productivity and sustainability.

It was estimated that the global population would touch 9 billion individuals, and the annual growth rate will be 0.75 percent by 2050. To feed this burgeoning human population only, it is required to produce a surplus of one billion tons of cereals by the end of 2050 [1]. It is well known that to achieve these targets new integrated approaches must be practiced with the conventional breeding programmes to accelerate the breeding cycle by reducing net time and cost per unit production [2, 3].

The primary objective of plant breeding is to increase crop yield [4], and the secondary objectives are quality improvement, development of photo & thermo-insensitive cultivars, tolerance to biotic and abiotic stresses, synchronous maturity, water and nutrient use efficiency, elimination of toxic substances, and different crop maturity groups [5, 6] for high agricultural output and sustainable development. The advanced understanding and developments in molecular genetics have significantly enhanced the efficiency of plant breeding to achieve the desired objectives in crop plants [7]. The efficient and effective application of molecular markers in crop improvement programmes improves the selection efficiency, degree of precision, and accelerates the breeding cycle to develop a new cultivar with a trait of interest [5].

Marker-assisted selection (MAS) can be defined as the manipulation of genomic regions that are involved in the desirable trait of interest through DNA markers [7], and their potential use in crop improvement begins a new era of molecular breeding [8]. The MAS has an edge over the visual phenotypic selection because the trait of interest is linked with a molecular marker which increases the selection efficiency of the targeted trait [9].

The fundamental aim of any crop improvement programme is the selection of effective plants with a trait of interest. In conventional plant breeding, there are more chances to skip the trait of interest and delays the time to develop new cultivars with desirable traits. Whereas, MAS has shown its utility in crop plants for improvement of various traits by reducing the environmental effect and by increasing selection efficiency for a trait of interest [10]. However, the efficacy of MAS on selection may be impeded by genetic background [11], reliability and accuracy of QTLs [12], the insufficient linkage between the gene of interest (QTLs) and marker [13], relative high input cost, [14, 15] limited molecular markers and their narrow range of polymorphism and knowledge gap between plant breeders and molecular biologist [5].

Various markers such as morphological (trait-specific), proteinaceous (isoenzyme), cytological (chromosome-specific), and DNA markers have been utilized in plant breeding: however, DNA based markers are used extensively in MAS for various traits and crops by the plant breeders [16]. The basic requirements for effective MAS in plant breeding are- reliability of DNA marker, qualitative and quantitative assurance of genetic material (DNA), marker analysis procedures, genomic coverage of marker, level of polymorphism, genetic nature of marker such as co-dominance [5, 17, 18, 19].

Recent advances in molecular breeding such as the use of PCR based techniques [simple sequence repeats (SSRs), and insertion/deletion mutations (Indels)]; single nucleotide repeats (SNPs); Genomic sequencing (GS) and genotype by sequencing (GBS), etc. have extensively been used in crop improvement programme throughout the world [3, 19].

2. MOLECULAR MARKERS: ROAD FOR EASY AND RELIABLE SELECTION

Any fixed property of an individual showing the heritable variations is termed as a character or trait, whereas marker can be defined as any mark which inherits together with the trait of interest throughout generation [20, 21]. Markers are categorized into four main groups- morphological, biochemical, cytological and molecular (DNA based) markers [22].

Morphological markers are also known as naked eye marker or phenotypic marker, used for quality traits such as flower shape, size, color, seed structure, growth habit, and other agronomic traits in plants. These markers are eco-friendly; easy to use, and need not require any specific instrument; however, their number is limited in crop species and highly influenced by prevailing environmental conditions [22, 23, 24].

Biochemical markers, mostly isozymes, are the results of variation in enzymes (protein and amino acid sequences) encoded by various genes, but functionally they are the same [25]. They are the result end product of allelic variation of enzymes. They are co-dominance in inheritance, cost-effective, and easy to use. They have been widely used in plant breeding for the study of gene flow, population structure, and genetic diversity [26]. However, they are limited in number, show less polymorphism, and predominantly affected by plant tissue being used, growth stage, and method of their extraction [27].

Cytological markers are based on prevailing variation in number, shape, size, the position of chromosomes, and their banding pattern. Cytological analysis reveals the unique characteristics of chromosomes such as knob and satellite, and the number of nucleoli in the nucleus, etc. This variation shows a different pattern of euchromatin and heterochromatin in the chromosome [22], such as Giemsa stain recognizes G bands. They have been extensively utilized in plant breeding for the identification of linkage groups and physical mapping [9]. In contrast, molecular markers are defined as nucleotides polymorphism present between individuals as a result of deletion, duplication, insertion, substitution, point mutation and translocation, etc. [27] but do not affect the function of the gene.

Molecular markers do not inevitably target genes, instead, inherit as a 'flag' with the gene of interest during transmission of a trait from one generation to the next generation [28]. Molecular markers associated with the close proximity of genes of interest are known as gene tags *i.e.* linked with target gene [9]. The essential characteristic features of an ideal marker are co-dominance inheritance, high level of polymorphism, high reproducibility, whole-genome coverage, easy and fast to detect, neutral to environmental conditions, high resolution, low cost, and whole-genome coverage [22, 27, 29]. Different types of molecular markers have been developed, and are used in various crops. These molecular markers are mainly categorized into the following classes based on their method of detection.

2.1 Hybridization-based markers

DNA bands are captured where labeled probe *i.e.* DNA fragment of known sequence hybridizes with DNA fragment digested by restriction endonuclease enzyme. The restriction fragment length polymorphism (RFLP) was the first and last marker which was only based on the hybridization method [22].

2.2 PCR-based markers

The idea of polymerase chain reaction (PCR) was conceived by Kary Mullis in 1983, and invented the process in 1985 which is based on denaturation, annealing, and extension [30]. The PCR based markers use primer dependent PCR amplification and/or DNA hybridization followed by electrophoresis. Polymorphism is detected based on the presence or absence of an amplicon or based on the band size and mobility. The most commonly used PCR based markers are Randomly amplified polymorphic DNA (RAPD) [31], Amplified fragment length polymorphism (AFLP) [32], microsatellites or simple sequence repeats (SSRs) [33], sequence-related amplified polymorphism (SRAP) [34], inter simple sequence repeat (ISSR) [35], cleaved amplified polymorphic sequences [36], sequence characterized amplified region (SCAR) [37].

2.3 Sequence-based markers

Sequencing technique is characterized by the identification of nucleotide sequences and their order along with the DNA strand [38]. Sequence-based markers are designed as per a specific sequence of DNA in a pool of unknown DNA. The modern sequencing techniques are genotyping by sequencing (GBS) and next-generation sequencing (NGS), which help to develop a large array of polymorphism at the nucleotide level; however, the most commonly used marker are single nucleotide polymorphisms (SNPs) [39] and diversity array technology (DArTSeq), which are known to be more accurate and reliable [22, 40].

The historical development of molecular markers is also represented in the Table 1, which is adapted and modified from: Singh and Singh [41].

We have discussed several molecular marker systems; however, the most commonly used markers in plant breeding are RFLP, SSR, RAPD, AFLP, SCAR, and SNP [42]. The single-locus markers are RFLP, VNTR, SSLP, STMS, SSR, STS, SNP, CAPS, and SCAR whereas; multi-locus markers are RAPD, AP-PCR, ISSR, AFLP, M-AFLP, and S-SAP marker [43]. All these markers are used in plant breeding for germplasm characterization and protection, gene tagging, genome mapping, linkage map construction and analysis, evolution studies, parental selection, F1 hybrid testing, genetic purity test of seeds, genes or QTLs mapping etc. [44, 45].

Table 1. A chronology of the historical steps in molecular breeding.

Year	Events
1923	Sax reported a linkage map between quantitative (seed size) and qualitative trait (seed coat color) in common bean for the first time.
1961	Thodey described QTLs mapping in *Drosophila melanogaster*
1980	Linkage mapping in humans using RFLP (Restriction Fragment Length Polymorphism) was described for the first time by Botstein *et al.*
1985	Kary Mullis discovered the Polymerized Chain Reaction (PCR) which led to the designing of PCR based markers
1989	Olson et al. reported Sequence-tagged site (STS) markers
1990	Williams JGK *et al.* developed 'RAPD' (Random Amplified Polymorphic DNA)
1991	Williams MNV *et al.* reported 'CAPs' (Cleaved Amplified Polymorphic sequence)
1993	Development of Marker-assisted techniques: Paran and Michelmore developed 'SCAR' (Sequence Characterized Amplified Regions) and Zabeau and Vos developed 'AFLP' (Amplified Fragment Length Polymorphism) technique
2001	Li and Quiros developed 'SRAP' (Sequence Related Amplified Polymorphism) technique
2009	Collard and Mackill reported 'SCoT' (Start Codon Targeted Polymorphism)
2014	Singh AK *et al* described 'CAAT box-derived polymorphism marker'

3. MARKER ASSISTED SELECTION (MAS)

The direct phenotypic selection in plant breeding for crop improvement is labor-intensive, costly, and time-taking. This selection is also affected by target gene expression, their specific biological or environmental condition, and heritability of a trait. Phenotypic selection is less efficient for the quantitative traits that are frequently under the selection [46].

In MAS, the phenotypic selection is made with the help of genotypic markers. This technique helps to avoid difficulties and challenges that are occurred during the conventional crop breeding [47]. It is mostly used by plant breeders in their breeding programmes for the identification of desired dominant or recessive alleles throughout generations, also it helps to identify best genotypes from segregating generations [48]. The prerequisite for an efficient MAS program is reliable markers, quality of DNA extraction method, genetic maps, knowledge of marker-trait association, quick and efficient data processing, and availability of high throughput marker detection system [49]. Marker development pipeline adapted from [5] Collard and Mackill, 2008, in Figure 1 explain that how marker assisted selection imposed from development of population through various steps.

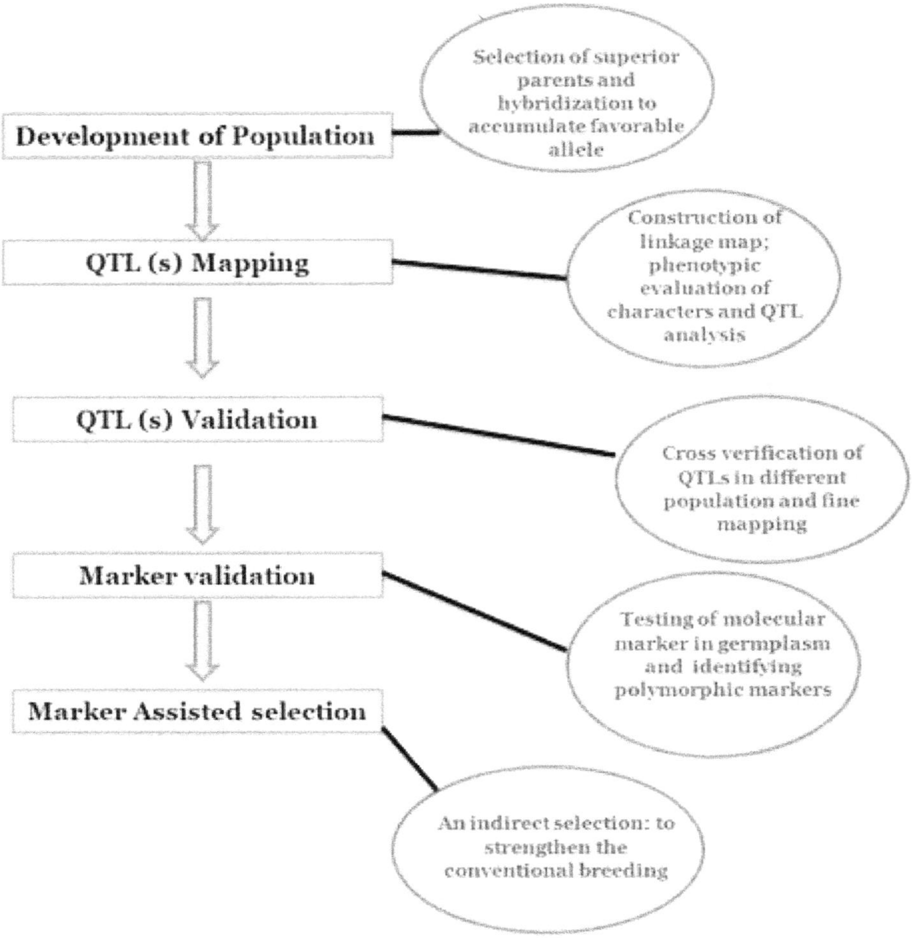

Figure 1. Marker development flow chart.

4. VARIATIONS OF MAS

There are different molecular approaches used under the umbrella of MAS, such as marker-assisted backcrossing (MABC), gene pyramiding, marker-assisted recurrent selection (MARS) and genomic selection (GS). These approaches have been utilized in plant breeding for the characterization of genetic material and selection of individuals in the early segregating generation, which fastens the breeding cycle with more accuracy [22].

4.1 Marker-assisted backcrossing (MABC)

Convention backcrossing is an age-old practice and is a very useful technique for the transfer of oligogenic traits from donor parents to recipient parents by recovering the whole genome of recipient parents except trait of interest after 6–7 generations of backcrossing. The MABC is a backcrossing technique and is assisted by molecular markers [50] to speed up the selection process and genome recovery of recipient parents. The MABC technique has been extensively used to remove the undesirable traits such as insect and disease susceptibility, and anti-nutritional factors etc. from high yielding popular varieties by introducing gene of interest or quantitative trait loci (QTLs) from donor parent [51].

The fundamental basis of MABC is the close association of marker with gene/s or QTLs. Recovery of recurrent parent genome is specified by using formula- $1-(1/2)^{m+1}$ (m is the number of generation of selfing or backcrossing). This technique has been used in different crops such as rice [52], wheat [53], barley [54], soybean [55], cotton [56], tomato [57], and pea [58], etc. There are three basic steps in the MABC technique *viz.* foreground selection, recombinant selection, and background selection.

Foreground selection is the first step of MABC, where the gene of interest from the donor parent is the primary target which is linked with the marker. The efficiency of foreground selection depends on marker-trait association, the physical distance between marker and gene of interest, genetic load or linkage drag, number of genes/QTLs/loci targeted to selection, etc. [59]. Linkage drag is undesirable for selection due to the negative effect of associated genes on targeted traits.

Recombinant selection is the second step of MABC, where selection is made for target gene in backcross progeny, and the recombination process is done between the gene of interest and linked flanking marker for reducing the effect of linkage drag [22].

Background selection is the third step of MABC, where the major target is the recovery of a large amount of recipient parental genome from backcross progeny by using molecular markers that are unlinked with the gene of interest [5]. The efficiency of background selection is determined by various factors such as the size of the population, the number of markers and targeted genes, and linkage drag, etc. It helps to speed up the recovery of the recipient parent genome with the trait of interest and also termed as 'complete line conversion' [60].

4.2 Marker-assisted gene pyramiding (MAGP)

Current breeding programs mainly focus on the development of lines governing complex traits such as biotic and abiotic stress. Modern MAS methods involve pyramiding of different genes to accomplish such goals referred to as MAGP. In MAGP, two or more than two genes at a time are selected for pyramiding. Different approaches have been utilized for pyramiding multiple genes/QTLs from donor parent to recipient parent. Some of them are recurrent selection, backcrossing, and multiple-parent crossing or complex crossing. The 3-4 desirable genes from other lines would be incorporated by convergent or stepwise backcrossing. The incorporation of more genes is usually carried through multiple crossing or recurrent selection. If we want to pyramid multiple genes/QTLs, marker-assisted convergent crossing (MACC) can be used [8, 61].

4.3 Marker-assisted recurrent selection (MARS)

Recurrent selection is an efficient technique used in plant breeding for the improvement of quantitative traits by continuous crossing and selection process. However, its efficiency of selection is adversely affected by environmental fluctuations which leads to delays breeding cycle. In MARS, molecular markers are used at each generation level for the targeted traits. Here, the selective crossing is done in selected individual plants at every crossing and selection cycle. The selection is made based on phenotypic data with marker scores. Thus, it increases the efficiency of recurrent selection and accelerates the breeding or selection cycle. The MARS has been extensively used for polygenic traits such as crop yield, biotic and abiotic stress tolerance, and considered as a forward breeding tool for augmenting multiple genes or QTLs [62].

4.4 Genomic selection (GS)

The genomic selection was developed by Hayes and Goddard [63] and is known as an advanced version of MAS. It can predict the genetic values of selected individuals which depend on genome estimated breeding values (GEBVs) by using high-density markers that are distributed throughout the genome. The GEBV prediction model combines genotypic data with phenotypic data with their pedigree and increases the prediction accuracy. The GS is mostly dependent on all the molecular markers which have both major and minor marker effect. Molecular markers are selected based on their whole genome coverage and all the QTLs should be in linkage disequilibrium with at least a single marker [23, 62, 63]. Two different types of populations are used in GS, such as training and testing population. The training population is related to the breeding population, and used to estimate the genomic selection model parameter. A testing population is a group of individuals in which genomic selection is carried out. The GEBV value is calculated by using molecular markers. Selection is based on GEBVs values, and no direct phenotypic selection is required [22, 64, 65, 66].

5. INNOVATIVE BREEDING SCHEMES OF MAS

Utilizing molecular markers, MAS has a broad spectrum application in plant breeding. Molecular markers can genotype all the accession present in germplasm. This potentiality permits the categorization of germplasm as well as reducing duplication. Here some of the innovative applications of MAS have been presented.

5.1 Combined marker-assisted selection

The MAS, along with phenotypic selection, increases genetic gain to unravel unidentified QTLs through QTL mapping compared to phenotypic screening or MAS alone [67]. The term 'combined MAS' was coined by Moreau et al., 2004 [68]. This approach not only reduces the population size but also increase selection efficiency. The combination of phenotypic selection and MAS also helps select traits where markers genotyping is economical compared to phenotypic screening [69]. With this view, this scheme explain that always a confirmation of MAS is necessary through phenotypic screening like in the case of QTL identified for Fusarium head blight resistance [70].

5.2 Marker-directed phenotyping

In most cases, there is a low level of recombination between QTL and marker is observed [13] which means we cannot believe 100% on markers for selecting desirable phenotypes. However, it will reduce the number of plants that are about to evaluate. This approach is mainly used for quality traits [71]; where phenotypic screening is costlier than marker genotyping [72]. The method is also known as tandem selection [71] and stepwise selection by [73]. One of the successful examples to explain this scheme is that rice primary QTL *sub 1*controls submergence tolerance, which assisted in breeding for the same [74].

5.3 Inbred or pureline enhancement and QTL mapping

This approach's main features are constructing the introgression library, evaluating the line for QTL detection, mapping, and further superior line used in the breeding program [41]. This scheme starts with hybridizing the two inbred line. One is the recurrent parent (agronomically superior having defects for one trait), and the other is the donor parent (have the desirable target gene). Further, the F_1 obtained from this cross is backcrossed again to the recurrent parent, and genome-wide markers have been utilized to select the genetic segment from the donor parent. To generate a set of NILs, F_1 is repeatedly backcrossed to the recurrent parent, and this set of NILs is known as the introgression line library. Therefore, this scheme seeks to introduce QTL from a suitable donor parent and simultaneously maps the QTL [75].

5.4 Advance backcross QTL analysis

It is designed to facilitate QTL introgression from unadapted germplasm like landraces and wild species into elite lines, simultaneously mapped for introgressed QTL [76]. This scheme is somewhat similar to the introgression line library, as discussed in section 5.3. However, the differences in the incorporation of phenotypic selection are in contrast to the introgression line library. Apart from this, several advantages like simplicity of mapping population in phenotype to the recurrent parent and reducing deleterious allele from donor parent, possibility of epistasis, andlinkage drag. After QTL mapping, only one or two generationsare needed for identifying QTL-NILs. In several crops like maize, tomato, soybean, cotton, rice, barley, and wheat, this approach is effectively used [9].

5.5 Single large scale MAS: a strategy applied at early generation

Single large scale MASwas proposed by Ribaut and Betran, 1999 [77], where marker-assisted selection is utilized at first segregating generation (F2 or F3). As the name describes, a single means one; large scale means up to three QTLs, explaining the most considerable phenotypic variance. The shortening of crop duration by reducing the breeding cycle prompted the idea of early generation MAS. Further plants having targeted gene/QTLs are selected whereas undesirable gene combination was discarded. Further, selected alleles were fixed in homozygous condition, and individual plants with undesirable genes would be discarded in early segregating generations. Thus, emphasis can be given on a few selected lines in the later stage, which reduces the wastage of resources and increases the selection efficiency [78].

5.6 Breeding by design

MAS's most ambitious objective is to improve plant type having the anticipated alleles at each locus participating in the control of all the traits [79]. Plant breeders will exploit known allelic variation to frame elite lines by accumulating multiple favorable alleles through this approach [80]. Therefore, the breeder can pre-plan the combination of genes he is looking for, and consequently, he can select the plant with the desired characteristics that will save expensive field testing.

5.7 Mapping As You Go (MAYG)

This method revised assessments of QTL allele effects for remapping new elite germplasm produced continuously over the selection cycle. In this approach, initial breeding crosses are utilized to estimate the QTL location and its impact. The information revealed from this estimation will be used in the mapping. This updated QTL information will be used in a new set of breeding cycles as the name suggest, mapping as you go, which means that the breeding cycle can be continued as long as desired. Overall an enhanced response has been achieved with frequent re-estimation of QTL compared to single QTL estimation at the initial level of this approach [41]. Hence, this method's advantage is that it ensures that the QTL estimate remains significant for the germplasm currently used in the breeding program [81].

5.8 Characterization of breeding material

Well-documented and characterized breeding material is a prerequisite for improving crop yield in plant breeding programs. The MAS could help to select desirable traits and have been exploited to identify cultivars/purity assessment, evaluate genetic diversity and selection of suitable donor parent, heterotic grouping, and identification of genomic regions for effective utilization in breeding programs [82, 83, 84].

6. ACHIEVEMENTS MADE THROUGH MAS

Several examples illustrate the achievement, made through marker-assisted selection; however, in Table 2, few paradigm crop-wise and trait wise have been presented.

Table 2. The paradigm of MAS in crops.

Traits	Gene/QTL	Maize Marker	Particulars	References
Yield	Six QTL	SNP	Cross between CLM495 and LPSC7F64	[85]
Maize earliness and yield	QTL	RFLP	QTLs on chromosomes number 5, 8 and 10	[86]
Sugarcane mosaic virus(SMV)	Scm1 and scm2	SCAR and CAPS	Fine mapping show present on chromosome number 6 (scm1) and 3 (scm2) in maize	[87]
Maize rough dwarf disease (MRDD)	QTL qMrdd	SSR	Conventional method coupled with MAS is used to introgress qMrdd8 from X178 into elite germplasm	[88]
Nothern corn leaf blight(NCLB)	Ht1, Ht2, Ht3, Ht4, HtM, HtP, HtNB, Htn1	RFLP	Ht 1 located on chromosome 2 and Ht2 located on chromosome 8.	[89]
European corn borer, sugarcane borer and southwestern corn borer	LIR4, 17, and 22 MQTL	SNP	LIR MQTL present on chromosome number 1 and contain QTL for cell wall acidic constituents, fiber components and diferulates.	[90]

Traits	Gene/QTL	Maize Marker	Particulars	References
European corn borer and Mediterranean corn borer	42 SIR MQTL	SNP	Highest SIR MQTL present on chromosome number 2 and 5. cross-linking between fiber and hydroxycinnamate against mechanical damage by insects.	[90]
maize weevil and the Mediterranean corn borer	KIR MQTL, KIR3, 15, and 16	SNP	Highest KIR MQTL presents on chromosome 4 and 10. Provide resistance to kernel damage and associated post-harvest loss and contaminations.	[90]
Drought resistance	Major QTL	—	Major QTLs on chromosome number 1, 2, 8 and 10	[91]
QPM	o2 allele	SSR	QPM hybrids accumulate essential amino acids (lysine, tryptophan) in the endosperm.	[92]
Rice				
Yield	Yld1.1 and yld2.1	SSR	Feasibility of SSR marker associated genes (yld1.1 and yld2.1) in screening rice HYVs.	[93]
Bacterial blight	Xa21	RFLP	Seedling and adult stage resistance against blight	[94]
Traits	Gene/QTL	Marker	Particulars	References
Bacterial blight, Rice Blast	Pi9, Xa23	PCR based primer	Rice blast and bacterial blight resistance.	[95]
Rice Blast	Pi9, Pi2	PCR based primer	Hybrid of Hui 316 (restorer line) and Pi9, Pi2 respectively impart blast resistance.	[96]
Brown plant hopper (BPH)	Bph3, bph4, Bph13(t), bph19(t), and Qbph-9	SSR	Phenotypic variations associated with BPH infestation varies from 17 to 20% concerning BPH biotypes.	[97]

Traits	Gene/QTL	Maize Marker	Particulars	References
Submergence tolerance beyond SUB1	5 QTL	SSR	5 QTL were found on Chromosome 1, 4, 8, 9, and 10	[98]
Barley				
Yield	QTL	PCR based marker	QTL present on Baronesse chromosome 2HL and 3HL fragments	[99]
Malting quality in barley	QTL1, QTl2	PCR based marker	QTL1 is located on chromosome number 1 and QTL2 located on chromosome number 4.	[71]
Fusarium head blight (FHB) along with agronomic traits	Additive and Epistatic QTLs	SSR and DArT markers	Multi-QTL analysis for the improvement of FHB resistance and agronomic traits using recombinant inbred population.	[100]
Drought tolerance	Yield and biomass associated QTLs	SSR	QTL alleles introgression ensured yield potential and biomass stability under multiple environments.	[101]
Wheat				
Drought tolerance	QTLs for photosynthesis, water content, cell membrane stability)	SSR	QTL present on 2A chromosome	[102]
Pulses				
Bacterial blight resistance (Common bean)	QTL	STS, RAPD, SCAR, AFLP	Multi-QTL loci analysis based on linkage maps can predict the phenotypic variation up to a large extent.	[103]
Resistance to Fusarium wilt (Chickpea)	QTLFoc02, QTLFoc5	SSR	Genetic distance is 10 cM	[104]
Ascochyta blight resistance	QTLAR3	SSR	Genetic distance is 24 cM	[105]
Powdery mildew resistance (Mungbean)	QTLs	RAPD, CAP, AFLP	Genetic distance is 1.3	[106]

Traits	Gene/QTL	Maize Marker	Particulars	References
Traits	Gene/QTL	Marker	Particulars	References
Common bean (Drought resistant)	QTLs (yield components, pod harvest index)	SNP	QTL present on chromosome 1, 3, 4, 7, 8, and 9	[107]
Chickpea (salinity)	48 QTLs (days to 50% flowering and maturity and days after sowing)	SSR and SNP	QTL present on CaLG05 and CaLG07 Chromosome loci	[108]
Drought (Chickpea)	93 QTLs (plant height, days to flowering and days to maturity	SSR	QTL present on LG3 and LG4 Chromosome loci	[109]
Salinity (Cowpea)	1 QTL (pod length and seed size)	SSR	QTL present on LG1 Chromosome loci	[110]
Al tolerance (Soybean)	2 QTLs (root extension)	RFLP and SSR	QTL present on Gm08 and Gm16 Chromosome loci	[111]
Salinity (Soybean)	1 QTL (salt-tolerant)	SSR and SNP	QTL present on 3 Chromosome	[112]
Drought (Soybean)	7 QTLs (canopy wilting trait)	SSR	QTL present on Gm12 Chromosome loci	[113]
Pea (Frost)	161 QTLs	SSR and SNP	QTL present on 1, 2, 3, 4, 5, 6, and 7 Chromosome	[114]
Pea (Drought)	10 QTLs	SSR	QTL present on LGI, LGIII, and LGIV Chromosome loci	[115]

Apart from the improvement in specific traits through an indirect selection via MAS, there are varieties that are released through MAS also presented in Table 3.

Table 3. Varieties developed through Marker Assisted Selection [41].

Varieties	Gene	Remark
Pusa Basmati 1 (IPB1) variety	QTL (xa13)-Chromosome 8 and QTL (xa21)-Chromosome 11	Bacterial leaf blight resistance from IRBB55
Improved Sambha Mahsuri (Improved BPT 5204)	xa5, xa13 and xa21	Bacterial leaf blight resistance
Vivek QPM9	opaque-2 from Vivek Hybrid Maize	High tryptophan, lysine and iron content
Improved Pusa RH10	xa13, xa21, pi54 and piz5	Bacterial leaf blight resistance and blast resistance

7. CONCISE SUMMARY

Molecular marker technology has traveled more than 30 years since the identification of the first marker *i.e.* RFLP, and reached its peak by using SNP or DArT. Molecular marker can assist in the selection process with phenotypic selection and speed up the pace of the breeding cycle. In recent times modern technologies such as NGS *i.e.* low cost with high throughput, GS, and GBS have been used in plant breeding but could not achieve the desired goal. The most probable reason is inaccurate phenotyping, and this problem can be alleviated by using modern throughputs phenotyping techniques such as camera or computer or sensor-based techniques in phenomics. Edge cutting technologies such as CRISPR/Cas and genome editing can be used for precise modification in the genome as per the need of human beings for their welfare.

AT A GLANCE

Burgeoning the human population with its required food demand created a burden on ever-decreasing cultivated land and our food production systems. This situation prompted plant scientists to breed crops in a short duration with specific traits. Marker-assisted selection (MAS) has emerged as a potential tool to achieve desirable results in plants with the help of molecular markers and improves the traits of interest in a short duration. The MAS has comprehensively been used in plant breeding to characterize germplasm, diversity analysis, trait stacking, gene pyramiding, multi-trait introgression, and genetic purity of different cereals, pulses, oilseeds, and fiber crops, etc. Mapping studies pointed out several marker-trait associations from different crop species, which specifies the potential application of MAS in accelerating crop improvement. This chapter presents an overview of molecular markers, their genesis, and potential use in plant breeding.

ABBREVIATIONS

AFLP	Amplified fragment length polymorphism
AP-PCR	Arbitrarily primed polymerase chain reaction
CAPS	Cleaved amplified polymorphic sequence
DArT	Diversity array technology
DNA	Deoxyribonucleic acid
GBS	Genotyping by sequencing
GS	Genomic selection
InDels	Insertions and Deletions
ISSR	Inter-simple sequence repeat
MABC	Marker-assisted backcrossing
MAS	Marker assisted selection

NGS	Next generation sequencing
NILs	Near isogenic line(s)
PCR	Polymerase chain reaction
QTL	Quantitative trait locus
RAPD	Random amplified polymorphic DNA
RFLP	Restriction fragment length polymorphism
SCAR	Sequence characterized amplified region
SCoT	Start codon-targeted
SNP	Single nucleotide polymorphism
SRAP	Sequence related amplified polymorphism
SSAP	Sequence-specific amplification polymorphism
S-SAP	Sequence-specific amplification polymorphism
SSLP	Simple sequence length polymorphism
SSR	Simple sequence repeats
STMS	Sequence-tagged microsatellite site
STS	Sequence-tagged site
VNTR	Variable number of tandem repeats

REFERENCES

1. Alexandratos N, Bruinsma J. World agriculture towards 2030/2050: the 2012 revision
2. Kage U, Kumar A, Dhokane D, Karre S, Kushalappa AC. Functional molecular markers for crop improvement. Critical reviews in biotechnology. 2016 Sep 2;36(5):917-30
3. Platten JD, Cobb JN, Zantua RE. Criteria for evaluating molecular markers: comprehensive quality metrics to improve marker-assisted selection. PloS one. 2019 Jan 15;14(1):e0210529
4. hand S, Chandra K, Khatik CL. Varietal Release, Notification and Denotification System in India. InPlant Breeding-Current and Future Views 2020 Oct 19. IntechOpen
5. ollard BC, Mackill DJ. Marker-assisted selection: an approach for precision plant breeding in the twenty-first century. Philosophical Transactions of the Royal Society B: Biological Sciences. 2008 Feb 12;363(1491):557-72
6. Bhargava A, Srivastava S. Toward Participatory Plant Breeding. InParticipatory Plant Breeding: Concept and Applications 2019 (pp. 69-86). Springer, Singapore
7. Ribaut JM, Hoisington D. Marker-assisted selection: new tools and strategies. Trends in Plant Science. 1998 Jun 1;3(6):236-9

8. Gupta PK, Langridge P, Mir RR. Marker-assisted wheat breeding: present status and future possibilities. Molecular Breeding. 2010 Aug 1;26(2):145-61
9. Jiang GL. Molecular markers and marker-assisted breeding in plants. Plant breeding from laboratories to fields. 2013 May 22:45-83
10. Tester M, Langridge P. Breeding technologies to increase crop production in a changing world. Science. 2010 Feb 12;327(5967):818-22
11. Charcosset A, Moreau L. Use of molecular markers for the development of new cultivars and the evaluation of genetic diversity. Euphytica. 2004 Jun 1;137(1):81-94
12. Melchinger AE, Utz HF, Schön CC. Quantitative trait locus (QTL) mapping using different testers and independent population samples in maize reveals low power of QTL detection and large bias in estimates of QTL effects. Genetics. 1998 May 1;149(1):383-403
13. Sharp PJ, Johnston S, Brown G, McIntosh RA, Pallotta M, Carter M, Bariana HS, Khatkar S, Lagudah ES, Singh RP, Khairallah M. Validation of molecular markers for wheat breeding. Australian Journal of Agricultural Research. 2001;52(12):1357-66
14. Brennan JP, Rehman A, Raman H, Milgate AW, Pleming D, Martin PJ. An economic assessment of the value of molecular markers in plant breeding programs. 2005
15. Brumlop S, Finckh MR. Applications and potentials of marker assisted selection (MAS) in plant breeding. Final report of the F+ E project "Applications and Potentials of Smart Breeding"(FKZ 350 889 0020) On behalf of the Federal Agency for Nature Conservation December. 2010 Dec
16. Madina MH, Haque ME, Dutta AK, Islam MA, Deb AC, Sikdar B. Estimation of genetic diversity in six lentil (Lens culinaris Medik.) varieties using morphological and biochemical markers. International Journal of Scientific & Engineering Research. 2013;4:819-25
17. Mackill DJ, Ni J. Molecular mapping and marker-assisted selection for major-gene traits in rice. In Rice genetics IV 2001 (pp. 137-151)
18. Mohler V, Singrün C. General considerations: marker-assisted selection. In Molecular Marker Systems in Plant Breeding and Crop Improvement 2004 (pp. 305-317). Springer, Berlin, Heidelberg
19. Bohar R, Chitkineni A, Varshney RK. Genetic molecular markers to accelerate genetic gains in crops
20. Singh AK, Gopalakrishnan S, Singh VP, Prabhu KV, Mohapatra T, Singh NK, Sharma TR, Nagarajan M, Vinod KK, Singh D, Singh UD. Marker assisted

selection: a paradigm shift in Basmati breeding. Indian Journal of Genetics and Plant Breeding. 2011 May 1;71(2):120

21. He J, Zhao X, Laroche A, Lu ZX, Liu H, Li Z. Genotyping-by-sequencing (GBS), an ultimate marker-assisted selection (MAS) tool to accelerate plant breeding. Frontiers in plant science. 2014 Sep 30;5:484

22. Nadeem MA, Nawaz MA, Shahid MQ , Doğan Y, Comertpay G, Yıldız M, Hatipoğlu R, Ahmad F, Alsaleh A, Labhane N, Özkan H. DNA molecular markers in plant breeding: current status and recent advancements in genomic selection and genome editing. Biotechnology & Biotechnological Equipment. 2018 Mar 4;32(2):261-85

23. Eagles HA, Bariana HS, Ogbonnaya FC, Rebetzke GJ, Hollamby GJ, Henry RJ, Henschke PH, Carter M. Implementation of markers in Australian wheat breeding. Australian Journal of Agricultural Research. 2001;52(12):1349-56

24. Karaköy T, Baloch FS, Toklu F, Özkan H. Variation for selected morphological and quality-related traits among 178 faba bean landraces collected from Turkey. Plant Genetic Resources. 2014 Apr 1;12(1):5

25. Tanksley S, Orton T. Isozymic variation and plant breeders' rights. Iso Plant Genet Breed. 1983;1:425

26. Mateu-Andres I, De Paco L. Allozymic differentiation of the Antirrhinum majus and A. siculum species groups. Annals of botany. 2005 Feb 1;95(3):465-73

27. Mondini L, Noorani A, Pagnotta MA. Assessing plant genetic diversity by molecular tools. Diversity. 2009 Sep;1(1):19-35

28. Nakaya A, Isobe SN. Will genomic selection be a practical method for plant breeding?. Annals of botany. 2012 Nov 1;110(6):1303-16

29. Collard BC, Jahufer MZ, Brouwer JB, Pang EC. An introduction to markers, quantitative trait loci (QTL) mapping and marker-assisted selection for crop improvement: the basic concepts. Euphytica. 2005 Jan 1;142(1-2):169-96

30. Mullis K, Faloona F, Scharf S, Saiki RK, Horn GT, Erlich H. Specific enzymatic amplification of DNA in vitro: the polymerase chain reaction. InCold Spring Harbor symposia on quantitative biology 1986 Jan 1 (Vol. 51, pp. 263-273). Cold Spring Harbor Laboratory Press

31. Williams JG, Kubelik AR, Livak KJ, Rafalski JA, Tingey SV. DNA polymorphisms amplified by arbitrary primers are useful as genetic markers. Nucleic acids research. 1990 Nov 25;18(22):6531-5

32. Vos P, Hogers R, Bleeker M, Reijans M, Lee TV, Hornes M, Friters A, Pot J, Paleman J, Kuiper M, Zabeau M. AFLP: a new technique for DNA fingerprinting. Nucleic acids research. 1995;23(21):4407-14

33. Zane L, Bargelloni L, Patarnello T. Strategies for microsatellite isolation: a review. Molecular ecology. 2002 Jan;11(1):1-6

34. Uzun AY, Yesiloglu T, Aka-Kacar Y, Tuzcu O, Gulsen O. Genetic diversity and relationships within Citrus and related genera based on sequence related amplified polymorphism markers (SRAPs). Scientia horticulturae. 2009 Jul 2;121(3):306-12

35. Semagn K, Bjørnstad Å, Ndjiondjop MN. An overview of molecular marker methods for plants. African journal of biotechnology. 2006;5(25)

36. Weiland JJ, Yu MH. A cleaved amplified polymorphic sequence (CAPS) marker associated with root-knot nematode resistance in sugarbeet. Crop science. 2003 Sep;43(5):1814-8

37. Yang L, Fu S, Khan MA, Zeng W, Fu J. Molecular cloning and development of RAPD-SCAR markers for Dimocarpus longan variety authentication. SpringerPlus. 2013 Dec 1;2(1):501

38. França LT, Carrilho E, Kist TB. A review of DNA sequencing techniques. Quarterly reviews of biophysics. 2002 May 1;35(2):169

39. Davey JW, Hohenlohe PA, Etter PD, Boone JQ ,Catchen JM, Blaxter ML. Genome-wide genetic marker discovery and genotyping using next-generation sequencing. Nature Reviews Genetics. 2011 Jul;12(7):499-510

40. Wenzl P, Carling J, Kudrna D, Jaccoud D, Huttner E, Kleinhofs A, Kilian A. Diversity Arrays Technology (DArT) for whole-genome profiling of barley. Proceedings of the National Academy of Sciences. 2004 Jun 29;101(26):9915-20

41. Singh BD, Singh AK. Marker-assisted plant breeding: principles and practices. New Delhi: Springer; 2015 Jun 26

42. Agarwal M, Shrivastava N, Padh H. Advances in molecular marker techniques and their applications in plant sciences. Plant cell reports. 2008 Apr 1;27(4):617-31

43. Barcaccia G. Molecular markers for characterizing and conserving crop plant germplasm. InMolecular techniques in crop improvement 2010 (pp. 231-254). Springer, Dordrecht

44. MirMohammadiMaibody SA, Golkar P. Application of DNA Molecular Markers in Plant Breeding. Journal of Plant Genetic Research. 2019 Sep 10;6(1):1-30

45. Ahmar S, Gill RA, Jung KH, Faheem A, Qasim MU, Mubeen M, Zhou W. Conventional and Molecular Techniques from Simple Breeding to Speed Breeding in Crop Plants: Recent Advances and Future Outlook. International Journal of Molecular Sciences. 2020 Jan;21(7):2590

46. Watson A, Hickey LT, Christopher J, Rutkoski J, Poland J, Hayes BJ. Multivariate Genomic Selection and Potential of Rapid Indirect Selection with Speed Breeding in Spring Wheat. Crop Science. 2019 Sep;59(5):1945-59

47. Tabor HK, Risch NJ, Myers RM. Candidate-gene approaches for studying complex genetic traits: practical considerations. Nature Reviews Genetics. 2002 May;3(5):391-7

48. Francia E, Tacconi G, Crosatti C, Barabaschi D, Bulgarelli D, Dall'Aglio E, Valè GI. Marker assisted selection in crop plants. Plant Cell, Tissue and Organ Culture. 2005 Sep 1;82(3):317-42

49. Dar AA, Mahajan R, Sharma S. Molecular markers for characterization and conservation of plant genetic resources. Indian J. Agric. Sci.. 2019 Nov 1;89(11):3-11

50. Holland JB. Implementation of molecular markers for quantitative traits in breeding programs—challenges and opportunities. InNew Directions for a Diverse Planet: Proceedings for the 4th International Crop Science Congress. Regional Institute, Gosford, Australia, www.cropscience.org.au/icsc2004 2004 Sep 26

51. Ribaut JM, Sawkins MC, Bänziger M, Vargas M, Huerta E, Martinez C, Moreno M. Marker-assisted selection in tropical maize based on consensus map, perspectives, and limitations. Resilient Crops for Water Limited Environments. 2004:267

52. Hasan MM, Rafii MY, Ismail MR, Mahmood M, Rahim HA, Alam MA, Ashkani S, Malek MA, Latif MA. Marker-assisted backcrossing: a useful method for rice improvement. Biotechnology & Biotechnological Equipment. 2015 Mar 4;29(2):237-54

53. Yadawad A, Gadpale A, Hanchinal RR, Nadaf HL, Desai SA, Biradar S, Naik VR. Pyramiding of leaf rust resistance genes in bread wheat variety DWR 162 through marker assisted backcrossing. Indian J Genet Pl Br. 2017 May 1;77:251-7

54. Xu Y, Zhang XQ ,Harasymow S, Westcott S, Zhang W, Li C. Molecular marker-assisted backcrossing breeding: an example to transfer a thermostable β-amylase gene from wild barley. Molecular Breeding. 2018 May 1;38(5):63

55. Rawal R, Kumar V, Rani A, Gokhale SM. Genetic elimination of off-flavour generating lipoxygenase-2 gene of soybean through marker assisted backcrossing and its effect on seed longevity. Plant Breeding and Biotechnology. 2020 Jun 1;8(2):163-73

56. Wu J, Zhang M, Zhang X, Guo L, Qi T, Wang H, Tang H, Zhang J, Xing C. Development of InDel markers for the restorer gene Rf1 and assessment of

their utility for marker-assisted selection in cotton. Euphytica. 2017 Nov 1;213(11):251

57. Osei MK, Prempeh R, Adjebeng-Danquah J, Opoku JA, Danquah A, Danquah E, Blay E, Adu-Dapaah H. Marker-Assisted Selection (MAS): A Fast-Track Tool in Tomato Breeding. InRecent Advances in Tomato Breeding and Production 2018 Nov 5. IntechOpen

58. Mannur DM, Babbar A, Thudi M, Sabbavarapu MM, Roorkiwal M, Sharanabasappa BY, Bansal VP, Jayalakshmi SK, Yadav SS, Rathore A, Chamarthi SK. Super Annigeri 1 and improved JG 74: two Fusarium wilt-resistant introgression lines developed using marker-assisted backcrossing approach in chickpea (Cicer arietinum L.). Molecular Breeding. 2019 Jan 1;39(1):2

59. Hospital F. Selection in backcross programmes. Philosophical Transactions of the Royal Society B: Biological Sciences. 2005 Jul 29;360(1459):1503-11

60. Ribaut JM, Jiang C, Hoisington D. Simulation experiments on efficiencies of gene introgression by backcrossing. Crop Science. 2002 Mar;42(2):557-65

61. Luo Y, Sangha JS, Wang S, Li Z, Yang J, Yin Z. Marker-assisted breeding of Xa4, Xa21 and Xa27 in the restorer lines of hybrid rice for broad-spectrum and enhanced disease resistance to bacterial blight. Molecular breeding. 2012 Dec 1;30(4):1601-10

62. Eathington SR, Crosbie TM, Edwards MD, Reiter RS, Bull JK. Molecular markers in a commercial breeding program. Crop Science. 2007 Dec;47:S-154

63. Hayes BJ, Goddard ME. Prediction of total genetic value using genome-wide dense marker maps. Genetics. 2001 Apr 1;157(4):1819-29

64. Ingvarsson PK, Street NR. Association genetics of complex traits in plants. New Phytologist. 2011 Mar;189(4):909-22

65. Crossa J, Pérez-Rodríguez P, Cuevas J, Montesinos-López O, Jarquín D, de los Campos G, Burgueño J, González-Camacho JM, Pérez-Elizalde S, Beyene Y, Dreisigacker S. Genomic selection in plant breeding: methods, models, and perspectives. Trends in plant science. 2017 Nov 1;22(11):961-75

66. Voss-Fels KP, Cooper M, Hayes BJ. Accelerating crop genetic gains with genomic selection. Theoretical and Applied Genetics. 2019 Mar 1;132(3):669-86

67. Lande R, Thompson R. Efficiency of marker-assisted selection in the improvement of quantitative traits. Genetics. 1990;124(3):743-56

68. Moreau L, Charcosset A, Gallais A. Experimental evaluation of several cycles of marker-assisted selection in maize. Euphytica. 2004; 137(1):111-8

69. Randhawa HS, Mutti JS, Kidwell K, Morris CF, Chen X, Gill KS. Rapid and targeted introgression of genes into popular wheat cultivars using marker-assisted background selection. PloS one. 2009; 4(6):e5752

70. Zhou WC, Kolb FL, Bai GH, Domier LL, Boze LK, Smith NJ. Validation of a major QTL for scab resistance with SSR markers and use of marker-assisted selection in wheat. Plant breeding. 2003 ;122(1):40-6

71. Han F, Romagosa I, Ullrich SE, Jones BL, Hayes PM, Wesenberg DM. Molecular marker-assisted selection for malting quality traits in barley. Molecular Breeding. 1997 Dec 1;3(6):427-37

72. Akhtar S, Bhat MA, Wani SA, Bhat KA, Chalkoo S, Mir MR, Wani SA. Marker assisted selection in rice. Journal of Phytology. 2010 Dec 19

73. Langridge P, Chalmers K. The principle: identification and application of molecular markers. InMolecular marker systems in plant breeding and crop improvement 2004 (pp. 3-22). Springer, Berlin, Heidelberg

74. Mackill DJ, Collard BC, Neeraja CN, Rodriguez RM, Heuer S, Ismail AM. QTLs in rice breeding: examples for abiotic stresses. InRice Genetics V 2007 (pp. 155-167)

75. Stuber CW, Polacco M, Lynn M. Synergy of empirical breeding, marker-assisted selection, and genomics to increase crop yield potential. Crop Science. 1999 Nov;39(6):1571-83

76. Pillen K, Zacharias A, Leon J. Advanced backcross QTL analysis in barley (Hordeum vulgare L.). Theoretical and Applied genetics. 2003 Jul 1;107(2):340-52

77. Ribaut JM, Betrán J. Single large-scale marker-assisted selection (SLS-MAS). Molecular Breeding. 1999 Dec 1;5(6):531-41

78. Eathington SR, Dudley JW, Rufener GK. Usefulness of Marker-QTAL ssociations in Early Generation Selection. Crop Science. 1997 Nov;37(6):1686-93

79. Peleman JD. Breeding by design: exploiting genetic maps and molecular markers through marker-assisted selection. The Handbook of Plant Genome Mapping: Genetic and Physical Mapping. 2005

80. Peleman JD and Van der Voort JR. Breeding by design. Trends in plant science. 2003 Jul 1;8(7):330-4

81. Podlich DW, Winkler CR, Cooper M. Mapping as you go: An effective approach for marker-assisted selection of complex traits. Crop Science. 2004 Sep;44(5):1560-71

82. Yashitola J, Thirumurugan T, Sundaram RM, Naseerullah MK, Ramesha MS, Sarma NP, Sonti RV. Assessment of purity of rice hybrids using microsatellite and STS markers. Crop Science. 2002 Jul;42(4):1369-73

83. Xu Y, Beachell H, McCOUCH SR. A marker-based approach to broadening the genetic base of rice in the USA. Crop science. 2004 Nov;44(6):1947-59

84. Reif JC, Melchinger AE, Xia XC, Warburton ML, Hoisington DA, Vasal SK, Beck D, Bohn M, Frisch M. Use of SSRs for establishing heterotic groups in subtropical maize. Theoretical and Applied genetics. 2003 Sep 1;107(5):947-57

85. Cerrudo D, Cao S, Yuan Y, Martinez C, Suarez EA, Babu R, Zhang X, Trachsel S. Genomic selection outperforms marker assisted selection for grain yield and physiological traits in a maize doubled haploid population across water treatments. Frontiers in plant science. 2018 Mar 20;9:366

86. Bouchez A, Causse M, Gallais A, Charcosset A. Marker-assisted introgression of favorable alleles at quantitative trait loci between maize elite lines. Genetics. 2002 Dec 1;162(4):1945-59

87. Dussle C, Quint M, Xu M, Melchinger A, Lübberstedt T. Conversion of AFLP fragments tightly linked to SCMV resistance genes Scmv1 and Scmv2 into simple PCR-based markers. Theoretical and Applied Genetics. 2002 Dec 1;105(8):1190-5

88. Xu Z, Hua J, Wang F, Cheng Z, Meng Q , Chen Y, Han X, Tie S, Liu C, Li X, Wang Z. Marker-assisted selection of qMrdd8 to improve maize resistance to rough dwarf disease. Breeding Science. 2020:19110

89. Galiano-Carneiro AL, Miedaner T. Genetics of resistance and pathogenicity in the maize/Setosphaeria turcica pathosystem and implications for breeding. Frontiers in plant science. 2017 Aug 29;8:1490

90. 90.Badji A, Otim M, Machida L, Odong T, Kwemoi DB, Okii D, Agbahoungba S, Mwila N, Kumi F, Ibanda A, Mugo S. Maize combined insect resistance genomic regions and their co-localization with Cell Wall constituents revealed by tissue-specific QTL meta-analyses. Frontiers in plant science. 2018 Jul 5;9:895

91. Prasanna BM. Molecular markers for maize improvement inAsia. InAsian Regional Maize Workshop, 10. Maize for Asia-Emerging Trends and; Technologies. Proceedings of The Asian Regional Maize Workshop; Makassar, Indonesia; 20-23 October, 2008 2009 (p. 202). CIMMYT

92. Hossain F, Muthusamy V, Pandey N, Vishwakarma AK, Baveja A, Zunjare RU, Thirunavukkarasu N, Saha S, Manjaiah KM, Prasanna BM, Gupta HS. Marker-assisted introgression of opaque2 allele for rapid conversion of elite hybrids into quality protein maize. Journal of genetics. 2018 Mar 1;97(1):287-98

93. Liang F, Deng Q , Wang Y, Xiong Y, Jin D, Li J, Wang B. Molecular marker-assisted selection for yield-enhancing genes in the progeny of "9311× O. rufipogon" using SSR. Euphytica. 2004 Jan 1;139(2):159-65

94. Chen H, Wang S, Zhang Q. New gene for bacterial blight resistance in rice located on chromosome 12 identified from Minghui 63, an elite restorer line. Phytopathology. 2002 Jul;92(7):750-4

95. Ni D, Song F, Ni J, Zhang A, Wang C, Zhao K, Yang Y, Wei P, Yang J, Li L. Marker-assisted selection of two-line hybrid rice for disease resistance to rice blast and bacterial blight. Field Crops Research. 2015 Dec 1;184:1-8

96. Tian D, Guo X, Zhang Z, Wang M, Wang F. Improving blast resistance of the rice restorer line, Hui 316, by introducing Pi9 or Pi2 with marker-assisted selection. Biotechnology & Biotechnological Equipment. 2019 Jan 1;33(1):1195-203

97. Shabanimofrad M, Yusop MR, Ashkani S, Musa MH, Adam NA, Haifa I, Harun AR, Latif MA. Marker-assisted selection for rice brown planthopper (Nilaparvatalugens) resistance using linked SSR markers. Turkish Journal of Biology. 2015 Sep 18;39(5):666-73

98. Gonzaga ZJ, Carandang J, Sanchez DL, Mackill DJ, Septiningsih EM. Mapping additional QTLs from FR13A to increase submergence tolerance in rice beyond SUB1. Euphytica. 2016 Jun 1;209(3):627-36

99. Schmierer DA, Kandemir N, Kudrna DA, Jones BL, Ullrich SE, Kleinhofs A. Molecular marker-assisted selection for enhanced yield in malting barley. Molecular Breeding. 2004 Dec 1;14(4):463-73

100. Lv C, Song Y, Gao L, Yao Q , Zhou R, Xu R, Jia J. Integration of QTL detection and marker assisted selection for improving resistance to Fusarium head blight and important agronomic traits in wheat. The Crop Journal. 2014 Feb 1;2(1):70-8

101. Merchuk-Ovnat L, Barak V, Fahima T, Ordon F, Lidzbarsky GA, Krugman T, Saranga Y. Ancestral QTL alleles from wild emmer wheat improve drought resistance and productivity in modern wheat cultivars. Frontiers in plant science. 2016 Apr 15;7:452

102. Malik S, Malik TA. Genetic mapping of potential QTLs associated with drought tolerance in wheat. JAPS: Journal of Animal & Plant Sciences. 2015 Aug 1;25(4)

103. Tar'An B, Michaels TE, Pauls KP. Mapping genetic factors affecting the reaction to Xanthomonas axonopodis pv. phaseoli in Phaseolus vulgaris L. under field conditions. Genome. 2001 Dec 1;44(6):1046-56

104. Cobos MJ, Winter P, Kharrat M, Cubero JI, Gil J, Millan T, Rubio J. Genetic analysis of agronomic traits in a wide cross of chickpea. Field Crops Research. 2009 Mar 15;111(1-2):130-6

105. Iruela M, Pistón F, Cubero JI, Millán T, Barro F, Gil J. The marker SCK13 603 associated with resistance to ascochyta blight in chickpea is located in a region of a putative retrotransposon. Plant cell reports. 2009 Jan 1;28(1):53-60

106. Chen HM, Liu CA, Kuo CG, Chien CM, Sun HC, Huang CC, Lin YC, Ku HM. Development of a molecular marker for a bruchid (Callosobruchus chinensis L.) resistance gene in mungbean. Euphytica. 2007 Sep 1;157(1-2):113-22

107. Mukeshimana G, Butare L, Cregan PB, Blair MW, Kelly JD. Quantitative trait loci associated with drought tolerance in common bean. Crop Science. 2014 May;54(3):923-38

108. Pushpavalli R, Krishnamurthy L, Thudi M, Gaur PM, Rao MV, Siddique KH, Colmer TD, Turner NC, Varshney RK, Vadez V. Two key genomic regions harbour QTLs for salinity tolerance in ICCV 2× JG 11 derived chickpea (Cicer arietinum L.) recombinant inbred lines. BMC plant biology. 2015 Dec 1;15(1):124

109. Hamwieh A, Imtiaz M, Malhotra RS. Multi-environment QTL analyses for drought-related traits in a recombinant inbred population of chickpea (Cicer arientinum L.). Theoretical and Applied Genetics. 2013 Apr 1;126(4):1025-38

110. Chankaew S, Isemura T, Naito K, Ogiso-Tanaka E, Tomooka N, Somta P, Kaga A, Vaughan DA, Srinives P. QTL mapping for salt tolerance and domestication-related traits in Vigna marina subsp. oblonga, a halophytic species. Theoretical and applied genetics. 2014 Mar 1;127(3):691-702

111. Abdel-Haleem H, Carter TE, Rufty TW, Boerma HR, Li Z. Quantitative trait loci controlling aluminum tolerance in soybean: candidate gene and single nucleotide polymorphism marker discovery. Molecular breeding. 2014 Apr 1;33(4):851-62

112. Ha BK, Vuong TD, Velusamy V, Nguyen HT, Shannon JG, Lee JD. Genetic mapping of quantitative trait loci conditioning salt tolerance in wild soybean (Glycine soja) PI 483463. Euphytica. 2013 Sep 1;193(1):79-88

113. Abdel-Haleem H, Carter TE, Purcell LC, King CA, Ries LL, Chen P, Schapaugh W, Sinclair TR, Boerma HR. Mapping of quantitative trait loci for canopy-wilting trait in soybean (Glycine max L. Merr). Theoretical and Applied Genetics. 2012 Sep 1;125(5):837-46

114. Klein A, Houtin H, Rond C, Marget P, Jacquin F, Boucherot K, Huart M, Rivière N, Boutet G, Lejeune-Hénaut I, Burstin J. QTL analysis of frost damage in pea suggests different mechanisms involved in frost tolerance. Theoretical and Applied Genetics. 2014 Jun 1;127(6):1319-30

115. Iglesias-García R, Prats E, Fondevilla S, Satovic Z, Rubiales D. Quantitative trait loci associated to drought adaptation in pea (Pisum sativum L.). Plant molecular biology reporter. 2015 Dec 1;33(6):1768-78

CHAPTER-12

SINGLE NUCLEOTIDE POLYMORPHISMS (SNPS): AN UP-TO-DATE APPROACH FOR EVALUATING PLANTS FOR FAVORABLE CHARACTERISTICS

1. INTRODUCTION: AN OVERVIEW

In the ever-evolving landscape of plant breeding, the advent of modern genomic tools has brought forth revolutionary advancements, among which Single Nucleotide Polymorphisms (SNPs) stand as a beacon of precision and efficiency. Serving as molecular signposts within the genetic code, SNPs have emerged as a powerful tool for screening plants with unparalleled accuracy, particularly in the quest for desirable traits.

Plant breeders usually screen a large number of plants for traits of economic value as determined by the breeding goal which may include breeding for resistance, biofortification to increase some micronutrients and gene pyramiding. The larger volume of plants screened at early stages of a breeding program can be laborious, capital intensive and time consuming. Germplasm screening is usually an initial step for a number of breeding programs. The aim of screening large collection of plants is for narrowing on those with desired characteristic for advancement to the next stage. It is extremely important to get it right from the beginning in other to meet breeding objectives for crop improvement.

Recently, molecular breeding has brought about a revolution in plant breeding and has been widely applied in plant improvement programs. Molecular breeding also called marker assisted breeding or marker assisted selection (MAS) is the method of using molecular or DNA markers that are closely linked to a phenotype to aid selection for such trait in a breeding scheme. A number of molecular markers have been developed and successfully used in selection due to their association with a phenotype of interest in different plant species. Thus, they have been applied in plant germplasm screening for desirable traits. Because selection is based on target genes, they provide a higher rate of accuracy during screening, reduce time and labor, therefore bringing about reduced cost.

Several types of molecular markers have been applied to different areas of plant breeding, firstly developed markers include Restriction fragment length polymorphism (RFLPs) and Randomly amplified polymorphic DNA (RAPDs). These gave way to the popular techniques Amplified fragment polymorphism (AFLP) and Simple sequence repeats (SSRs) due to their ease in detection and automation. Simple sequence repeats have been extensively applied to screen for resistant germplasms, biotic and abiotic stress and variety identification in potato, groundnut rice [1, 2, 3]. SSR markers are generally PCR-based, their technicality is simple and relatively cheap. The disadvantage of these markers is that they require polyacrylamide gel electrophoresis to achieve a high resolution of allele fragments, which is laborious to perform. They give information about a single locus per assay although multiplexing of more than two markers is possible, it is relatively expensive. The continued automation in sequencing technology has resulted in a shift from first generation DNA based markers to the use of functional and gene targeted single nucleotide polymorphism (SNP) markers. Therefore, this chapter covers ways of discovering new SNPs and their conversion to molecular marker. Also, different methods of SNPs genotyping and their application in some aspects of crop improvement are documented. The discovery of SNP markers is cost effective, it is highly multiplexable and they are availability of high throughput technologies for SNP genotyping. Because SNPs are based on target genes, they are highly reliable in cultivar identification and germplasm screening.

2. SINGLE NUCLEOTIDE POLYMORPHISM MARKERS

Single nucleotide polymorphism is a change in a single base pair at a specific locus involving two alleles where the rare allele frequency is >1% (Figure 1). It is an individual nucleotide base difference between two DNA sequences. It represents the site where the DNA sequence shows a difference by one base. They are categorized based on nucleotide substitution as transition which is an interchange of pyrimidines (C/T) and purines (A/G) or transversion which is an interchange of purine base for pyrimidine (G/C, A/T, A/C and G/T). A nucleotide base represents the basic unit of inheritance therefore SNPs present a powerful tool as molecular marker. SNPs which can be categorized as insertion or deletions, may be found in the coding or non-coding sequences of crop plants. Some individuals of a species may be heterozygous at a SNP locus or ambiguous as seen in Figure 1 (Y refers to C and T nucleotides being present at one locus), this means they possess both nucleotides on the same position on that gene. Such individuals will display an intermediate phenotype when screened phenotypically.

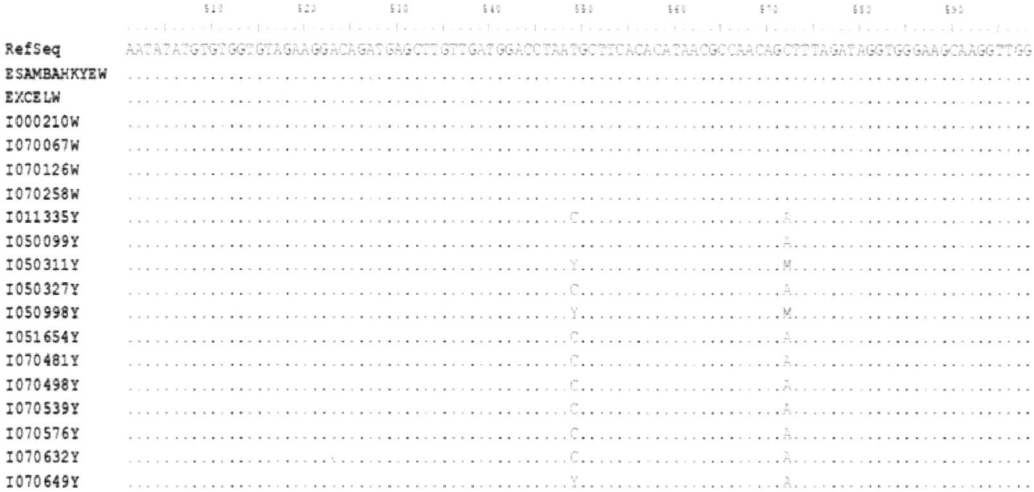

Figure 1. Multiple sequence alignment of nucleotide sequences to a reference sequence (RefSeq) reveal SNPs of PSY2 gene (Y)C/T and (M)A/C in yellow and white root accessions of cassava. Consensus to reference sequence are plotted as dots.

When SNPs or insertion/deletions found in coding sequences result in non-conservative amino acid changes, it can cause variation in the phenotype of individuals of same species (Figure 2). But the translation of non-coding sequences containing SNPs that results in conservative amino acid change will not significantly affect the phenotype of the individuals. Considering the figures above, Figure 1 had SNP present at two different positions (549 and 572) for C/T and A/C nucleotides of the PSY2 gene in cassava, respectively. Upon translation to amino acid as seen in Figure 2, the SNP at position 549 was synonymous because it caused no amino acid changes therefore may not be an effective marker in marker assisted selection with respect to yellow and white cassava roots. But the SNP located at position 572 of the gene caused non-synonymous changes in their amino acid sequence as individuals with the A nucleotide gave Alanine while those with the C nucleotide gave aspartic acid. Individuals with white root carrying the C nucleotide gave Alanine while those with yellow root carrying the A nucleotide gave Aspartic acid. Thus, this SNP can be effectively utilized as molecular marker in selecting for root color of cassava even before it gets to the stage of developing roots for phenotypic evaluation. The presence of SNPs in regulatory and coding regions of genes can cause significant phenotypic effect on function of protein and how genes are expressed. This permits the association of genotypic and phenotypic variations which has been successfully exploited in cultivar identification and genetic diversity analysis. Though the association of traits of economic importance may be less than 100%, they can still be successfully utilized in marker assisted selection and gene isolation. Given its precision in germplasm identification, it can also be an efficient tool for plant germplasm screening.

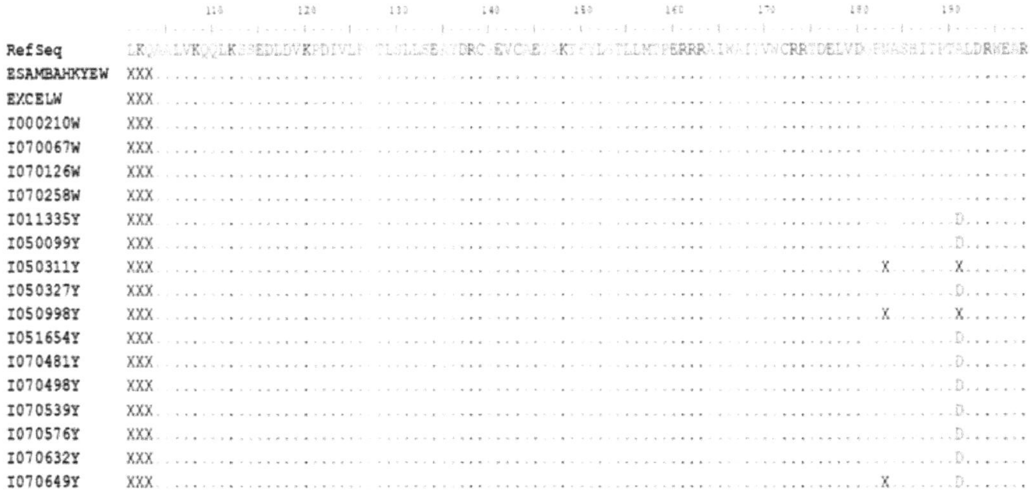

Figure 2. Multiple sequence alignment of amino acid sequences showing a non-synonymous mutation at the position of the A/C nucleotide polymorphism of PSY2 gene in cassava.

The translation of individuals with the A nucleotide gave Alanine (A) while those with the C nucleotide gave Aspartic acid (D), X shows ambiguous position. Consensus sequences are plotted as dot with reference to the reference sequence (RefSeq).

Single nucleotide polymorphism is now a highly preferred genetic marker due to the increase in amount of sequence information and the determination of gene function due to genomic research. Their widespread abundance in the genome and the development of new SNP genotyping platforms has made them the preferred marker for plant germplasm screening or characterization and identification of functional genes for traits of interest [4, 5, 6]. SNPs are easily automated with high throughput techniques and are being used for large segregating populations.

A lot of techniques and methods have been applied to identify SNPs and use those that can successfully discriminate between traits of interest for marker development. SNP markers can be identified by carrying out locus specific PCR. Here, locus specific PCR primers are synthesized from genomic sequences that are known and available in the public databases or previously sequenced data. The primers are used to amplify DNA samples from several individuals of a plant species. The resultant PCR amplicons are sequenced and aligned. Alignment is searched for availability of SNPs which are base changes within a particular locus. Depending on the informativeness of such SNP after characterization, it can be further evaluated for its effectiveness as a marker for germplasm screening or marker assisted selection. This method of SNP discovery can only be used if there is an existing information concerning the sequence to be amplified. This method was used by Udoh et al. [7] to identify SNPs causing non-synonymous changes in amino acid of phytoene synthase gene in cassava linked to expression of yellow color in roots of some cassava varieties. Also, Harjes et al. [8] identified SNPs in the regulatory regions of lycopene epsilon cylase genes causing accumulation of carotenoids in maize.

The availability of whole genome sequences and expressed sequence tags (ESTs) databases has allowed for non-gel-based approach to SNP discovery. Unigenes or EST sequences of interest can be analyzed de novo or exported to other convenient computer software proprams for alignment and SNP searches. Alignment of genomic sequences may identify SNPs in both coding and non-coding regions of the genome but ESTs are preferred because they are coding sequences and SNPs identified here can affect gene expression thus can be evaluated further for downstream applications. This approach to discover SNP is relatively easy and cost effective although the authenticity of sequences used may not be guaranteed because they were mined from public databases.

Also, high throughput automated next generation sequencing (NGS) platforms such as Illumina Genome Analyzer, Roch/454 FLX and ABI SOLiD can generate lots of SNPs when used for whole genome sequencing, RNA sequencing, methylated DNA sequencing and exome capture procedures. SNPs generated through these platforms can be between different varieties of plant species or between the same unigenes. Although these platforms are relatively expensive to utilize, but prices are gradually easing with increasing patronage. This method has been used to discover thousands of good quality SNPs in four pea recombinant inbreed lines [9]. Nevertheless, limitations exist with regards to accuracy, sensitivity and reproducibility of reads generated. A major concern using the NGS platform is the need to use very good assemblers to organize reads for SNP calling; examples of some assemblers include Genome Analysis Toolkit (GATK) [10], SOAPsnp [11, 12] and freebayes [13]. Different SNP callers have been compared in searching for a more versatile tool [14, 15, 16, 17, 18]. In a study for SNP discovery using RNA-sequence data, a combination of SNP callers Trinity-GATK gave 100% accuracy in peach and mandarin RNA-sequencing [19].

The versatility of SNPs has also led to their widespread use in phylogenetics to study the relatedness of organisms through the use of molecular sequencing data resulting in the identification and accurate classification of organisms. It has also been applied in phytogeography for determining the distribution of plant species. A major advantage of the single-base resolution of SNPs is that it allows better detection of 'perfect' markers, which are causally linked to agronomic traits. Another high-throughput method used for detecting SNPs is the GBS (genotyping by sequencing) which utilizes a range of techniques including those of reduced-representation sequencing and whole genome resequencing. Generally, this method identifies SNPs that are broadly distributed throughout the genome of organisms by fragmenting the genome using restriction enzymes, fragments are ligated to adapters and amplified. Amplified products are sequenced and aligned to a reference genome to call SNPs. GBS is gradually leading a transition from population genetics to population genomics, so that high-throughput marker recognition in plant population is affordable. A lot of commercial crops have been studied using GBS to aid breeding processes in Rice [20, 21, 22], Maize [23, 24, 25], potato [26, 27, 28]. Although GBS was initially developed as a reduced-representation sequencing (RRS) approach using restriction enzymes to decrease genome complexity before sequencing [29, 30, 31, 32]. Whole genome re-sequencing approaches was applied to allow higher genomic resolution. Since the creation of GBS, it has undergone continuous development, based on reduced-representation sequencing or whole genome resequencing methods. Combined with phenotypic data, GBS procedures provide a powerful basis for rapid mapping and identification of SNPs in genes underlying agronomic traits, which can then be utilized as efficient molecular markers for crop germplasm improvement.

Notwithstanding, the fact that a lot of next generation SNP genotyping techniques have been developed, they are all within the same bracket with regards to limitations of cost, complexity and accuracy. Important quantitative trait loci and SNPs associated with desirable agronomic traits have been employed to improve productivity of crops. Whole genome resequencing of *G.max* and *G. soja by Ramakrishna et al.* [33], *identified S*NPs and InDels and seven genes that hold a probable role in the determination of seed permeability. The expression differences of these genes at different stages of water imbibition wasanalyzed and two genes were identified that revealed preliminary, but a relevant association with soybean seed permeability trait. Genome-wide association study was performed by Do et al. [34] to map genomic regions for salt tolerance in a diverse panel of 305 soybean accessions using a SNP dataset derived from the SoySNP50K iSelect BeadChip [35]. The analysis revealed a major locus for salt tolerance on Chr. 3 confirmed by a number of significant SNPs, of which three gene-based SNP markers, Salt-20, Salt14056 and Salt11655, had the highest association with the studied trait.

In *Arachis* species Clevenger et al. [36] re-sequenced 20 genotypes and selected genome-wide SNPs to develop large-scale SNP genotyping array which will be very useful for further genetic and breeding applications in *Arachis*. In Maize, Unterseer et al. [37] developed a high-density maize SNP array comprising 616,201 variants (SNPs and indels) and used to design commercially available Affymetrix®

Axiom® Maize Genotyping Array. The array is composed of 609,442 SNPs and 6759 indels. Among these were 116,224 variants in coding regions and 45,655 SNPs of the Illumina® MaizeSNP50 BeadChip for study comparison. The Array although optimized for European and American climate is suitable for a broad range of applications because of the stringent quality filter criteria implemented. Cereals like *Zea mays* and *Oryza sativa* have extensively been studied for SNP diversity using diverse germplasm.

2.1 Single nucleotide polymorphism genotyping methods

Single nucleotide polymorphism is an individual nucleotide base difference between two DNA sequences. When SNPs occur within a gene, they may play a more direct role on the trait by affecting the gene's function and such SNPs can be exploited as molecular markers. Molecular markers enable precise identification of genotypes without the confounding effect of the environment [38], because selection is based on molecular determination and not the morphological expressions observed. A more informative marker gives a high polymorphic information content result [39]. SNP markers for chickpea and pigeon pea were evaluated and found to show 100% consistency and polymorphic information content values between 0.02 to 0.5 [22, 40]. SNP markers can be used for association studies, conservation genetics, germplasm screening or characterization, genetic diversity analysis and are fast becoming the preferred marker system in marker assisted breeding programs.

In the last 10 years, the rapid transformation in sequencing technology have enormously affected crop genotyping procedures. These new procedures enhanced rapid, high-throughput genotyping of whole crop population and gives opportunity to advance use of molecular tools in plant breeding. There is an urgent need in crop improvement programs to speed up crop production through marker assisted selection or introduce alleles that confers plants with resistance to pest and diseases, abiotic stress adaptation and high yield potential. Elite cultivars, store very useful genetic information that needs to be introgressed. Molecular marker approaches have been used in analyzing and identifying alleles associated with desirable agronomic traits in diverse germplasm pool of legumes and cereals [41].

Some SNPs genotyping methods that are easy to use and accurate and can specifically genotype SNP markers at specific loci for a collection of plant population are presented below.

2.1.1 Tetra ARMS allele specific PCR

Tetra ARMS (tetra-primer amplification refractory mutation system) allele specific PCR is a versatile, rapid and economical SNP detection tool. Other contemporary SNP genotyping tools include allele specific PCR, high resolution melting analysis, PCR single stranded conformation polymorphism, PCR-primer introduced restriction analysis and real-time PCR-based genotyping. It involves a single PCR step followed by gel electrophoresis. Tetra ARMS allele specific PCR utilizes four primers including outer forward, outer reserve, inner forward and inner reverse primers. The outer forward or outer reverse primer combination generates the outer fragment of the SNP locus and acts as an internal control for the PCR. The inner forward or outer reverse and outer forward or inner reverse primer combination yield allele-specific amplicons depending on the genotype of the sample used. The placing of the inner primers is not the same as those of the corresponding outer primer to produce amplicons with different sizes and easily visualized on gel and distinction is made accordingly [42, 43, 44].

A study by Ehnert et al. [45] using tetra ARMS allele specific PCR method described three common single nucleotide polymorphisms in the *PADI4* gene involved in diverse post-translational modifications of proteins in eukaryotes. The SNPs which are thought to affect PAD4 expression activity are rs874881, rs11203366 and rs11203367. Hypercitrullination as a result of increased PAD expression or activity, is associated with autoimmune diseases like rheumatoid arthritis, lupus, Alzheimer's disease, ulcerative colitis, multiple sclerosis, and certain cancers. SNP markers identified by genotyping-by-sequencing were used to distinguish four varieties of sweet potato through tetra-primer ARMS-PCR method. It was shown that three variety-specific fragments (164 bp and 241 bp of SNP 04-27457768 and 292 bp of SNP 03-16195623) were amplified in the 'Beniharuka', 'Pungwonmi', and 'Annobeni' sweet potato varieties [43]. Some SNP markers developed by Angiolillo et al. [46] was used to resolve the issue of nomenclature for 65 olive samples as the markers were able to discriminate 77% of the olive cultivars. SNP markers developed in this study were used to assess the genetic variability and diversity of widely cultivated olive cultivars important for oil production. To show the reliability of the tetra-primer ARMS-PCR technique and its potential for use in low- to moderate-throughput situations Chiapparino et al. [47] unambiguously assayed five SNPs in a set of 132 varieties of cultivated barley.

With this technique, they is almost always a need for trouble shooting to standardize the procedure especially at initial steps of the protocol to adapt it to the genotype investigated. This really reduced its wide spread application, despite the fact that it is economical and precise in SNP genotyping. In other to improve the ARMS-PCR procedure, several modifications have been suggested to optimize its usage. Two improvements were suggested by Tanha et al. [42]; one is to equalize outer primer and inner primer strength by adding a mismatch at 2 positions of outer primers and the second is to equal annealing temperature which should be a little higher than melting temperature. This resulted in the improvement of expected result and specificity. Another study by Alyethodi et al. [44] suggest that the use of Strand displacement polymerase rather than conventional Taq polymerase resulted in the generation of amplicons by 25 cycles in the PCR reaction while Taq polymerase needed a minimum of 35 cycles. Also, reaction with Strand displacement polymerase did not require PCR enhancers like dimethyl sulfoxide, thus it was time saving and efficient.

2.1.2 KASP assay for SNPs genotyping

Another robust and easy to use SNP genotyping method is Kompetitive allele-specific PCR (KASP) genotyping assay based on competitive allele specific polymerase chain reaction, developed by LGC genomics (www.lgcgroup.com). It is widely applied in plant breeding because of its reduced cost in genotyping large number of samples. It allows for biallelic scoring of SNPs and insertion/deletions at specific loci and can be conveniently used for small number of SNPs. Here, the SNP-specific KASP assay mix and the universal KASP master mix are added to the DNA samples, followed by thermal cycling. The bi-allelic discrimination is carried out by competitive binding of two allele-specific forward primers. Each of the primers are labeled with fluorescence resonant energy transfer cassettes a FAM dye and an HEX dye [48, 49].

This method is a PCR-based homogenous fluorescent SNP genotyping set up which is cost-effective to run and more reliable than other SNP genotyping techniques. Since the introduction of KASP, it has been developed and used to genotype rice, wheat, soybean, cumber, chickpeas and many other crops. It has been employed in the enhancement and production of efficient markers in Chinese cabbage. The authors re-sequenced 4 Chinese cabbage and carried out SNP survey in the genome. They established KASP-SNP resource and converted 258 SNP variations into KASP molecular markers. These molecular markers discovered in Chinese cabbage will be invaluable for germplasm identification and cabbage research around the world [50]. Also, Khanal et al. [51] reported flanking sequences of 162 putative SNPs, none of them have been previously evaluated to determine whether they performed as intended. Therefore, a subset of 31 putative SNPs that represent the entire nematode genome were designed to form a residual emission fluorescence KASP.

With KASP primers, biallelic scoring of SNPs at specific loci is possible. Cotton (*Gossypium hirsutum*) leaf was mapped using the underlying gene of okra leaf using KASP assay. The sequences of okra leaf gene, GhOKRA, has been link to other plant species and is involved in regulating leaf morphology in plants. SNP markers located on the okra gene (GhOKRA) was successfully establish to be the best candidate gene responsible for okra leaf trait in upland cotton [52]. SNP marker analysis can be used for genetic diversity analysis, create genetic maps, and marker assisted selection of crops. To use these technologies, one must first identify and validate putative SNPs. Many technologies exist for SNP genotyping but KASP performs well when it comes to adaptability, efficiency and cost-effectiveness. It is efficient in determining the alleles at specific locus within genomic DNA [53].

For crop improvement purposes, the maize breeders at International Maize and Wheat Improvement Center developed 16 marker-assisted recurrent select (MARS) populations. The parents of these MARS populations were initially genotyped along with over 450 maize inbred and advanced breeding lines using the GoldenGate assay [54].

2.2 Application of SNP in crop improvement

SNP markers are used for the improvement of crops in a number of ways which include to select disease resistant crops, high yielding varieties, plants that can withstand biotic and abiotic stress and many more. The development of SNP markers has become a regular process, especially for crops with reference genome and this new order has influenced the application of SNP markers in plant breeding. For more than two decades, researchers experienced so much throwback in a bid to develop markers linked to discovered QTLs. However, the revolution in sequencing technology, brought about easy identification of SNP markers underlying genes in a QTL. SNP markers were used to characterize natural variation of sorghum grain nutrients composition in a global sorghum panel and genome wide association study was use to map QTL responsible for this variation. It was discovered by Rhodes et al. [55] that protein, fat and starch all had strong correlation across years, but protein was the most significant. Also, protein had the highest narrow sense-heritability. Further investigation showed that there is a strong negative correlation between starch and protein and fat, and strong positive correlation between protein and fat.

SNP markers have been frequently used in marker assisted selection due to its abundance in the genes of all species. In breeding for resistance to root-knot nematode (*Melodogyne incognita*) that infects soybean by Dubiela et al. [56], SNP markers were identified for *M. incognita* in soybean using a microarray panel. The markers were used to identify susceptible soybean which was confirmed by phenotypical evaluation of plants on the field. This marker assisted selection protocol helped in the identification of resistant varieties. Also, Potato blight is considered as one of the most destructive disease of potato, which is caused by a fungus, *Phytophthora infestans*. Globally, there are 2 dominant mating type strains: type A1 and A2. SNP markers were used to determine potato late blight of susceptible and resistant genotypes. Using genome wide association studies, Nay et al. [57] identified SNP markers that co-segregated with resistance loci for Angular leaf spot (ALS) in common bean and haplotypes. The discovered markers will increase breeding efficiency for ALS resistance and allow researchers to react faster to future changes in pathogen pressure and composition.

The allelic, high number of loci that can be multiplexed and possibility of automation of SNP markers makes them very useful for cultivar identification. Grapevine cultivar identification was carried out using over 300 SNPs in its genome. Re-sequencing method was used in the selection of 11 genotypes, 48 SNPs spread across all grapevine chromosome providing enough information content for genetic cultivar identification [58].

The quality of a crop is highly dependent on the number of micronutrients it contains. Genome wide study was used to identify SNPs associated with micronutrient (Fe, Se and Zinc) concentration in pea (*Pisum sativum*). The SNP was very useful in identification of seed mineral concentration in pea and the loci were mapped. For Fe concentration, 5 SNPs were identified and each marker was distinct phenotypically. Markers identified for Zn was 5 chromosomal and 3 non-chromosomal SNP markers while eight was identified for Se. It was also stated by Dissanayaka et al. [59] that genome wide association studies can be effectively used for reliable marker assisted selection scheme.

3. CONCISE SUMMARY

Germplasm screening is usually a first step in plant breeding programs. It aims to reduce the large collection of plants and narrow down on those that fit the breeding objectives in view. It is usually laborious and time consuming, but the use of molecular markers can substantially aid this process. Several molecular markers including simple sequence repeats (SSRs) have previously been utilized in plant germplasm screening but SNPs markers enable selection based on target genes that code for specific trait of interest.

Advances in sequencing technologies has given rise to SNP markers and now the markers of choice in genetic studies because they are robust, widely distributed throughout the genome of plants and highly multiplexable. SNPs represent difference in a single nucleotide in the genome and those that are linked to a phenotype as a result of nonsynonymous amino acid changes can be reliably used as molecular markers.

A number of techniques and methods have been applied to discover new SNPs including non-gel methods where SNPs are mined from multiple sequence alignment in databases. SNPs have also been generated using next generation sequencing platforms like Illumina genome analyzer and Roch/454 FLX. Discovered SNPs are developed into user-friendly SNPs markers and used for genotyping. A number of SNP markers have been validated for marker assisted selection. In a study by Burow et al. [60], SNP markers were developed from sequences of brown midrib (bmr) trait of sorghum and used to accurately identify bmr6 or bmr12 individuals at the seedling stage. This validation was for a group of sorghum germplasm and a genetic population. Also, fifteen KASP SNP markers for bmr6 and bmr12 were developed and used for allele discrimination to select bmr individuals. Another study by Khanal et al. [61] employed KASP SNPs to determine the genetic variability present in 26 isolates of *Rotylenchulusreniformis* a plant parasitic nematode of cotton and soybean, this will be of benefit in resistant breeding programs. Also, Udoh et al. [62] developed KASP SNP markers for phytoene synthase2 gene associated with carotenoids in cassava. The validation SNP makers explained most of the proportion of phenotypic variation for carotenoids in a genetic gain cassava population.

In Panax species, Nguyen et al. [30] identified 1128 SNPs in coding gene sequences and developed 18 SNP markers from the chloroplast genic coding sequence region that can be used to distinguish all the seven Panax species from each other. Because SNPs markers are based on target gene, they are a highly reliable tool in identification of cultivars and germplasm screening.

AT A GLANCE

Single nucleotide polymorphism (SNP) represents a change in a single nucleotide within the genome. This can alter the phenotype of an individual within the same species if it occurs in a coding region of the gene. The change in nucleotide can produce desirable characteristic in plants and can become an object for selection. New SNPs have been discovered and subsequently converted to molecular markers using various non-gel based and next generation sequencing platforms. Considering that SNP markers are based on target genes, its abundance in the genome, high automation and multiplexability, has made it a marker of choice and an effective tool for screening plant germplasm for desirable traits. This chapter considers SNP as molecular marker, their discovery and different SNP genotyping methods was documented. A few case studies of SNP as allele specific markers and their association with traits of interest was considered. Thus, highlighting their efficacy as useful tool for marker assisted selection and plant germplasms screening.

APPENDICES AND NOMENCLATURE

AFLP	Amplified fragment length polymorphism
ARM	Amplification refractory mutation system
GBS	Genotyping-by-sequencing
KASP	Competitive allele-specific polymerase chain reaction assay
MAS	Marker assisted selection
NGS	Next generation sequencing
PCR	Polymerase chain reaction
SNPs	Single nucleotide polymorphism
SSR	Simple sequence repeats
QTL	Quantitative trait loci

REFERENCES

1. Slater AT, Schultz L, Lombardi M, et al. Screening for Resistance to PVY in Australian Potato Germplasm. 2020; 1-15
2. Bhawar PC, Tiwari S, Tripathi MK, et al. Screening of Groundnut Germplasm for Foliar Fungal Diseases and Population Structure Analysis Using Gene Based SSR Markers. 2020; 39: 75-84
3. Zhang K, Wu Z, Tang D, et al. Development and Identification of SSR Markers Associated with Starch Properties and β-Carotene Content in the Storage Root of Sweet Potato (Ipomoea batatas L.). Front Plant Sci 2016; 7: 1-21
4. Gray IC, Campbell DA, Spurr NK. Single nucleotide polymorphisms as tools in human genetics. Hum Mol Genet 2000; 9: 2403-2408
5. Zhou L, Vega FE, Tan H, et al. Developing Single Nucleotide Polymorphism (SNP) Markers for the Identification of Coffee Germplasm. Trop Plant Biol 2016; 9: 82-95
6. Chen X, Sullivan PF. Single nucleotide polymorphism genotyping: biochemistry, protocol, cost and throughput. Pharmacogenomics J 2003; 3: 77-96
7. Udoh LI, Adesoye A, Gedil M. Identification and Molecular Analysis of Pro - vitamin A Carotenoid Genes in Cassava (Manihot esculenta Crantz). 2017; 8: 38-44
8. Harjes CE, Rocheford TR, Bai L, et al. Natural genetic variation in lycopene epsilon cyclase tapped for maize biofortification. Science 2008; 319: 330-3
9. Boutet G, Carvalho SA, Falque M, et al. SNP discovery and genetic mapping using genotyping by sequencing of whole genome genomic DNA from a pea RIL population. BMC Genomics. Epub ahead of print 2016. DOI: 10.1186/s12864-016-2447-2
10. Poplin R, Ruano-Rubio V, DePristo M, et al. Scaling accurate genetic variant discovery to tens of thousands of samples. bioRxiv 2017; 201178
11. Li R, Li Y, Kristiansen K, et al. SOAP: Short oligonucleotide alignment program. Bioinformatics 2008; 24: 713-714
12. Li R, Li Y, Fang X, et al. SNP detection for massively parallel whole-genome resequencing. Genome Res 2009; 19: 1124-1132
13. Garrison E, Marth G. Haplotype-based variant detection from short-read sequencing
14. Sandmann S, De Graaf AO, Karimi M, et al. Evaluating Variant Calling Tools for Non-Matched Next-Generation Sequencing Data. Sci Rep 2017; 7: 1-12

15. Cornish A, Guda C. A Comparison of Variant Calling Pipelines Using Genome in a Bottle as a Reference. Biomed Res Int; 2015. Epub ahead of print 2015. DOI: 10.1155/2015/456479
16. Liu F, Zhang Y, Zhang L, et al. Systematic comparative analysis of single-nucleotide variant detection methods from single-cell RNA sequencing data. Genome Biol; 20. Epub ahead of print 19 November 2019. DOI: 10.1186/s13059-019-1863-4
17. Ni G, Strom TM, Pausch H, et al. Comparison among three variant callers and assessment of the accuracy of imputation from SNP array data to whole-genome sequence level in chicken. BMC Genomics; 16. Epub ahead of print 21 October 2015. DOI: 10.1186/s12864-015-2059-2
18. Bian X, Zhu B, Wang M, et al. Comparing the performance of selected variant callers using synthetic data and genome segmentation. BMC Bioinformatics 2018; 19: 429
19. Zhao Y, Wang K, Wang W, et al. A high-throughput SNP discovery strategy for RNA-seq data. 2019; 1-10
20. Yadav S, Sandhu N, Singh VK, et al. Genotyping-by-sequencing based QTL mapping for rice grain yield under reproductive stage drought stress tolerance. 2019; 1-12
21. Ali J, Aslam UM, Tariq R, et al. Exploiting the Genomic Diversity of Rice (Oryza sativa L.): SNP-Typing in Introgression-Breeding Populations. 2018; 9: 1-10
22. Hiremath PJ, Kumar A, Penmetsa RV, et al. Large-scale development of cost-effective SNP marker assays for diversity assessment and genetic mapping in chickpea and comparative mapping in legumes. Plant Biotechnol J 2012; 10: 716-32
23. Wang N, Yuan Y, Wang H, et al. Applications of genotyping - by - sequencing (GBS) in maize genetics and breeding. Sci Rep 2020; 1-12
24. Su C, Wang W, Gong S, et al. High Density Linkage Map Construction and Mapping of Yield Trait QTLs in Maize (Zea mays) Using the Genotyping-by-Sequencing (GBS) Technology Genome Wide Identification of SNPs Using. 2017; 8: 1-14
25. Paulsmeyer MN, Brown PJ, Juvik JA. Discovery of Anthocyanin Acyltransferase1 (AAT1) in Maize Using Genotyping-by-Sequencing (GBS). 2018; 8: 3669-3678
26. Technology FSS, Yu X, Zhang M, et al. An SNP-Based High-Density Genetic Linkage Map for Tetraploid Potato Using Specific Length Amplified
27. SALIHU BZ, FALUSI OA, ADEPOJU AO, et al. Assessment of Genetic Diversity of Promising Castor Been (*Ricinus communis* L.) Genotypes in Nigeria. Not Sci Biol 2019; 11: 467-474

28. Wadl PA, Olukolu BA, Branham SE, et al. Genetic Diversity and Population Structure of the USDA Sweetpotato (Ipomoea batatas) Germplasm Collections Using GBSpoly. 2018; 9: 1-13
29. Elshire RJ, Glaubitz JC, Sun Q, et al. A Robust , Simple Genotyping-by-Sequencing (GBS) Approach for High Diversity Species. 2011; 6: 1-10
30. Nguyen VB, Linh Giang VN, Waminal NE, et al. Comprehensive comparative analysis of chloroplast genomes from seven Panax species and development of an authentication system based on species-unique single nucleotide polymorphism markers. J Ginseng Res 2020; 44: 135-144
31. Shaibu AS, Sneller C, Motagi BN, et al. Genome-Wide Detection of SNP Markers Associated with Four Physiological Traits in Groundnut (Arachis hypogaea L.) Mini Core Collection. Agronomy 2020; 10: 192
32. Meger J, Ulaszewski B, Vendramin GG, et al. Using reduced representation libraries sequencing methods to identify cp DNA polymorphisms in European beech (Fagus sylvatica L)
33. Ramakrishna G, Kaur P, Nigam D, et al. Genome-wide identification and characterization of InDels and SNPs in Glycine max and Glycine soja for contrasting seed permeability traits. 2018; 1-15
34. Do TD, Vuong TD, Dunn D, et al. Identification of new loci for salt tolerance in soybean by high-resolution genome- wide association mapping. 2019; 1-16
35. Song Q, Hyten DL, Jia G, et al. Development and Evaluation of SoySNP50K , a High- Density Genotyping Array for Soybean. 2013; 8: 1-12
36. Clevenger J, Chu Y, Chavarro C, et al. Genome-wide SNP Genotyping Resolves Signatures of Selection and Tetrasomic Recombination in Peanut. Mol Plant 2017; 10: 309-322
37. Unterseer S, Bauer E, Haberer G, et al. A powerful tool for genome analysis in maize : development and evaluation of the high density 600 k SNP genotyping array. 2014; 1-15
38. Meuwissen THE, Hayes BJ, Goddard ME. Prediction of Total Genetic Value Using Genome-Wide Dense Marker Maps
39. Species R, Lebedev VG, Subbotina NM, et al. Transferability and Polymorphism of SSR Markers Located in Flavonoid Pathway Genes in Fragaria and
40. Saxena RK, Penmetsa RV, Upadhyaya HD, et al. Large-scale development of cost-effective single-nucleotide polymorphism marker assays for genetic mapping in pigeonpea and comparative mapping in legumes. DNA Res 2012; 19: 449-461
41. Dwivedi SL, Scheben A, Edwards D, et al. Assessing and Exploiting Functional Diversity in Germplasm Pools to Enhance Abiotic Stress Adaptation and Yield in Cereals and Food Legumes. 2017; 8: 1-30

42. Tanha HM, Naeini MM, Rahgozar S, et al. Modified Tetra-Primer ARMS PCR as a Single-Nucleotide Polymorphism Genotyping Tool. Epub ahead of print 2015. DOI: 10.1089/gtmb.2014.0289
43. Park H, Kim S, Nie H, et al. Molecular identification of sweet potato accessions using ARMS-PCR based on SNPs. 2020; 2818: 124-130
44. Alyethodi RR, Singh U, Kumar S, et al. RESEARCH NOTE T - ARMS PCR genotyping of SNP rs445709131 using thermostable strand displacement polymerase. BMC Res Notes 2018; 1-5
45. Ehnert S, Linnemann C, Braun B, et al. One-Step ARMS-PCR for the Detection of SNPs — Using the Example of the PADI4 Gene
46. Angiolillo A, Lucentini L, Pilla F, et al. SNP-based markers for discriminating olive (Olea europaea L.) cultivars. Epub ahead of print 2006. DOI: 10.1139/g06-068
47. Chiapparino E, Lee D, Donini P. NOTE / NOTE Genotyping single nucleotide polymorphisms in barley by tetra-primer ARMS – PCR. Epub ahead of print 2014. DOI: 10.1139/g03-130
48. Neelam K, Brown-Guedira G, Huang L. Development and validation of a breeder-friendly KASPar marker for wheat leaf rust resistance locus Lr21. Mol Breed 2013; 31: 233-237
49. Allen AM, Barker GLA, Berry ST, et al. Transcript-specific, single-nucleotide polymorphism discovery and linkage analysis in hexaploid bread wheat (Triticum aestivum L.). Plant Biotechnol J 2011; 9: 1086-1099
50. Yang S, Yu W, Wei X, et al. An extended KASP-SNP resource for molecular breeding in Chinese cabbage (Brassica rapa L. ssp. pekinensis). PLoS One 2020; 15: 1-17
51. Khanal C, Kularathna MT, Ray JD, et al. Single Nucleotide Polymorphism Analysis Using KASP Assay Reveals Genetic Variability in Rotylenchulusreniformis. 2019; 1835-1842
52. Zhu QH, Zhang J, Liu D, et al. Integrated mapping and characterization of the gene underlying the okra leaf trait in Gossypium hirsutum L. J Exp Bot 2016; 67: 763-774
53. Graves H, Rayburn AL, Gonzalez-Hernandez JL, et al. Validating DNA Polymorphisms Using KASP Assay in Prairie Cordgrass (Spartina pectinata Link) Populations in the U.S. Front Plant Sci; 6. Epub ahead of print 2016. DOI: 10.3389/fpls.2015.01271
54. Semagn K, Babu R, Hearne S, et al. Single nucleotide polymorphism genotyping using Kompetitive Allele Specific PCR (KASP): Overview of the technology and its application in crop improvement. Molecular Breeding 2014; 33: 1-14

55. Rhodes DH, Hoffmann L, Rooney WL, et al. Genetic architecture of kernel composition in global sorghum germplasm. BMC Genomics 2017; 18: 1-8

56. Dubiela C, Montecelli TDN, Lazzari F, et al. Development and validation of SNP assays for the selection of resistance to Meloidogyne incognita in soybean. Crop Breed Appl Biotechnol 2019; 19: 102-109

57. Nay MM, Mukankusi CM, Studer B, et al. Haplotypes at the Phg-2 Locus Are Determining Pathotype-Specificity of Angular Leaf Spot Resistance in Common Bean. Front Plant Sci 2019; 10: 1-11

58. Cabezas JA, Ibáñez J, Lijavetzky D, et al. A 48 SNP set for grapevine cultivar identification. BMC Plant Biol; 11. Epub ahead of print 2011. DOI: 10.1186/1471-2229-11-153

59. Dissanayaka DN, Gali KK, Jha AB, et al. Genome-wide association study to identify single nucleotide polymorphisms associated with Fe, Zn, and Se concentration in field pea. Crop Sci 2020; 60: 2070-2084

60. Burow G, Chopra R, Sattler S, et al. Deployment of SNP (CAPS and KASP) markers for allelic discrimination and easy access to functional variants for brown midrib genes bmr6 and bmr12 in Sorghum bicolor. Mol Breed 2019; 39: 115

61. Khanal C, Kularathna MT, Ray JD, et al. Single Nucleotide Polymorphism Analysis Using KASP Assay Reveals Genetic Variability in Rotylenchulusreniformis. Plant Dis 2019; 103: 1835-1842

62. Udoh LI, Gedil M, Parkes EY, et al. Candidate gene sequencing and validation of SNP markers linked to carotenoid content in cassava (Manihot esculenta Crantz). Mol Breed; 37. Epub ahead of print 2017. DOI: 10.1007/s11032-017-0718-5

CHAPTER-13

ENHANCING FIBER QUALITY IN UPLAND COTTONS THROUGH ASSOCIATION MAPPING

1. INTRODUCTION: AN OVERVIEW

In the pursuit of advancing agricultural practices and improving the quality of cotton fibers, researchers are increasingly turning to innovative approaches such as association mapping. Upland cotton, a predominant and economically significant cotton variety, serves as a crucial raw material for the textile industry. Recognizing the paramount importance of fiber quality in meeting industry standards and consumer expectations, this study delves into the application of association mapping as a powerful tool for unraveling the genetic factors influencing fiber traits in Upland cotton varieties. By exploring the intricate genetic architecture underlying fiber quality, this research aims to pave the way for targeted breeding programs and the development of high-performing cotton varieties that meet the evolving demands of the textile market.

Cotton is a crop of immense importance as being a dominant source of fiber and oil from cottonseed all over the world [1]. The improvement of cotton fiber quality has become more important because of changes in spinning technology and ever-increasing demands of fiber. Cotton is grown in more than 80 countries, and contributes to the world economy as a raw material for textile industry [2].

Gossypium" genus is made up of about 52 species of which 47 are diploid and 7 are as allotetraploids [3, 4, 5, 6, 7]. Of all the species of the genus, two most common diploids are *G. arboreum* L., *G. herbaceum* L., while *G. hirsutum* L., and *G. barbadense* L. are considered as the most commercially valuable tetraploids. *G. hirsutum*, is characterized by high yield, moderate fiber quality and wide adaptability contributes for 95% of overall cotton production [8]; while *G. barbadense* (Pima, and Egyptian) increases superior fiber quality [9, 10].

Efforts for broadening the genetic base of *Gossypium* genus have not generated successful outcomes due to the complex and large genetic architecture of its genome. Moreover, owing to its developmental barriers, genetic studies have not yet been able to produce the required traits in cotton [11]. Association among markers and characters can be used for fastening the breeding program. The hereditary variation present among the gene pool land races can be exploited by applying the mapping based on linkage disequilibrium. It will speed up the cotton breeding through identification of markers among trait of interest and ensure molecular breeding. Single reproducibility of genetic marker which govern a specific appearance on sequence of nucleotides can be analyzed with genome wide association [12, 13]. Association mapping relies upon the magnitude of different pair of genes for population analysis. Moreover, this mapping shows powerful connection between required character and a genetic marker while nonrandom combination between two quantitative trait loci or markers manifests linkage disequilibrium [8]. The valuable information about the origin of an individual is determined with the degree and the size of the population [13, 14]. Many loci relating to polygenic characters have been determined via genetic maps and linkage disequilibrium (LD) was measured in humans through diverse analysis methods [15, 16]. Population based polygenic characters mapping for desired traits became a widely used technique thanks to the innovations in omics and availability of advanced bioinformatic tools for analyzing genetic variations [17]. The ultimate benefits of this technique includes the ability to work with a large number of loci, producibilty of highly saturated maps, its speed and its low cost [18].

2. FIBER QUALITY

Single cell elongation of ovule in cottonseed outer layer forms a natural fiber known as "trichome" which contains about 89–100% cellulose. [19, 20, 21, 22]. As little as, 30% of lint primordia have the ability to be differentiated as mature fibers forming about 20,000 of it within a single ovule [23, 24]. The ideal cotton fiber should be white like frozen vapor, durable like iron, attractive like silk and stretched as a wool [25]. Nonetheless it is hard to include all these qualities within a breeding program for cotton production, but efforts have been made to obtain the most desired ones. Fiber quality is an array of quantitative traits (length, fineness, strength, uniformity and elongation) that enhance yarn value during spinning [26, 27, 28]. Fiber quality is a difficult association of physiology and genetic make-up of plant within a growing season of cotton [29, 30].

Fiber quality enhancement through genetics is the ultimate objective of breeding strategy in cotton. Cotton scientists have been involved in fiber quality improvement for a long time due to the increase in demand for multiple products from cotton. The critical goals of all cotton related techniques are fiber yield and quality, and the precise parameters which contribute its economic value on global level. Spinning automation renders fiber improvement according to interests of textile sector, as a result fiber quality measurements for breeders are considered. As an instance, prevailing spinning automation highly signify strength instead of fiber length and fineness [31]. Moreover, fiber quality improvement is a demanding task as it is determined after harvesting of crop.

The main goal of all genetic improvement is to increase yield. The intensity of improvement for lint production has deteriorated since the 1980s [32, 33, 34]. Nonetheless, genetic diversity has increased at the start of 21st century [35, 36].

3. MARKER ASSISTED SELECTION

Due to the inverse relationship between seedcotton yield and fiber quality, and the complicated involvement of multiple genes in traits demand breeders to evolve varieties through more useful methods. In the past textile industry flourished principally via selection of new recombinants among germplasm entries with traditional breeding approaches [37, 38]. Elite grown cotton genotypes have narrow genetic base, therefore it has been thought that germplasm should be used for improvement of traits. Some of popular characters such as disease and insect resistance have been enhanced by introgression [39]. The advent of DNA markers paved the way for plant breeders to fasten breeding process through fast, authentic and substitutive techniques instead of the traditional methods for the selection to develop both agronomic and economic characters of plants [40].

Molecular marker is a specific DNA portion with a known position on the chromosome [41], or a gene whose phenotypic expression is frequently easily distinguished and used to detect an individual [42, 43]. DNA markers are having the property of polymorphism which can be used for the differentiation of homozygotes and heterozygotes [44]. Marker assisted selection has a great amount of advantages over conventional breeding, reviewed by many researchers [45, 46, 47]. Plant breeders utilize DNA markers for selection of desirable traits on molecular basis in spite of observing them phenotypically [48], furnishing the basis for using the molecular assisted selection [49, 50, 51]. Molecular markers are desired for improving traits in many essential crops; rice [52], wheat [53], maize [54, 55] and barley [56, 57]. Cotton is an important cash crop at global level and marker assisted selection has not got desired goals because of compatibility barriers through historic domestication and insufficient polymorphism [58, 59, 60].

Molecular characterization is the way to transfer required traits into modern genotypes [45, 61, 62, 63, 64]. Quantitative trait loci (QTLs) allow gene pyramiding for yield and fiber quality through evolution of linkage maps. Association mapping using linkage disequilibrium on genome wide level is the most valuable strategy among scientists for searching QTLs in crop sciences. The association among trait of interest and germplasm entries is observed using population construction information and linkage disequilibrium (LD) with association mapping [65]. LD mapping is highly popular thanks to the sophistication of mathematical methods and accessibility of large number of DNA markers.

The traits controlled by multiple genes such as fiber quality can be studied more precisely with linkage maps after the availability of new genomic data of *Gossypium* spp. like *Gossypium raimondii* Ulbrich [66, 67], *Gossypium arboreum* L. [68] and *Gossypium hirsutum* L. [69, 70]. [71] revealed that tetraploid species derived from crossing of two diploid species *Gossypium arboreum* L. (A genome) and *Gossypium raimondii* Ulbrich (D genome) about 1–2 million years ago. Moreover, it may pave the way for fiber improvements as higher number of QTLs assigned to the Dt sub-genome compared to At sub-genome in hawian cotton [72, 73, 74].

Many researchers have observed QTLs for seedcotton yield and its components [9, 70, 75, 76, 77, 78, 79]. But, mostly filial generations were used for QTLs. Quantitative trait loci are highly effected by low heritability and more experimental error which are high in such plant materials, hence it is need of the day that a useful way should be employed for the development of stable populations for overcoming these obstacles. The accuracy of QTL determination relies upon allelic frequency among QTL of the desired character and related marker [80]. Molecular breeding methods designed with the information obtained through quantitative trait loci analysis in association mapping creates valuable genetic variation from stable populations [81].

4. ASSOCIATION MAPPING OF FIBER TRAITS USING GENOTYPING BY SEQUENCING (GBS)

Molecular markers are highly favored for linkage map development because they are polymorphic, easily transferred to next generation with Mendelian ratio and do not show epistasis. Molecular breeding with highly saturated maps having QTLs connected with economic traits through impactful genetic markers provides a good source for cotton improvement [64]. Genomic analysis in many crop species including cotton has been done using populations derived from hybridization of only two ancestors; which is major drawback for omics information. Therefore, there has been hindrance in applying QTL information gained from such populations to accomplishing breeding objectives, as, in these populations, the genetic aspects are the same owing to the share of genetically similar backgrounds.

The foundation of association mapping is on hypothesis about occurrence of markers as a panel in which the alleles are found almost adjacent to the required traits with co-segregation and thought to be in linkage disequilibrium. Germplasm entries are used for determining QTLs of interest using genome wide association mapping [82]. There are many agents including type of copulation, gene flow frequency and population structure can affect such mapping approach [18]. Association mapping allows to overcome drawbacks found in bi-parental mapping from traditional methods which include using populations which are found as well-established genotypes, detects only the required gene and identify high polymorphism [83, 84, 85]. This methodology also urges to use knowledge based on linkage disequilibrium instead of linkage mapping.

Marker assisted breeding involves recent approaches of genomics combined with traditional breeding procedures for improving traits in crop sciences. For this reproducibility is essential among genetic markers. Morphological characters grading and genotyping with molecular markers is accomplished [86]. Molecular markers are very effective for identifying and overcoming problems for transfer of traits from other species such as segregation distortion [87]. Genetic markers are effective for determining genetic variation in Gossypium gene pool. [88] classified DNA markers into groups: 1) non-hybridization based; which include Amplified Fragment Length Polymorphism (AFLP), Simple Sequence Repeats (SSR), Sequence Repeat Amplified polymorphism (SRAP), İnter-Simple Sequence Repeats (ISSR), Expressed Sequence Tag (EST-SSR), Single Nucleotide Polymorphism (SNPs) etc. Numerous linkage maps have been developed in allotetraploid cotton employing diverse mapping populations and different DNA markers techniques [76, 89, 90, 91, 92, 93, 94]. Numerous SSRs and SNPs have been evolved in cotton [95, 96, 97, 98, 99]. Saturated genetic maps development through loci information of SSR and SNPs in cotton paves the way for ascertaining quantitative traits related to breeder objectives [100, 101, 102, 103, 104] Nonetheless, association analysis and very fine mapping is not possible owing to less information from these maps. It is need of the day that highly saturated mapping should be devised in cotton for overcoming the sequencing drawbacks and fastening the variety development.

Availability of microsatellites (SSR) and single-nucleotide polymorphisms (SNPs) have fastened genome mapping owing to their wider applicability in diverse populations derived from discrete genetic backgrounds [93, 95, 99, 105, 106, 107]. Thanks to advances in genotyping and SNPs calling tools; broadening of genetic base is being explored excessively in plants owing to availability of valuable loci information [108, 109, 110, 111, 112, 113, 114].

Single nucleotide polymorphisms are distinct points of nucleotides on chromosomes between two genotypes differentiated by a single base [64]. [115] speculated that each SNP is found after 100-300 bp in any genome while revealed that such genetic markers are highest in occurrence than any other marker and manifest higher degree compared to microsatellites. SNPs can be formed rapidly with economical cost owing to availability of high-throughput tools for genotyping [116]. Assessment of gene expression [117, 118], genome wide association [68, 119] and SNPs detection has been carried among the individuals having different sizes of genomes and also polyploid species having limited genetic variation like cotton [10, 120] and wheat [121] through low-cost high-throughput genotyping tools. SNPs have been explored and genotyped among different species via diverse ways [10, 120, 121, 122].

Genotyping-by-sequencing (GBS) is powerful and easy approach which paves the way for the discovery of numerous SNPs concurrently among large number of genotypes [123]. Restriction enzymes with methyl sensitivity are used to mark the flanking restriction sites in the genome for the development of reduced representation of the genome via GBS [121, 122]. GBS method is much easier, requires lower amount of DNA and library preparation is achieved in just two steps on plates, circumvents DNA fragment analysis preceded by PCR amplification of pooled library in contrast to reduced representation libraries (RRL) and restriction site associated DNA (RAD) [122]. The discovery and verification of reproducibility is not required in this procedure and can be applied in any species having polymorphism or mapping population with diverse size [124]. A number of SNPs has been discovered in many species using GBS like maize [122], wheat, barley [121], sorghum [125], rice [126], soybean [127], oat [128] and cotton [10, 79, 129, 130].

Association mapping furnishes saturated map of desired trait in contrast to pair of genes harboring a required character [131]. Therefore, verification of QTLs is compulsory for mapping. Association mapping is the way to examine genetic variation of required characters; integrates the variation of the desired characters through reproducibility of the alleles and genetic markers are selected connected to economic traits using linkage disequilibrium extent [132]. Moreover, LD elaborates the ancestral pattern through information among populations and ecology [133, 134].

LD based association mapping has been applied by using different strategies for determining genetic diversity contributing source pattern and design of population [135, 136]. Grouping of population individuals with combined genetic distance among the entries established via LD [137, 138, 139]. LD extent among natural population is not contributed by linked loci but non-homologous chromosomes are also involved, accountable to selection, behavior of population and hybridization. Owing to which immense care should be considered for analyzing such relations. Reproducibility in a sequence controlling a specific character is the property of this mapping [140]. Moreover, considerable concern is prevailed among association studies and linkage mapping relating to depth and precision of QTLs, the magnitude of knowledge and evaluating procedures [132].

In spite of the fact, statistical analysis is not appropriate with LD derived tools. Natural population partitioned into distinct categories with model-based procedures [141]. Bayesian modeling is used widely for assessing the probability of a genotype related to a specific population category through allele repetition. With this technique the genotypes are allotted to particular population which can be interspersed into statistical methods for association mapping with population organization. The population framework is analyzed by using STRUCTURE software [135] which has been used for association studies in many plants. Various studies have been conducted in cotton for different aspects in cotton through association mapping like seedcotton yield and components [142, 143, 144], salt tolerance [145], architecture of plant, earliness [146] and protein and oil contents [147] and fiber quality [8, 60, 132, 148, 149, 150].

In-contrast to genetic mapping in populations developed from hybridization of parents using conventional ways are not saturated, labor intensive, always in danger, high investment for development and more work after evaluating numerous genotypes of gene pool [84]. Nonetheless, association mapping use LD and overcomes the requirement of bi-parental populations by utilizing the extent of genetic variation present within the available stable populations like cultivars, accessions developed with the time and maintained as gene pool. Association mapping on whole genome has been studied in Arabidopsis [151], rice [152] for observing loci connected to economical characters. Association studies allow the development of highly saturated maps via determination of QTLs related to economic characters at whole genome level in permanent mapping populations.

Abdurakhmonov et al. [60] used association analysis for observing association among fiber traits in cotton among germplasm entries for utilizing the genetic variation in marker-based breeding. Linkage disequilibrium based association mapping determined in the germplasm having diverse genotypes from all over the world. 95 SSR were screened among all germplasm entries for ascertaining QTLs at whole genome level associated with fiber properties. They found about 11–12% LD among all SSRs. They also observed significant population orientation among all entries. They employed mixed linear model and general linear model using kinship and population structure and as a whole determined 6 & 13% pair of primers related to fiber quality. They concluded that the markers selected in this study can be used for refinement of fiber using hidden sources of genetic variability.

Genetic variation, population behavior and LD based association analysis for fiber conducted in germplasm under two different climatic zones [85]. The upland gene pool containing 335 elite entries screened with 202 SSRs. Mean of LD prolonged to 25 cM at whole genome level among all genotypes at 0.01 probability. They found that LD dropped to about 5 cM at ($r2 > 0.2$) showing potential for association among genotypes for yield contributing characters. They performed mixed linear model and population analysis for observing association contributing to permutation significance and population pattern. As a whole developed many common markers for fiber traits among genotypes in both locations. They revealed that mixed linear model associations ranged from 7 to 43% having strong to very strong relation to fiber properties as confirmed by Bayes factor which will be a very effective source for association analysis of yield improvement in marker based breeding techniques.

Wang et al. [153] found association among yield and fiber characters in using mixed linear model in pima cotton germplasm entries. They observed 72 loci, out of which 46 were connected to fiber while 26 related to cotton. They concluded that marker-associations among fiber characters are of vital value for enhancing quality.

Fang et al. [154] used multi-parents population for observing association among yield and fiber quality traits. They revealed that common and new QTLs deducted in this study can be used for overcoming problems in fiber quality enhancement. They screened 1582 polymorphic microsatellites among 275 RILs in first set developed from diverse parents for screening QTLs connected to fiber. 131 QTLs found for fiber quality sharing characters via association analysis with TASSEL while same QTLs verified in second set of 275 RILs with 270 SSR. The distinction showed that 54 new QTLs and 77 QTLs are in accordance to previous studies.

Genetic map constructed using RIL developed from transference of superior fiber quality from *G. barbadense* (TM-1) to *G. hirsutum* cv. NM24016 and relationship determined among yield components and fiber. 429 SSR and 412 GBS-based single nucleotides were involved in the development of map which spanned to about half length of upland cotton genome [10]. They revealed that all makers are distributed randomly among all loci of the genome. The yield components and fiber characters showed extreme phenotypic expression under multiple locations. They found 28 QTLs which are useful from breeding perspectives for agronomic and fiber properties.

Cai et al. [8] used 99 upland cotton genotypes to ascertain the association for fiber traits. The relationship among fiber components determined with 97 polymorphic microsatellites. The genomic regions associated with fiber were 107 including 70 in 2 or more than 2 zones and 37 found in just one. It was revealed that most of the associations were reliable as verified from earlier findings for fiber quality. They also observed genomic regions related with 2 or more characters and assumed that such regions derived from the genotypes which are having minor allele frequency less than five, from local sources or acclimatized in china. They concluded that fiber traits can be renovated by using such loci from diverse resources.

Islam et al. [123] carried GBS for observing SNPs which can be used for improving economic traits in cotton gene pool. RILs and 11 contrasting parents were used in the study with two separate methods were applied for determining SNPs with variant allele frequency of >0.1. SNPs quality control performed and calling done with available G. raimondii Ulbrich genome. As a whole 1071 and 1223 SNPs observed among At and Dt genomes respective. Moreover these SNPs were found in coding region usually in higher frequency. GBS was conducted in germplasm consisting of 154 accessions for the verification of 111 of total SNPs and the SNPs verified in all parents and none of the genotype was found with same SNP. They revealed that SNPs can be determined in *G. hirsutum* with ease and genetic improvement can be done after getting true SNPs.

Association among fiber traits conducted in germplasm collection of Hawaiian cotton consisting of 503 genotypes [132]. They used 494 microsatellites at whole genome and as a whole 179 replicable SSRs were screened among genotypes under diverse climatic conditions. Population pattern and LD used for observing association among various fiber traits with mixed linear model via TASSEL program. The QTLs were selected among markers and phenological characters with association values. 426 alleles were evolved and germplasm was differentiated into seven subgroups upon the basis of hybridization, climate and topographical pattern. 216 polymorphic loci were associated with fiber contributing characters having mean of 2.7% and showed phenotypic variation from 0.58–5.12%. LD decreased significantly to 0-5 cM and observed 13 QTLs which are same to earlier findings and 3 connected to similar character while 7 QTLs were corresponded to fiber formation. They concluded that novel alleles identified based association mapping based LD for fiber quality can be applied in breeding cultivars for tagging genes of interest.

GBS carried in a population evolved using various parents for overcoming the inverse relation among yield and fiber traits [155]. They assumed that GBS will serve as a valuable source for the development of high saturated map with the development of large frequency of SNPs. Association analysis via mixed linear model in TASSEL observed among fiber traits in four separate climates with 5071 SNPs developed from GBS and 223 SSRs from 547 RILs. One QTL cluster related to fiber traits including length, short fiber content, strength and uniformity found and verified on locus A07. They also studied the ultimate genes connected to fiber traits and revealed that SNP (CFBid0004) formed from deletion of 10 bp GhRBB1_A07 is directly associated with fiber traits among RIL and 104 approved American varieties. Moreover, GhRBB1_A07 can be used in MAS for the improvement of fiber traits among germplasm entries.

Sun et al. [150] studied the genetic architecture of major fiber traits in cotton germplasm using association mapping under different climatic zones. The mixed linear model association analysis showed that fiber length, strength and uniformity had 16, 10 and 7 SNPs respectively while G. raimondii 7th chromosome had two main genomic locations and fiber length contributing four genes were also observed. Moreover population structure showed that populations from low peaks were having less genetic variation among accessions compared to high peaks. The valuable allelic frequency was more in genotypes from less elevation in-contrast to high. They concluded that the desired allelic number among genotypes can be used for enhancement of fiber.

Association was observed for plant ideotype, heat tolerance, yield contributing traits and fiber quality among germplasm collection under different climatic conditions for consecutive three years at whole genome [156]. The genetic stock associations were observed using SNPs. Fiber characters were found to be low to highly heritable as value ranged from 0.26–0.89 for boradsense heritability as compared to yield components having 0.14–0.43. Phylogenetic analysis showed that the genotypes were developed from diverse parents having multiple characters from breeding perspectives. They pointed that less number of informative markers can be used for association mapping studies as LD value found upto 5Mbp which decreased to 2Mbp at r2 ≥ 0.2. 17 significant SNPs connected fiber length while 50 SNPs for fineness were observed using mixed linear model. The results revealed that associations among most of the characters at whole genome were non-significant as numerous SNPs impact on phenotype was found lower than 5% and assumed this to be due to low reproducibility of markers among cotton or SNP Chip less coverage in the germplasm.

Sun et al. [150] used association analysis in germplasm containing wide variation among genotypes at multiple locations for fiber quality traits. Illumnia SNP array was used for genome-wide study for quality analysis. They found 10,511 SNPs which were distributed over all loci and 46 SNPs associated with fiber quality with significance. They observed two QTLs for strength and length on At07 and Dt11.

5. CONCISE SUMMARY

Fiber quality enhancement through genetics is the ultimate objective of breeding strategy in cotton. Cotton scientists have been involved in fiber quality improvement for a long time due to the increase in demand for multiple products from cotton. Furthermore, conventional ways would be tiresome and stagnant. Hence, the modern plant improvement methods should be integrated. Molecular characterization is the way to transfer required traits into modern genotypes. Genotyping-by-sequencing (GBS) is powerful and easy approach which paves the way for the discovery of numerous SNPs concurrently among large number of genotypes. Quantitative trait loci (QTLs) allow gene pyramiding for yield and fiber quality through evolution of linkage maps. Molecular breeding with highly saturated maps having QTLs connected with economic traits through impactful genetic markers provides a good source for cotton improvement. Association mapping using linkage disequilibrium on genome wide level is the most valuable strategy among scientists for searching QTLs in crop sciences. It is need of the day that highly saturated mapping should be devised in cotton for overcoming the sequencing drawbacks and fastening the variety development.

AT A GLANCE

Improved fiber yield is considered a constant goal of upland cotton (Gossypium hirsutum) breeding worldwide, but the understanding of the genetic basis controlling yield-related traits remains limited. Dissecting the genetic architecture of complex traits is an ongoing challenge for geneticists. Two complementary approaches for genetic mapping, linkage mapping and association mapping have led to successful dissection of complex traits in many crop species. Both of these methods detect quantitative trait loci (QTL) by identifying marker–trait associations, and the only fundamental difference between them is that between mapping populations, which directly determine mapping resolution and power. Nowadays, the availability of genomic tools and resources is leading to a new revolution of plant breeding, as they facilitate the study of the genotype and its relationship with the phenotype, in particular for complex traits. Next Generation Sequencing (NGS) technologies are allowing the mass sequencing of genomes and transcriptomes, which is producing a vast array of genomic information with the development of high-throughput genotyping, phenotyping will be a major challenge for genetic mapping studies. We believe that high-quality phenotyping and appropriate experimental design coupled with new statistical models will accelerate progress in dissecting the genetic architecture of complex traits.

REFERENCES

1. Bardak, A., Bolek, Y. 2012. Genetic diversity of diploid and tetraploid cottons determined by SSR and ISSR markers. Turkish J. of Field Crops. 17(2): 139-144
2. Tan, Z., Fang, X., Tang, S., Zhang, J., Liu, D., Teng, Z., et al. (2014). Genetic map and QTL controlling fiber quality traits in upland cotton (Gossypium hirsutum L.). Euphytica 203, 615-628. doi: 10.1007/s10681-014-1288-9
3. Fryxell, P.A. 1979. The Natural History of the Cotton Tribe. Texas A&M University Press, College Station, TX, USA
4. Fryxell, P.A. 1992. A revised taxonomic interpretation of Gossypium L. (Malvaceae). Rheeda, 2(2): 108-165

5. Stewart J. McD. 1995. Potential for crop improvement with exotic germplasms and genetic engineering. In challenging the Future, World Cotton Research Conference, Brisbance Australia 14-17 February 1994. Constable CA and Forester NW (Eds.) CSIRO, Melbourne. pp.313-332

6. Grover, C.E., Gallagher, J.P., Jareczek, J.J., Page, J.T., Judall, J.A., Wendel, J.F. 2015. Re-evaluating the phylogeny of allopolyploid Gossypium L. Mol. Phylognet. Evol., 92: 45-52

7. Gallagher, J.P., Gover, C.E., Rex, K., Moran, M., Wendel, J.F. 2017. A new species of cotton from Wake Atoll. Gossypium stephensii (Malvaceae). Systematic Botany, 42:115-123

8. Cai, C., Ye, W., Zhang, T., Guo, W. 2014. Association analysis of fiber quality traits and exploration of elite alleles in upland cotton cultivars/accessions (G. hirsutum L.). J. of Integ. Plant Bio., 56: 51-62

9. Ulloa, M., Saha, S., Jenkins, J.N., Meredith, W.R.Jr., McCarty, J.C.Jr., Stelly, D.M. 2005. Chromosomal assignment of RFLP linkage groups harboring important QTLs on an intraspecific cotton (Gossypium hirsutum L.) joinmap. J. Here., 96(2): 132-144

10. Gore, M.A., Fang, D.D., Poland, J.A., Zhang, J., Percy, R.G., Cantrell, R.G., Thyseen, G., Lipka, A.E. 2014. Linkage map construction and quantitative trait analysis of agronomic and quality traits in cotton, The Plant Genome, 7(1): 1-10

11. Rahman, M., Ullah, I., Ashraf, M., Stewart JM, Zafar Y. 2008. Genotypic variation for drought tolerance in cotton. Agron. Sustain Dev., 28:439-447

12. Cerda, S., Cloutier, B.J.; S.; Quian, R.; Gajardo, H.A.; Olivos, M.; You, F.M. 2018. Genome-Wide Association Analysis of Mucilage and Hull Content in Flax (Linum usitatissimum L.) Seeds. Int. J. Mol. Sci. 19, 2870

13. Nordborg, M., Tavare, S. 2002. Linkage disequilibrium: what history had to tell us. Trend Genet., 18: 83-90

14. Slatkin, M. 2008. Linkage disequilibrium-understanding the evolution past and mapping the medical future. Nat. Rev. Genet. 9: 477-485

15. Weiss, K.M., Clark, A.G. 2002. Linkage disequilibrium and mapping of human traits. Trend Genet., 18: 19-24

16. Kruglyak, L. 2008. The road to genome-wide association studies. Nat. Rev. Genet., 9: 314-318

17. Zhu, C., Gore, M., Buckler, E., Yu, J. 2008. Status and prospects of association mapping in plants. The Plant Genome, 1(1): 5-20

18. Flint-Garcia, S.A., Thornsberry, J.M., Buckler, E.S. 2003. Structure of linkage disequilibrium in plants. Annual Review of Plant Bio., 54: 357-374

19. Basara, A.S., Malik, C.P. 1984. Development of cotton fiber. Inter Rev. Cyto., 65-113

20. Ryser, U. 1985. Cell wall biosynthesis in differentiating cotton fibers. Eur. J. Cell Bio. 39: 236-256

21. Delmer, P.D., Amor, Y. 1995. Cellulose biosynthesis. The Plant Cell, 7: 987-1000

22. Haigler, C.H., Zhang, D., Wilkerson, C.G. 2005. Biotechnological improvement of cotton fiber. Phsioyl. Plant., 124: 285-294

23. Berlin, J.D. 1986. The outer epidermis of the cottonseed. In: Mauney, I.R., Srewart I.M. (Ed.), Cotton physiology. The cotton foundation reference book series. Memphis, TN: The Cotton Foundation, : 375-414

24. Tiwari, S.C., Wilkins, T.A. 1995. Cotton (Gossypium hirsutum L.) seed trichomes expand via diffuse growing mechanism. Canadian Journal of Botany, 73: 746-757

25. Bradow, J.W., Davidonis, H.D. 2000. Quantification of fiber quality and the cotton production-processing interface: A physiologist perspective. The J. of Cotton sci., 4: 34-64

26. Dutt, Y., Wang, X.D., Zhu, Y.G., Li, Y.Y. 2004. Breeding for high yield and fiber quality in coloured cotton. Plant Breeding, 123:145-151

27. Ali, M.A., Khan, I.A., Nawab, N.N. 2009. Estimation of genetic divergence and linkage for fiber quality traits in cotton. J. Agri. Res., 47(3): 229-236

28. Shen, X., Cow, Z., Singh, R., Lubbers, E.L., Smith, C.W., Paterson, A.H., Chew, P.W. 2011. Efficacy QFL-chr1 a quantitative trait locus for fiber length on cotton (Gossypium sp.). Crop Sci., 51:2005-2010

29. Rahman, H., Murtaza, N., Shah, M.K.N. 2007. Study of cotton fiber traits inheritance under different temperature regimes. J. Agron. Crop Sci. 193:45-54

30. Ali, M.A., Khan, I.A., Awan, S.I., Ali, S., Niaz, S. 2008. Genetics of fiber quality traits in cotton (Gossypium hirsutum). Aus. J. of Crop Sci., 2: 10-17

31. Shen, X.L., Guo, W.Z., Zhu, X.F., Yuan, Y.L., Yu, J.Z., Kohel, R.J., Zhang, T.Z. 2005. Molecular mapping of QTL for fiber qualities in three diverse lines in upland cotton using SSR markers. Mol. Breed. 15:169-181

32. Meredith, W.R.Jr., Heitholt, J.J., Pettigrew, W.T., Rayburn, S.T. 1997. Comparsion of obsolete and modern cultivars at two nitrogen levels. Crop Sci., 37: 1453-1457

33. Meredith, W.R.Jr. 2002. Factors that contribute to lack of genetic progress. Proc. Beltwide Cotton Conf. Atlanta, GA 8-13 Jan 2002. National Cotton Coun. Am, Memphis, TN. Pp: 528-532

34. Bayles, M.B., Verhalen, L.M., Johnson, W.M., Barnes, B.R. 2005. Trends over time among cotton cultivars released by the Oklahoma Agricultural Experiment Station. Crop Sci., 45: 966-980

35. Kerby, T.A., Hugie, W.V. 2006. Yield and quality improvement of significant D&PL varieties during the last 25 years. In Proc. Beltwide Cotton Conf. San Antonio, TX 3-6 Jan 2006 National Cotton Council Memphis, TN. pp: 858-865

36. Kuraparthy, V., Bowman, D.T. 2013. Gains in breeding upland cotton for fiber quality. J. Cotton Sci.,

37. Flint-Garcia, S. A. (2013). Genetics and consequences of crop domestication. Journal of Agricultural and Food Chemistry, 61(35), 8267-8276

38. Zhang, B.L., Yang, Y.W., Chen, T.Z., Yu, W.G., Liu, T.L., Li, H., Fan, X., Ren, Y., Shen, D., Liu, L., Dou, D., Chang, Y. 2012. Island cotton GbVe1 gene encoding a receptor-like protein confers resistance to defoliating and non-defoliating isolates of Verticiliumdahliae. PLoS One, 7(12): 1-10

39. McCarty, J.C., Percy, R.G. 2001. Genes from exotic germplasm and their use in cultivar improvement in Gossypium hirsutum L. and G. barbadense L. pp. 65-80. In Jonkins J.J., Saha S. (ed.) Genetic improvement of cotton-emerging technologies. Sci. Publ. Enfield, NH

40. Tankley, S.D., Hewitt, J. 1988. Use of molecular markers in breeding for soluble solids in tomato- a re-examination. Theor. And App. Genet. 75: 811-823

41. Kumar, L.S. 1999. DNA markers in plant improvement: an overview, Biotechnology Advances, 17(2): 143-182

42. King R.C., Stansfield, W.D. 1990. A dictionary of genetics, Oxford University Press, New York, pp.188

43. Schulman, A. H. (2007). Molecular markers to assess genetic diversity. Euphytica, 158(3), 313-321

44. Roychowhury, R., Taoutaou, A., Khalid, R.K., Mohamed R.A.G., Jagatpati, T. 2014. Crop improvement in the ear of climate change, In: Roychowhury (eds.), I.K International Publication Ltd

45. Collard, B.C., Mackill, D.J. 2008. Marker-assisted selection: an approach for precision plant breeding in the twenty-first century. Philo. Trans. Of the Royal Soc. Of Lon., 363: 557-572

46. Kumpatla, S.P., Buyyarapu, R., Abdurakhmonov, I.Y., Mammadov, J.A. 2012. In: Abdurakhmonov I.Y., (ed.), Genomics-Assisted Plant Breeding in the 21st Century: Technological Advances and Progress, Plant Breeding, In Tech, Croatia, 131-184

47. Malik, W., Ashraf, J., Iqbal, M. Z., Ali Khan, A., Qayyum, A., Ali Abid, M., ... & Hasan Abbasi, G. (2014). Molecular markers and cotton genetic improvement: current status and future prospects. The Scientific World Journal, 2014

48. Helentjaris, T., Slocum, M., Wright, S., Schaefer, A., Nienhuis, J. 1986. Construction of genetic linkage maps in maize and tomato using restriction fragment length polymorphisms, Theoretical and Applied Genetics, 72(6):761-769

49. Welsh J., McClelland, M. 1990. Finger printing genomes using PCR with arbitrary primers, Nucleic Acids Research, 18(24): 7213-7218

50. Vos, P., Hogers, R., Bleeker, M., Reijans, M., Van, D.L.T., Hornes, M., Friters, A., Pot, J., Paleman, J., Kuiper, M., Zabeau, M. 1995. AFLP: a new technique for DNA fingerprinting, Nucleic Acids Research, 23(21): 4407-4414

51. Struss, D., &Plieske, J. (1998). The use of microsatellite markers for detection of genetic diversity in barley populations. Theoretical and Applied genetics, 97(1-2), 308-315

52. Mackill, D.J., Nguyen, H.T., Zhang, J. 1999. Use of molecular markers in plant improvement programs for rainfed lowland rice. Field Crops Res., 64, 177-185

53. Koebner, R.M.D., Summers R.W. 2003, 21st century wheat breeding: plot selection or plate detection? Trends in Biotechnology, 21(2): 59-63

54. Stuber, C.W., Polacco, M., Senior, M.L. 1999. Synergy of empirical breeding, marker-assisted selection, and genomics to increase crop yield potential. Crop Science, 39(6): 1571-1583

55. Tuberosa, R., Salvi, S., Sanguineti, M.C., Maccaferri, M., Giuliani, S., Landi, P. 2003. Searching for quantitative trait loci controlling root traits in maize: a critical appraisal. Developments in Plant and Soil Science, 101: 35-54

56. Thomas, W. 2003, Prospects for molecular breeding of barley. Annals of Applied Biology, 142(1): 1-12

57. Williams, K. J. (2003). The molecular genetics of disease resistance in barley. Australian Journal of Agricultural Research, 54(12), 1065-1079

58. Iqbal, M.J., Aziz, N., Saeed, N.A., Zafa, Y., Malik, K.A. 2001. Genetic diversity evaluation of some elite cotton cultivars by RAPD analysis. Theo. Applied Gene., 94:139-144. 13. Nordborg, M., Tavare, S. 2002. Linkage disequilibrium: what history had to tell us. Trend Genet., 18: 83-90

59. Rahman, M., Asif, M., Ullah, I., Malik, K.A., Zafar, Y. 2005. Overview of cotton genomic studies in Pakistan, Plant and Animal Genome Conference XIII, San Diego, USA

60. Abdurakhmonov, I.Y., Kohel, R.J., Yu, J.Z., Pepper, A.E., Abdullaev, A.A., Kushanov, F.N., Salakhutdinov, I.B., Buriev, Z.T., Saha, S., Scheffler, B.E., Jenkins, J.N., Abdukarimov, A. 2008. Molecular diversity and association mapping of fiber quality in exotic G. hirsutum L. germplasm. Genomics, 92: 478-487

61. Paterson, A.H., Tanksely, S.D., Sorells, S.E. 1991. DNA markers in plant improvement. Advances in Agronomy, 46: 39-90

62. Mohan, M., Nair, S., Bhagwatt, A., Karisma, T., Yano, M., Bhatia, C.R., Sasaki, T. 1997. Genome mapping, molecular markers and marker assisted selection in crop plants. Mol. Breed., 3: 87-103

63. Zhu, Y., Chen, H., Fan, J., Wang, Y., Li, Y., Chen, J., et al. (2000). Genetic diversity and disease control in rice. Nature 406, 718-722. doi: 10.1038/35021046

64. Bolek, Y., Hayat K., Bardak, A., Azhar, M.T. 2016. Molecular breeding of cotton.: Adurakhmonov, I.Y. Cotton Research. Intech Publishers, in Cotton Research, ed I. Y. Abdurakhmonov (InTech): 123-166

65. Thornsberry, J.M., Goodman, M.M., Doebley, J., Kresovich, S., Nielson, D., Buckler, E.S. 2001. Dwarf8 polymorphisms associate with variation in flowering time. Nature Genet., 28: 286-289

66. Wang, K., Wang, Z., Li, F., Ye, W., Wang, J., Song, G., Yue, G., Cong, L., Shang, H., Zhu, Z., Zou, C., Li, Q., Yuan, Y., Lu, C., Wei, H., Gou, C., Zheng, Z., Yin, Y., Zhang, X., Liu, K., Wang, B., Song, C., Shi, N., Kohel, R.J., Percy, R.J., Yu, Z.Y., Zhu, Y., Wang, J., Yu, S. 2012. The draft genome of diploid cotton Gossypium raimondii Ulbrich. Nat. Genet., 44: 1098-1103

67. Paterson, A.H., Wendel, J.F., Gundlach, H., Guo, H., Jenkins, J., et al. 2012. Repeated polyploidization of Gossypium genomes and the evolution of spinnable cotton fibers. Nature, 492: 423-427

68. Li, F., Fan. G., Wang, K., Sun, F., Yuan, Y., Guoli. Song., Qin, Li., Zhiying, Ma., Cairui, Lu., Zou, C., Chen, W., Liang, X., Shang, H., Liu, W., Shim C., Xiao, G., Goum C., Ye1, W., Xu. X., Zhang, X., Wei1, H., Li, Z., Zhang, G., Wang, J., Liu, K., Kohel, R.J., Percy, R.J., Yu, J.Z., Zhu, Y., Wang, J., Yu, S. 2014. Genome sequence of the cultivated Gossypium arboreum. Nat. Genet., 46:567-572

69. Li, F., Fan, G., Lu, C., Xiao, G., Zou, C., Kohel, R. J., ... & Liang, X. (2015). Genome sequence of cultivated Upland cotton (Gossypium hirsutum TM-1) provides insights into genome evolution. Nature biotechnology, 33(5), 524-530

70. Zhang, T., Hu, Y., Jiang, W., Fang, L., Guan, X., et al. 2015. Sequencing of allotetraploid cotton (Gossypium hirsutum L. acc. TM-1) provides a resource for fiber improvement. Nat. Biotechnol., 33: 531-537

71. Chen, Z., Scheffler, B., Dennis, E., Triplett B., Zhang, T., et al. 2007. Towards sequencing of (Gossypium) genomes. Plant Physiol., 145: 1303-1310

72. Jiang, C.X., Chee, P.W., Draye, X., Morrell, P.L., Smith, C.W., Paterson, A.H. 2000. Multi-locus interactions restrict gene introgression in interspecific populations of polyploid Gossypium (cotton). Evolution., 54: 798-814

73. Paterson, A.H., sarangi, Y., Menz, M., Jiang, C.X., wright, R.J. 2003. QTL analysis of genotype x environment interactions affecting cotton fiber quality. Theor. Appl. Genet., 106: 384-396

74. Rong, J., Feltus, F.A., Waghmare, V.N., Pierce, G.J., Chee, P.W., Draye, X., et al. 2007. Meta-analysis of polyplois cotton QTL shows unequal distributions of sub-genomes to a complex network of genes and gene clustered implicated in lint fiber development. Genetics, 176: 2577-2588

75. Shappley, Z.W., Jenkins, J.N., Zhu, J., McCarty, J.C. 1998. Quantitative trait loci associated with agronomic and fiber traits of upland cotton. J. Cotton Sci., 4:153-163. 86. Lande, R., Thomson, R. 1990. Efficiency of marker assisted selection in the improvement of quantitative traits. Genetics, 124:743-756

76. Ulloa, M., Meredith Jr, W. R., Shappley, Z. W., & Kahler, A. L. (2002). RFLP genetic linkage maps from four F 2.3 populations and a joinmap of Gossypium hirsutum L. Theoretical and Applied Genetics, 104(2-3), 200-208

77. He, D.H., Lin, Z.X., Zhang, X.L., Nie, Y.C., Guo, X.P., Feng, C.D., Stewart, Mc.D.J. 2005. Mapping QTLs contributing to yield and analysis of genetic effects in tetraploid cotton. Euphytica, 144(1-2):141-149

78. Fang, D.D., Jenkins, J.J., Deng, D.D., McCarty, J.C., Li, P., Wu, J. 2014. Quantitative trait loci analysis of fiber quality using a random-mated recombinant inbred line population in upland cotton (Gossypium hirsutum L.). BMC Genomics, 15-397

79. Said, J.I., Lin, Z., Zhang, X., Song, M., Zhang, J.A. comprehensive meta QTL analysis for fiber quality, yield, yield related and morphological traits, drought tolerance, and disease resistance in tetraploid cotton. BMC Genomics, 2013, 14:776

80. Mackay, I., Powell, W. 2007. Methods for linkage disequilibrium mapping in crops. Trends plant sci., Trends Plant Sci., 12: 57-63

81. Breseghello, F., Sorrells, M.E. 2006. Association mapping of kernel size and milling quality in wheat (Triticum aestivum L.) cultivars. Genetics, 172: 1165-1177

82. Nordborg, M., Borevitz, J.O., Bergelson, J., Berry, C.C., Chory J., Hagenblad, J., Kreitman, M., Maloof, J.N., Noyes, T., Oefner, P.J., Stahl, E.A., Weigel, D. 2002. The extent of linkage disequlibrium in Arabdopsis thaliana. Nature Genetics, 30: 190-193

83. Abdurakhmonov, I. Y., & Abdukarimov, A. (2008). Application of association mapping to understanding the genetic diversity of plant germplasm resources. International journal of plant genomics, 2008

84. Abdurakhmonov, I.Y., Buriev, Z.T., Saha, S., Pepper, A.E., Musaev, J.A., Almatov, A., Shermatov, S.E., Kushanov, F.N., Mavlonov, G.T., Reddy, U.K., Yu, J.Z., Jenkins., J.N., Kohel, R.J., Abdukarimov, A. 2007. Microsatellite Markers Associated with Lint Percentage Trait in Cotton, Gossypium hirsutum. Euphytica, 156(1-2): 141-156

85. Abdurakhmonov, I. Y. Saha, S., Jenkins, J.N., Buriev, Z.T., Shermatov, S.E., Scheffler, B.E., Pepper, A.E., Yu, J.Z., Kohel, R.J., Abdukarimov A. 2009. Linkage disequilibrium based association mapping of fiber quality traits in G. hirsutum L. variety germplasm. Genetica, 136, 401-417

86. Lande, R., Thomson, R. 1990. Efficiency of marker assisted selection in the improvement of quantitative traits. Genetics, 124:743-756

87. Chee P., Draye, X., Jiang, C.X., Decaanini L., Delmonte, T.A. Bredjauer, R., Smith, C.W., Paterson, A.H. 2005. Molecular dissection of interspecific variation between Gossypium hirsutum and Gossypium barbadense (cotton) by a backcrossself approach: I. Fiber elongation. Theor. Appl. Genet. 111: 757-763

88. Kumar, P., Gupta, V.K., Misra, A.K., Modi, D.R., Pandey, B.K. 2009. Potential of molecular markers in plant biotechnology. Plant Omics Journal, 2: 141-162

89. Reinisch, A.J., Dong, J.M., Brubaker, C.L., Stelly, D.M., Wendel, J.F., Paterson, A.H. 1994. A detailed RFLP map of cotton, Gossypium hirsutum x Gossypium barbadense: chromosome organization and evolution in a disomic polyploid genome. Genetics, 138: 829-847

90. Blenda, A., Fang, D.D., Rami, J.F., Garsmeur, O., Feng, L., Lacape, A. 2012, High density consensus genetic map of tetraploid cotton that integrates multiple component maps through molecular marker redundancy. J. Genetics, 91(3): 22-25

91. Rong, J.K., Abbey, C., Bowers J.E., Brubaker, C.L., et al. 2004. A 3347-locus genetic recombination map of sequence-tagged sites reveals types of genome organization, transmission and evolution of cotton Gossypium. Genetics, 166: 389-417

92. Mei, M., Syed, N.H., Gao, W., Thaxton, P.M., Smith, C.W. Stelly, D.M., Chen, Z.N. 2004. Genetic mapping and QTL analysis of fiber-related traits in cotton (Gossypium). Theor. Appl. Genet. 108:280-291

93. Nguyen, T.B., Giband, M., Brottier, P., Risterucci, A.M., Lacape, J.M. 2004. Wide coverage of the tetraploid cotton genome using newly developed microsatellite markers. Theo. and App, Gene. 109:167-175

94. Han, Z.G., Guo, W.Z., Song, X.L., Zhang, T.Z. 2004. Genetic mapping of EST-derived microsatellites from the diploid Gossypium arboreum in allotetraploid cotton. Molecular Genetics and Genomics, 272: 308-327

95. Guo, W., Cai, C., Wang, C., Han, Z., Song, X., Wang, K., Niu, X., Wang, C., Lu, K., Shi, B., Zhang, T. 2007. A microsatellite-based gene-riched linkage map of reveals genome structure, function, and evolution in Gossypium. Genetics, 176: 527-541

96. Lacape, J. M., Jacobs, J., Arioli, T., Derijcker, R., Forestier-Chiron, N., Llewellyn, D., ... &Viot, C. (2009). A new interspecific, Gossypium hirsutum× G. barbadense, RIL population: towards a unified consensus linkage map of tetraploid cotton. Theoretical and applied genetics, 119(2), 281-292

97. Blenda, A., Fang, D.D., Rami, J.F., Garsmeur, O., Luo, F., Lacape, J.M. 2012. A high-density consensus genetic map of tetraploid cotton that integrates multiple component maps through marker redundancy check. PLoS ONE, 7: e45739

98. Fang, D. D., & Yu, J. Z. (2012). Addition of 455 microsatellite marker loci to the high-density Gossypium hirsutum TM-1× G. barbadense 3-79 genetic map. J Cotton Sci, 16, 229-248

99. Yu, J.Z., Fang, D.D., Kohel, R.J., Ulloa, M., Hinze, L.L., Percy, R.G., Zhang, J., Chee, P., Scheffler, B.E, Jones, D.C. 2012. Development of a core set of SSR markers for the characterization of Gossypium germplasm. Euphytica, 187: 203-213

100. Zhu, H., Han, X., Lv, J., Zhao, L., Xu, X., Zhang, T., & Guo, W. (2011). Structure, expression differentiation and evolution of duplicated fiber developmental genes in Gossypium barbadense and G. hirsutum. BMC plant biology, 11(1), 40

101. Marathi, B., Guleria, S., Mohapatra, T., Parsad, R., Mariappan, N., Kurungara, V. K., ... & Singh, A. K. (2012). QTL analysis of novel genomic regions associated with yield and yield related traits in new plant type based recombinant inbred lines of rice (Oryza sativaL.). BMC plant biology, 12(1), 137

102. Lacape, J. M., Gawrysiak, G., Cao, T. V., Viot, C., Llewellyn, D., Liu, S., ... &Palaï, O. (2013). Mapping QTLs for traits related to phenology, morphology and yield components in an inter-specific Gossypium hirsutum× G. barbadense cotton RIL population. Field Crops Research, 144, 256-267

103. Li, X., Yuan, D., Zhang, J., Lin, Z., & Zhang, X. (2013). Genetic mapping and characteristics of genes specifically or preferentially expressed during fiber development in cotton. PLoS One, 8(1), e54444

104. Lacape, J. M., Jacobs, J., Arioli, T., Derijcker, R., Forestier-Chiron, N., Llewellyn, D., ... &Viot, C. (2009). A new interspecific, Gossypium hirsutum× G. barbadense, RIL population: towards a unified consensus linkage map of tetraploid cotton. Theoretical and applied genetics, 119(2), 281-292

105. Reddy, O. U. K., Pepper, A. E., Abdurakhmonov, I., Saha, S., Jenkins, J. N., Brooks, T., ... & El-Zik, K. M. (2001). New dinucleotide and trinucleotide microsatellite marker resources for cotton genome research

106. Van Deynze, A., Stoffel, K., Lee, M., Wilkins, T. A., Kozik, A., Cantrell, R. G., ... & Stelly, D. M. (2009). Sampling nucleotide diversity in cotton. BMC plant biology, 9(1), 125

107. Xiao, J., Wu, K., David, D.F., Stelly, D.M., Yu, J., Roy, G.C. 2009. New SSR markers for use in cotton (Gossypium spp.) improvement. J. Cotton Sci., 13: 75-157

108. McCouch, S.R., Zhao, K., Wright, M., Tung, C.W., Ebana, K., Thomson, M., Reynolds, A., Wang, D., DeClerck, G., Ali, M.L., McClung, A., Eizenga, G., Bustamanteet, C. 2010. Development of genome-wide SNP assays for rice. Breed Sci., 60, 524-535

109. Davey, J.W., Hohenlohe, P.A., Etter, P.D., Boone, J.Q., Catchen, J.M., Blaxter, M.L. 2011. Genome-wide genetic marker discovery and genotyping using next-generation sequencing. Nat. Rev. Genet. 12: 499-510

110. Feuillet, C., Leach, J.E., Rogers, J., Schnable, P.S., Eversole, K. 2011. Crop genome sequencing: lessons and rationales. Trends Plant Sci., 16: 77-88

111. Morrell, P. L., Buckler, E. S., & Ross-Ibarra, J. (2012). Crop genomics: advances and applications. Nature Reviews Genetics, 13(2), 85-96

112. Poland, J.A., Rife, T.W. 2012. Genotyping-by-sequencing for plant breeding and genetics. Plant Genome, 5: 92-102

113. Cai, C., Zhu, G., Zhang, T., and Guo, W. (2017). High-density 80 K SNP array is a

114. Huang, B. E., Clifford, D., & Cavanagh, C. (2013). Selecting subsets of genotyped experimental populations for phenotyping to maximize genetic diversity. Theoretical and applied genetics, 126(2), 379-388

115. Gupta, P.K., Roy, J.K., Prasad, M., 2001. Single nucleotide polymorphisms: a new paradigm molecular marker technology and DNA polymorphism detection with empahais on their use in plants. Curr. Sci., 80:524:535

116. Maughan, P.J., Yourstone, S.M., Jellen, E.N., Udall, J.A. 2009. SNP discovery via genome reduction, barcoding and 454-pyrosequencing in amaranth. Plant Genome, 2: 260-270

117. Harper, A.L., Trick, M., Higgins, J., Fraser, F., Clissold, L., Wells, R., Hattori, C., Werner, P., Bancroft I. 2012. Associative transcriptomics of traits in the polyploid crop species Brassica napus. Nature Biorechnol. 30: 798-802

118. Naoumkina, M.G., Thyssen, G., Fang, D.D., Hinchliffe, D.J., Florane, C., et al. 2014. The Li2 mutation results in reduced subgenome expression bias in elongation fibers of allotetraploid cotton (Gossypium hirsutum L). PLoS One, e90830

119. Xu, Y., Xu, C. & Xu, S. Prediction and association mapping of agronomic traits in maize using multiple omic data. Heredity 119, 174-184 (2017). https://doi.org/10.1038/hdy.2017.27

120. Byers, R. L., Harker, D. B., Yourstone, S. M., Maughan, P. J., & Udall, J. A. (2012). Development and mapping of SNP assays in allotetraploid cotton. Theoretical and Applied Genetics, 124(7), 1201-1214

121. Poland, J.A., Brown, P.J., Sorrells, M.E., Jannink, J.L. 2012. Development of high-density genetic maps for barley and wheat using a novel two-enzyme genotyping-by-sequencing approach. PLoS One, 7:32253 Genomics 18:654

122. Elshire, R.J., Glaubitz, J.C., Sun, Q., Poland, J.A., Kawamoto, K., Buckler, E.S., Mitchell, S.E. 2011. A robust, simple genotyping-by- sequencing (GBS) approach for high diversity species. PLoS ONE, 6: e19379

123. Islam, M.S., Thyssen, G.N., Jenkins, J.N., Fang, D.D. 2015. Detection, validation, and application of genotyping-by-Sequencing based single nucleotide polymorphisms in upland cotton. The Plant Genome, 8:1-10

124. Schnable, P.S., Liu, S., Wu, W. 2013. Genotyping by next-generation sequencing. U.S. Patent Appl. No. 13/739,874

125. Ma, X. F., Jensen, E., Ale

126. Spindel, J., Wright, M., Chen, C., Cobb, J., Gage, J., Harrington, S., Lorieux, M., Ahmadi, N., McCouchet, S. 2013. Bridging the genotyping gap: Using genotyping by sequencing (GBS) to add high-density SNP markers and new value to traditional bi-parental mapping and breeding populations. Theor. Appl. Genet., 126:2699-2716

127. Sonah, H., Bastien, M., Iquira, E., Tardivel, A., et al. 2013. An improved genotyping by sequencing (GBS) approach offering increased versatility and efficiency of SNP discovery and genotyping. PLoS ONE, 8(1): 1-9

128. Huang, Y.F., Poland. J.A., Wight, C.P., Jackson, E.W., Tinker, N.A. 2014. Using genotyping-bysequencing (GBS) for genomic discovery in cultivated oat. PLoS One, 9: e102448

129. Islam, M., Zeng, L., Thyseen, G., Delhom, C., Kim, H. 2016b. Mapping by sequencing in cotton (Gossypium hirsutum) line MD52ne identified genes for fiber strength and its related quality attributes. Theor. Appl. Genet., 129: 1071-1086

130. Hayat, K.B. 2018. Association analysis and Mapping of Fiber quality in upland cotton. Ph. D Thesis submitted to Kahramanmaras Sutcu Imam University/ Turkey

131. Yang, X.H., Yan, J.B., Zheng, Y.P., Yu, J.M., Li, J.S. 2007. Reviews of association analysis for quantitative traits in plants. Acta Agron Sin., 33(4):523-30

132. Nie X., Huang, C., You, C., Li, W., Zhao, W et al. 2016. Genome-wide SSR-based association mapping for fiber quality in nation-wide upland cotton inbreed cultivars in China. BMC Genomics, 17:1-16. 131. Yang, X.H., Yan, J.B., Zheng, Y.P., Yu, J.M., Li, J.S. 2007. Reviews of association analysis for quantitative traits in plants. Acta Agron Sin., 33(4):523-30

133. Gould, S. J., Johnston, R.F. 1972. Geographic variation. Annu. Rev. Ecol. Syst. 3: 457-498

134. Roesti, M., Moser, D., Berner, D. 2013 Recombination in the three-spine stickleback genome-patterns and consequences. Mol. Ecol., 22: 3014-3027

135. Pritchard, J.K., Stephens, M., Donnelly, P. 2000. Inference of population structure using multilocus genotype data. Genetics, 155, 945-959

136. Peleg, Z., Fahima, T., Abbo, S., et al. 2008. Genetic structure of wild emmer wheat populations as reflected by transcribed versus anonymous SSR markers. Genome, 51: 187-195

137. Nei, M. 1972. Genetic distances between populations. American Naturalist, 106, 283-292

138. Rogers, J.S. 1972. Measures of genetic similarity and genetic distance. Studies in genetics. VII. Univ. Texas Publ. 7213,145-153

139. Nei, M. 1978. Estimation of average heterozygosity and genetic distance from a small number of individuals. Genetics, 89, 583-590

140. Yan, J. B., Kandianis, C.B., Harjes, C.E., Bai, L., et al. 2010. Rare genetic variation at Zea mayscrtRB1 increases beta-carotene in maize grain. Nat. Genet., 42:322-327

141. Badigannavar, A., and Myers, G. O. (2015). Genetic diversity, population structure

142. Mei, H., Zhu, X., Zhang, T. 2013. Favorable QTL alleles for yield and its components identified by association mapping in Chinese upland cotton cultivars. PLoS One, 8: e82193

143. Zhang, T., Qian, N., Zhu, X., Chen, H., Wang, S., Mei, H., Zhang, Y. 2013. Variations and transmission of QTL alleles for yield and fiber qualities in upland cotton cultivars developed in china. PLoS One, 8(2): 1-12

144. Qin, H., Chen, M., Yi, X., Bie, S., Zhang, C., Zhang, Y., Lan, J., Meng, Y., Yuan. Y., Jiao, C. 2015. Identification of associated SSR markers for yield component and fiber quality traits based on frame map and upland cotton collections. PLoS One, 10(1): 1-16

145. Saeed, F., Farooq, J., Mahmood, A., Hussain, T., Riaz, M., Ahmad, S. 2014. Genetic diversity in upland cotton for cotton leaf curl virus disease, earliness and fiber quality. Pak. J. Agric. Res. 27, 226-236

146. Li, C., Zhang, J., Hu, G., Fu, Y., Wang, Q. 2016. Association mapping and favorable allele mining for node of first fruiting/sympodial branch and its height in Upland cotton (Gossypium hirsutum L.). Euphytica, 210:57-88

147. Liu, G.Z., Mei, H.X., Wang, S., Li, X.H., Zhu, X.F., Zhang, T.Z. 2015. Association mapping of seed oil and protein contents in upland cotton. Euphytica, 205:637-645

148. Zeng, L., Meredith, W. R., Gutiérrez, O. A., & Boykin, D. L. (2009). Identification of associations between SSR markers and fiber traits in an exotic germplasm derived from multiple crosses among Gossypium tetraploid species. Theoretical and applied genetics, 119(1), 93-103

149. Iqbal, M.A., Rahman, M. 2017. Identification of marker-trait associations for lint traits in cotton. Front Plant Sci., 8: 86

150. Sun, Z., Wang, X., Liu, Z., Gu, Q., Zhang, Y.,. 2017. Genome-wide association study discovered genetic variation and candidate genes of fibre quality traits in Gossypium hirsutum L. Plant Biotechnology Journal, 1-15

151. Atwell, S., Huang, Y.S., Vilhjalmsson, B.J., Willems, G., Horton, M. et al. 2010. Genome-wide association study of 107 phenotypes in Arabidopsis thaliana inbred lines. Nature, 465 (7298): 627-631

152. Huang, B. E., George, A. W., Forrest, K. L., Kilian, A., Hayden, M. J., Morell, M. K., & Cavanagh, C. R. (2012). A multiparent advanced generation inter-cross population for genetic analysis in wheat. Plant biotechnology journal, 10(7), 826-839

153. Wang, X., Yu, Y., Sang, J., Wu, Q., Zhang, X., Lin, Z. 2013. Intraspecific linkage map construction and QTL mapping of yield and fiber quality of Gossypium barbadense. Aust. J. Cop Sci., 7:1252-1261

154. Fang, H., Zhou, H., Sanogo, S., Lipka, A. E., Fang, D. D., et al. (2014). Quantitative trait locus analysis of Verticillium wilt resistance in an introgressed recombinant inbred population of Upland cotton. Molecular breeding,33(3), 709-720

155. Islam, A., Gregory, N., Thyssen, Johnie, N., Jenkins, J.J., 2016a. A MAGIC population-based genome-wide association study reveals functional association of GhRBB1_A07 gene with superior fiber quality in cotton. BMC Genomics, 17:903

156. Gapare, W., Conaty, W., Zhu, Q. H., Liu, S., Stiller, W., Llewellyn, D., & Wilson, I. (2017). Genome-wide association study of yield components and fibre quality traits in a cotton germplasm diversity panel. Euphytica,213(3), 66

CHAPTER-14

OPPORTUNITIES FOR MOLECULAR BREEDING IN COTTON, *GOSSYPIUM* SPP

1. INTRODUCTION

The landscape of cotton breeding is undergoing a transformative shift with the advent of molecular breeding techniques. Gossypium spp., encompassing various cotton species, stands at the forefront of this paradigm shift, offering immense potential for advancements through molecular breeding strategies. As the global demand for cotton continues to rise, driven by the textile industry's needs, the integration of molecular tools holds promise for enhancing breeding efficiency, precision, and the overall quality of cotton varieties. This exploration into the prospects for molecular breeding in Gossypium spp. aims to shed light on the opportunities and challenges that lie ahead, with the ultimate goal of fostering sustainable and resilient cotton crops that meet the demands of a dynamic agricultural landscape.

Cotton is one among many fiber-producing species, but it is the only major crop cultivated for quenching one of the basic human necessities i.e., clothing. The ancient Harappan civilizations that were discovered in the Indus valley suggested that the first use of cotton was around the 2nd millennium BC [1]. However, the discoveries of cotton fabric at Duweilah in Jordan indicate that cotton was used as early as 4th millennium BC, but the latest discoveries at Mehrgarh in Pakistan suggest that cotton fibers were used as early as 6th millennium BC [2]. Hundreds of years ago cotton was a chief source of clothing and in the future, it would continue to be, because of its unique unparalleled qualities such as comfort, safety and eco-friendly attributes. However, with the revolution in the textile industry, the synthetic fibers were dumped into the markets with the big tag line as "cost-effective" as these synthetic fibers can be manufactured at will with the desired fiber properties to meet the spinning demands. Synthetic fibers were assumed as a major threat to cotton cultivation but sooner than later when people realized the unsustainability, unsafe and less eco-friendly characters of the Petro-chemical based synthetic fibers, cotton is still the most preferred and produced fiber for clothing [3]. It's a big surprise that the majority of us would strictly prefer cotton-based clothing for newborn children but not synthetic fibers, which describes its safety and comfort. Nowadays, the reinforcement of natural fiber by the synthetic fibers has proved excellent in terms of improved properties of the new fiber synthesized [4]. Apart from the primary application as clothing, coarse cotton is widely used in hospitals as cotton swabs. Cotton linters (fiber<3.5 mm) are used in the paper industry along with other pulps to manufacture technical papers, art papers etc. Cotton oil extracted from seed is used in the cosmetics and paint industry, oil can also be used for consumption/cooking if the gossypol content is very low. Cottonseed meal is used as dairy feed. Apart from being economically important, the cotton fiber serves as a powerful single-celled model in studying cell wall and cellulose research. Around 100 million people are involved in cotton production with over 250 million deriving employment in transportation, ginning process, and several million people in textile manufacturing, agriculture inputs sector and cottonseed crushing, among others [5]. The cotton export value during 2017 was around 15.62 billion US$ (Rice export value; 24.99 billion US$, Wheat export value: 45.13 billion US$) [6]. Cotton has around 30% share in world textile fibers [3]. The global textile and apparel market were to the tune of 1.7 trillion US$ [7], indicating that cotton is a very important crop globally. The world population is booming and it is expected to be 8–10 billion people by 2050 [8] On the contrary, cotton productivity has also seen a rise with few but impactful breeding and molecular breeding interventions such as the introduction of hirsutum, early maturing types, introduction of intra and interspecific hybrids/derivatives and Genetically Modified (GM) crops (pest and herbicide tolerance). However, the trend of increase in productivity of cotton (1961:8.62 q/ha vs. 2018:21.90 q/ha) compared to principal crops such as wheat (1961:10 q/ha vs. 2018: 34.25 q/ha) and rice (1961, 18.69 vs. 2018: 46.78 q/ha) (Figure 1) is very low due to less international collaboration and lesser germplasm, technology exchange. In 2050, the cotton production is required to be 94.71 MT of seed cotton (33.15 MT of lint) [9]. To meet the projected demand with the same amount of land i.e., 33 mha, there is a need to boost productivity to 28.7 t/ha of seed cotton. After the era of transgenic introductions, there is no new technological breakthrough to push the stagnant yield plateau to higher peaks. To sustain future demand with available scanty cultivable land with uncertain climatic vulnerabilities, there is a need for a strong, focused and coordinated cotton research among the world cotton research community. There lie huge prospects for molecular breeding to break the yield stagnation. Here we attempt to review the cotton genomics research carried out till date and its ability to meet the future demands.

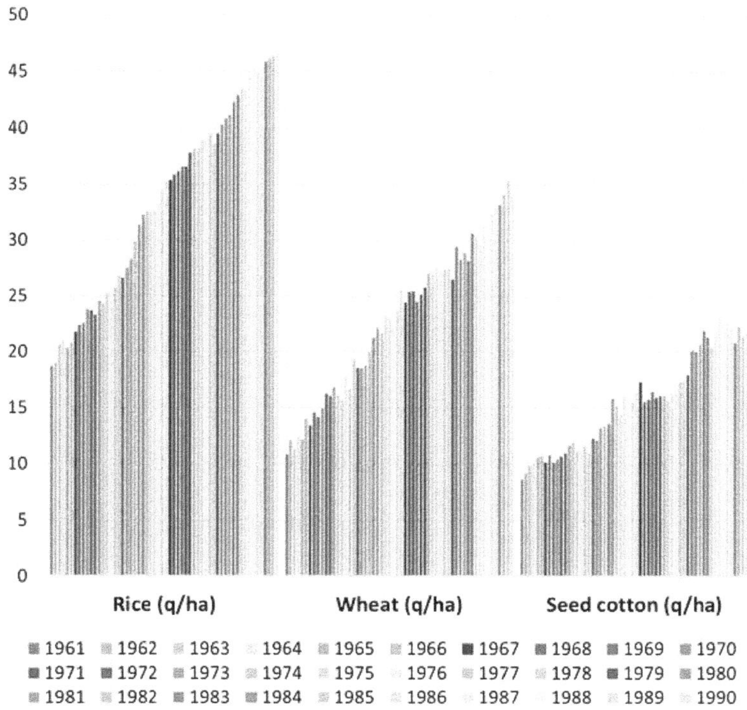

Figure 1. Productivity improvement comparison.

2. CURRENT SITUATION

Cotton is grown in around 105 countries and total cotton production during 2018–2019 was 71.02 million tons (Figure 2). India, China, United States of America, Brazil, Pakistan, Turkey, Uzbekistan, Australia, Greece and Benin are the top ten producers. India is the highest producer followed by China and United States. Australia has the highest productivity of 49.05 q/ha seed cotton followed by China (45.32 q/ha) and Brazil (39.76 q/ha) (Figure 3). In the last fifty years, the area is more or less the same with the productivity however showing an upward trend (Figure 2). The percent productivity improvement graph (Figure 4) shows us that there were two negative growth trends in the world during 1975 and 1995 but they were addressed with technological improvements. Currently, we are again witnessing a negative trend of cotton productivity growth around the world. India and USA together contribute 51% area of cotton cultivation, however the productivity (India: 13.9 t/ha & USA: 26.9 t/ha) (Figure 5) is low compared to other countries like Australia (49.05 q/ha seed cotton) and China (45.32 q/ha) [6], Improving cotton productivity in USA and India would change the outlook of cotton production sustainability. The world's two largest democracies, India & USA, have a mutual interest in promoting global security, stability and economic prosperity with cooperation at various levels. The co-operation can be extended to cotton research with other counties like Australia to make cotton productivity sustainable.

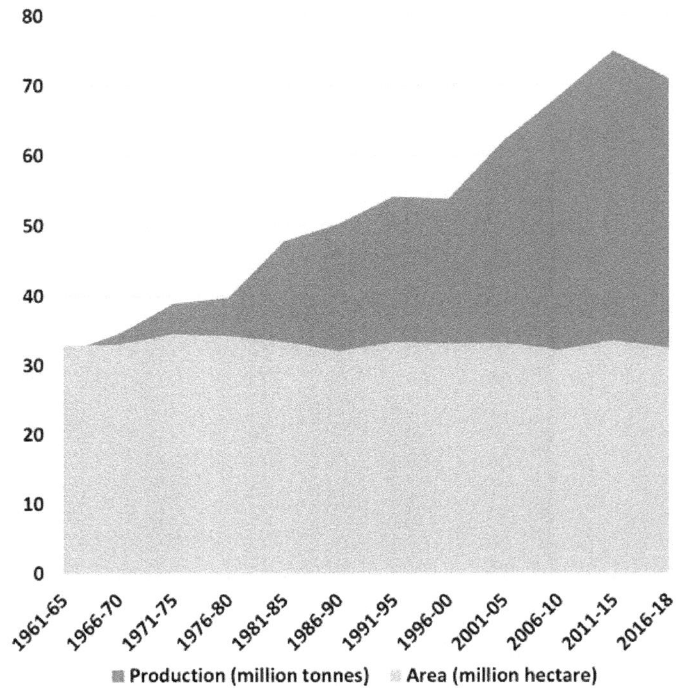

Figure 2. World area and production statistics.

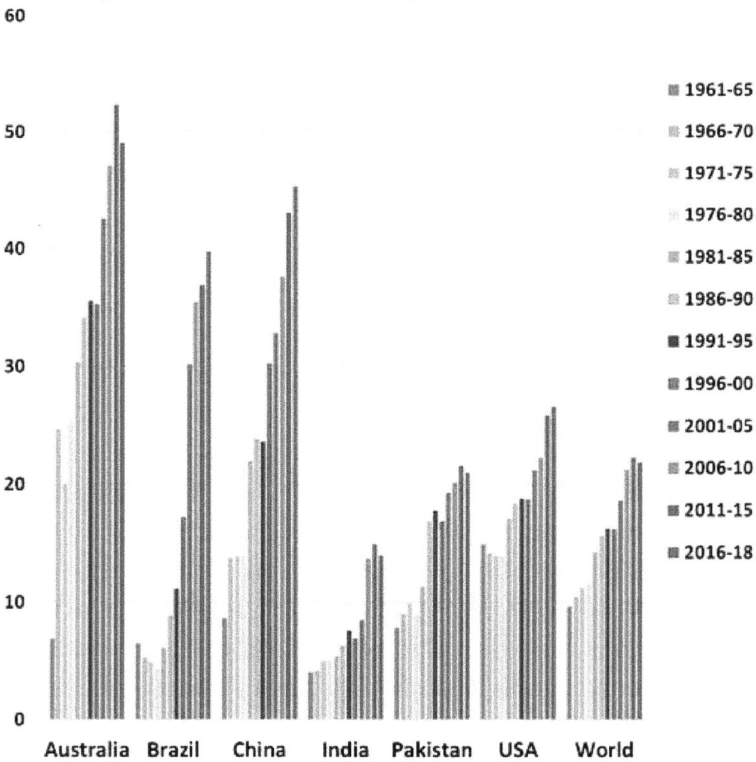

Figure 3. World cotton productivity dynamics (q/ha).

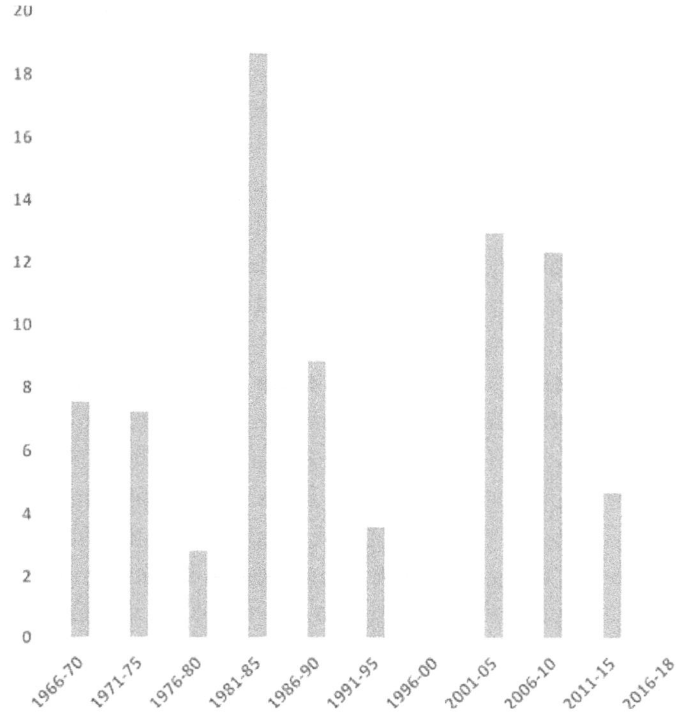

Figure 4. Percent cotton productivity growth.

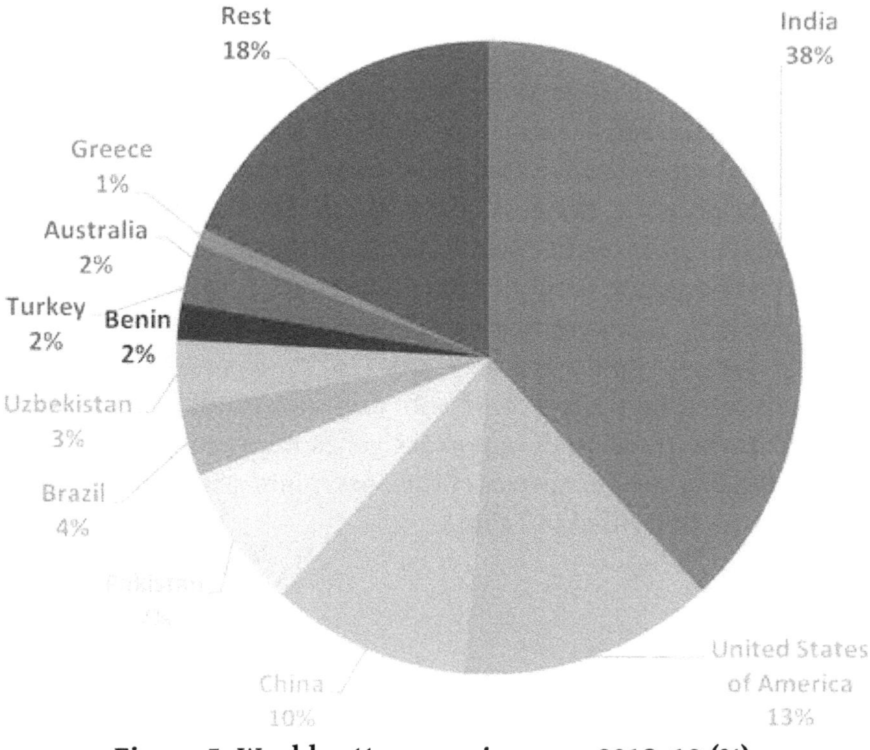

Figure 5. World cotton growing area 2018–19 (%).

3. GENETIC RESOURCE, ORIGIN, DISTRIBUTION AND USES

Genetic variability is the main driving force of all breeding programs. The basic requirement of varietal improvement/hybrid development/marker-assisted selection is the availability of genetic resources. Even basic molecular understanding requires the occurrence of special important morphological/ physiological characters upon which the studies are imposed. It is therefore necessary to understand and utilize within and between species diversity for crop improvement. The Cotton, belonging to family Malvaceae and genus Gossypium has high species diversity, which includes diploids and tetraploids with all the diploids sharing a common chromosome number i.e. 2n = 2x = 26. However, with 3-fold variation in DNA content per genome [10], they are classified into eight cytological groups (A, B, C, D, E, F, G, and H) [11, 12, 13]. All the tetraploids also share the same chromosome number i.e. 2n = 4x = 52 and have an AD genome. So, in total, the family Gossypium has around 51 recognized species, which includes 7 tetraploids and 44 diploids (Table 1). The genome A is thought to be originated in Asia/Africa, but the D genome is a derivative of A-genome formed by allopatric speciation i.e. due to trans-oceanic dispersion (Africa to Peru) of the A-genome. The modern-day tetraploids (AD-Genome) have originated from the trans-oceanic dispersion of A-genome to Peru followed by a polyploidization event with the native D-genome of Peru [14]. Cotton fiber is a single cell extension of the seed cell epidermis with deposition of cellulose. Only four species underwent the parallel selection pressure of domestication in America (*Gossypium hirsutum* and *G. barbadense*; tetraploids) and Africa-Asia (*G. arboreum* and *G. herbaceum*; diploids) and only these species produce the seed epidermal cell extension that is between 10 mm to 35 mm and hence are cultivated for lint purpose. The rest of the species produce lint less than 10 mm with varying shades of brown to white. Some of the recent updates in number of species in the Gossypium family include *G. trifurcatum* being tentatively placed in the B genome [15]. *G. lanceolatum* was proved to be a domesticated form of *G. hirsutum* and it does not hold a species status [16]. *G. stephensii* and *G. ekmanianum* are the two new tetraploids discovered with a species status [17, 18]. Wild species are considered as the treasure of important genes required to combat biotic and abiotic stress [19]. The various species of the Gossypium genus and their important traits of interest are presented in Table 1.

Table 1. Species diversity in cotton and their importance.

Sl. no.	Species	Genome	Ploidy/chromosome number	Origin	Habitat/Important traits
Primary Gene Pool					
1	G. hirsutum	AD1	2n = 4x = 52	Mexico	Cultivated
2	G. barbadense	AD2	2n = 4x = 52	South America	Cultivated, Verticillium wilt resistance
3	G. tomentosum	AD3	2n = 4x = 52	Hawaiian Islands	Wild, sucking pest tolerance, Drought/ Heat Resistance, Fiber strength
4	G. mustelinum	AD4	2n = 4x = 52	NE Brazil	Wild
5	G. darwinii	AD5	2n = 4x = 52	Galapagos Islands	Wild, Nematode resistance, Drought Resistance
6	G. ekmanianum	AD6	2n = 4x = 52	Dominican Republic	Wild
7	G. stephensii	AD7	2n = 4x = 52	Wake Atoll	Wild
Secondary Gene Pool					
1	G. herbaceum	A1	2n = 2x = 26	India	Cultivated, Drought resistance
2	G.arboreum	A2	2n = 2x = 26	Africa	Cultivated, Drought resistance
3	G. anomalum	B1	2n = 2x = 26	Africa	Wild, Fiber Length, Fiber Strength, Fiber fineness, Bollworm tolerance, Sucking pest tolerance, Bacterial Blight resistance, Drought Resistance, and Mite resistance.
4	G. triphyllum	B2	2n = 2x = 26	Cape Verde Islands	Wild, Jassid, Bollworm resistance, highly resistant to Bacterial blight
5	G. trifurcatum	B	2n = 2x = 26	Somalia	Wild
6	G. capitis-viridis	B3	2n = 2x = 26	Cape Verde Islands	Wild, Immune to bacterial blight

Sl. no.	Species	Genome	Ploidy/chromosome number	Origin	Habitat/Important traits
7	G. thurberi	D1	2n = 2x = 26	Sonora Desert	Wild, Fiber Strength, Bollworm tolerance, Fusarium wilt resistance, Frost resistance, Prolific boll bearing and high GOT.
8	G. armourianum	D2-1	2n = 2x = 26	Baja California	Wild, Bollworm tolerance, Sucking pest tolerance, Bacterial Blight resistance
9	G. harknessii	D2-2	2n = 2x = 26	Baja California	Wild, Verticillium wilt resistance, Fusarium wilt resistance, CMS male sterility source, Drought Resistance.
10	G. davidsonii	D3-d	2n = 2x = 26	Baja California	Wild, sucking pest tolerance, Resistance to salinity, and Bacterial Blight resistance.
11	G . klotzschianum	D3-k	2n = 2x = 26	Galapagos Islands	Wild, Sucking pest resistance
12	G. aridum	D4	2n = 2x = 26	Pacific slopes of Mexico	Wild, arborescent, CMS male sterility source, Drought resistance, High seed index
13	G. raimondii	D5	2n = 2x = 26	Pacific slopes of Peru	Wild, Fiber Length, Fiber Strength, Fiber finess, Bollworm tolerance, sucking pest tolerance, Bacterial Blight resistance, Drought Resistance, High GOT.

Sl. no.	Species	Genome	Ploidy/chromosome number	Origin	Habitat/Important traits
14	G. gossypioides	D6	2n = 2x = 26	South Central Mexico	Wild, Resistance to Leaf Hoppers
15	G. lobatum	D7	2n = 2x = 26	SW Mexico	Wild, arborescent, Resistance to bollworm
16	G. trilobum	D8	2n = 2x = 26	West-Central Mexico	Wild, CMS male sterility source, Glabrous leaves
17	G. laxum	D9	2n = 2x = 26	SW Mexico	Wild, arborescent
18	G. turneri	D10	2n = 2x = 26	NW Mexico	Wild
19	G . schwendimanii	D11	2n = 2x = 26	SW Mexico	Wild, arborescent
20	G. longicalyx	F1	2n = 2x = 26	East Central Africa	Wild, trailing shrub, Fiber Length, Fiber fineness, and Nematode resistance
Tertiary Gene Pool					
1	G. sturtianum	C1	2n = 2x = 26	Central Australia	Wild, Ornamental, Fiber Strength, Fusarium wilt resistance, Cold and Frost resistance and Insensitive to photoperiod
2	G. robinsonii	C2	2n = 2x = 26	Western Australia	Wild
3	G. stocksii	E1	2n = 2x = 26	Arabian Peninsula and the Horn of Africa	Wild, Fiber Length, Fiber Strength, Drought Resistance
4	G. somalense	E2	2n = 2x = 26	Horn of Africa and Sudan	Wild
5	G. areysianum	E3	2n = 2x = 26	Arabian Peninsula	Wild, Fiber Length, Fiber Strength, Drought Resistance

Sl. no.	Species	Genome	Ploidy/chromosome number	Origin	Habitat/Important traits
6	G. incanum	E4	2n = 2x = 26	Arabian Peninsula	Wild
7	G. benadirense	E	2n = 2x = 26	Somalia, Ethiopia, Kenya	Wild
8	G. bricchettii	E	2n = 2x = 26	Somalia	Wild
9	G. vollensenii	E	2n = 2x = 26	Somalia	Wild
10	G. bickii	G1	2n = 2x = 26	Central Australia	Wild
11	G. australe	G2	2n = 2x = 26	North Trans Australia	Wild, high GOT, Drought Resistance
12	G. nelsonii	G3	2n = 2x = 26	Central Australia	Wild
13	G. costulatum	K	2n = 2x = 26	North Kimberleys of W Australia	Wild, decumbent
14	G. populifolium	K	2n = 2x = 26	N Kimberleys, Australia	Wild
15	G. cunninghamii	K	2n = 2x = 26	The northern tip of NT, Australia	Wild
16	G. pulchellum	K	2n = 2x = 26	N Kimberleys, Australia	Wild
17	G. pilosum	K	2n = 2x = 26	NW Australia	Wild
18	G. anapoides	K	2n = 2x = 26	N Kimberleys, Australia	Wild
19	G. enthyle	K	2n = 2x = 26	N Kimberleys, Australia	Wild
20	G. exgiuum	K	2n = 2x = 26	N Kimberleys, Australia	Wild, prostrate,
21	G. londonderriense	K	2n = 2x = 26	N Kimberleys, Australia	Wild

Sl. no.	Species	Genome	Ploidy/chromosome number	Origin	Habitat/Important traits
22	G. marchantii	K	2n = 2x = 26	Australia	Wild, decumbent
23	G. nobile	K	2n = 2x = 26	N Kimberleys, Australia	Wild
24	G.rotundifolium	K	2n = 2x = 26	N Kimberleys, Australia	Wild, prostrate

G.arboreum and *G. herbaceum* species known as old-world cotton were majorly grown in the Indian sub-continent. The Indus valley discoveries prove that cotton was grown as early as in 6 millennium BC, with the use of cotton being mentioned in Rig Veda (15 century BC) and Manu's Dharmashastra (800 BC) [14]. From the Indian sub-continent cotton has spread to Mesopotamia, Egypt and Nubia. During the first century, cotton was introduced to Europe by the Arab traders. The East India Company that colonized India (1757) and started ruling were the biggest importers of raw cotton and they used to sell the finished goods to India and the world. *G. arboreum var neglecta* grown in Bengal was known to produce lint that could be spun to 480 counts yarn and made into Muslins which was a result of both a beautiful skill set and the cotton germplasm that were available. Garments produced here were called "webs of woven wind" [20]. However the East India Company that wanted to sell only their finished cotton garments, chopped off the thumbs of weavers and with the weavers lost, the germplasm too vanished from the world forever [21]. Though the polyploidization event of tetraploids happened in Peru, the *Gossypium hirsutum* and *G. barbadense* also called new world cottons originated in Mexico and Peru, respectively, from where it spread to both South and North America. The arrival of European colonists hastened the spread of the new world cotton to the rest of the world [22]. The East Indian Company also bought and introduced early maturing and high yield *G. hirsutum* cotton to India with many unsuccessful attempts being made (1790 in Bombay & Madras, 1840–42 in Deccan, Konkan and Hubli, 1853 in Punjab). However, the most significant development in terms of the spread of cotton was the introduction of Cambodia variety in Tamil Nadu region [23]. At present, *G. hirsutum* is cultivated in 95% cotton area due to its high yield ability. *G. barbadense* is the best source for fiber quality improvement of *G. hirsutum* as *G. barbadense* is known to produce lint that can be spun to 80–120 counts yarn [24]. However, owing to their low yields, *G. arboreum*, *G. herbaceum,* and *G. barbadense* are not grown widely. Maintaining germplasm and utilizing within-species variation are big challenges in varietal development as it will be an expensive proposition. The germplasm maintained elsewhere in different countries can be efficiently utilized in breeding programs. The list of germplasm preserved is mentioned in Table 2. Commendable efforts are needed in pre-breeding to utilize the rare alleles/genes present in wild species. In the principal crops like rice and wheat, the IRRI and CIMMYT, respectively, are taking up the pre-breeding work and the genetic materials are being supplied to breeders around the world. Some notable works concerning cotton pre-breeding include developing populations involving genes/segments/whole chromosomes from wild species. RHMBHMTUP-C4 a random mated population was developed involving *G. hirsutum*, *G. barbadense*, *G. mustelinum* and *G. tomentosum* [31]. RMBUP-C4 was developed from crossing three elite hirsutum lines with 18 chromosome substitution lines from *G. barbadense* [32]. There is huge scope for pre-breeding work in cotton to combat biotic and abiotic stresses.

Table 2. Country-wise list of germplasm maintained.

Country	G. hirsutum	G. barbadense	G. herbaceum	G. arboreum	Other species	Total	Reference
India	8851	536	565	2053	330	12,335	[25]
Uzbekistan	13,241	3019	1495	1185	31	18,971	[26]
United States of America	6302	1584	194	1729	502	10,311	[27]
China	7752	633	18	433	32	8868	[28]
Russia	4503	1057	336	365	15	6261	[29]
Brazil	1660	1509	19	219	889	4296	[29]
France	2173	483	50	69	294	3069	[29]
Australia	1573	99	39	211	22	1944	[30]

4. COTTON CONVENTIONAL BREEDING

Conventional breeding is the base for any trait improvement and without proper knowledge of conventional breeding techniques advanced molecular breeding techniques would surely lead to costly mistakes. Until the advent of molecular marker technologies, conventional breeding was the sole method for genetic improvement. Some of the popular conventional interventions in cotton were the development of determinate growth types. Development of early maturing types (resistant to boll weevils), the inclusion of morphological traits such as fergo bracts, glabrous leaves, nectariless, high gossypol for resistance to boll weevils and bollworms then followed.

Australian Conventional Breeding: American bollworm, Bacterial blight and Verticillium wilt were the major problems in Australia. Okra leaf types were used in an Australian breeding program by Norm Thomson to develop a variety called

Siokra1-1 which was the first okra leaf type along with bacterial blight resistance in 1985. Dr. Peter Reid released Verticillium wilt resistance variety which was popular outside Australia [33].

Indian Conventional Breeding: Introduction of *G. hirsutum* during the 1970s and development of first intra-hirsutum hybrid cotton (H4) in India by C T Patel in 1970 [34] and Development of first interspecific hybrid (*G. hirsutum* x *G. barbadense*) in cotton (Varalaxmi) by B.H. Katarki in 1972 [35] led to utilizing hybrid vigour for higher productivity in the Indian subcontinent. Morphological traits such as fergo bracts, glabrous leaves, nectariless, and antibiosis [36, 37, 38, 39, 40] were included in the breeding program for bollworm tolerance. Wild species were used widely used through introgression breeding in developing novel varieties like Badnawar-1, Khandwa-1, and Khandwa-2 from *G. hirsutum* x *G. tomentosum* cross, Arogya and PKV081 using *G. hirsutum* x *G. anomalum* cross, Devitej using *G. hirsutum* x *G. herbaceum* cross, SRT-1, Deviraj and Gujarat 67 using *G. hirsutum* x *G. arboreum* cross. MCU2 and MCU5 from *G. hirsutum* x *G. barbadense* cross [41].

US Conventional breeding: Boll weevils were a major threat to cotton cultivation and elimination of boll weevils by developing early maturing short-staple types was one of the significant interventions [20]. To reduce the negative association between yield and fiber quality, exotic germplasm was used in breaking the association and varieties like MD51ne with higher fiber quality were developed [42]. Development of sub okra, smooth leaf, and nectariless for reducing tarnished plant bug populations [43] was another achievement. Some of the private companies like Delta Pine made huge advances in cotton productivity improvement in cotton. They released mechanical harvest suitable variety called Delta pine smooth leaf which had around 25% US cotton area by 1963, Deltapine-16 with improved disease resistance and better fiber quality had around 28% area in the US by 1972, Delta pine Acala 90 premium quality cotton released was used as parental germplasm in development of many other varieties globally, all these interventions of Delta pine improved cotton production of United States significantly.

Uzbekistan Conventional Breeding: Turkestan Breeding Station established in 1992 with a major emphasis on the collection of cotton germplasm under the leadership of Dr. Zaitsev and Dr. Mauer was a major milestone in realizing the huge germplasm collection of present-day Uzbekistan. Early maturing types AK-Djura and Dehkam by Dr. Zaitsev were an important contribution. Utilizing *G. barbadense* as a resistant source to fusarium wilt and large boll types led to the release of 35-1 and 35-2 cultivars. Termez-14 high yielding cultivar developed by Dr. Ibragimov was another breakthrough. Development of Verticillium wilt resistance variety C-6524 by Dr. Alexander Avtonomov and Dr. Vadim Avtonomov had occupied more than two hundred thousand hectares for fifteen years till 2004 [44].

All these interventions along with production technologies have improved world productivity from 9.65 q/ha in 1960 to 16.20 q/ha in 2000 in the span of 40 years [6]. Conventional breeding methods are effective in transferring the traits but take considerably higher time, resources and uncertainty in the transfer of the trait. A breeding program involving large entries involves more samples needed to be tested for traits like fiber quality/oil content in cotton. We need to put a lot of resources and time to wait for the maturity of cotton. On the contrary, the Bt cotton technology, one of the spin-off technologies of genomics and molecular breeding played a significant role along with other production technologies and has achieved 5 q/ha improvement in just five years (Figure 4). Thus, cotton genomics/marker-assisted selection has huge potential in reducing a considerable amount of time, resource and assist conventional breeding in achieving future demands.

5. ADVANCES IN COTTON GENOME SEQUENCING

Having a reference genome is a boon since it is possible to characterize gene/gene families that are species-specific and which are further amenable to functional genomics work [45]. Since the publication "Toward Sequencing Cotton (Gossypium) Genomes" in 2007 by Chen and Co-workers [10], the framework was laid down for genome sequencing of cotton. The initial framework was to get first sequence of D-genome (G. raimondii) followed by A and then AD genome. [46] developed the first assembly in cotton (D-Genome). Now, as many as nine assemblies of G. hirsutum, four assemblies of G. barbadense, three assemblies of G. arboreum, three assemblies of G. raimondii and one each assembly of wild species such as G. australe, G. darwinii, G. longicalyx, G. mustelinum, G. tomentosum and G. turneri have been documented in CottonGen. The assembly statistics of various Gossypium genomes assembled are presented in Table 3. The reference genomes of cotton were used to produce a total of 17,224,361 SNPs that are documented in CottonGen by various researchers, the gene annotations and physical maps provided are valuable information for cotton scientists for various studies like the development of linkage maps, GWAS, validating the linkage maps, expression studies, development of guide RNAs in gene editing, genome-wide characterization studies of gene families etc. Since the sequencing cost is reducing day by day there are huge prospects for developing newer assemblies to catch all the variation (within and between species) for identifying rare alleles/genes that would help us to sustain future demands in cotton improvement.

Table 3. Genome assembly statistics of various cotton species.

Reference	[47]	[48]	[48]	[49]	[50]	[51]	[52]	[53]
Species	G. hirsutum	G. hirsutum	G. hirsutum	G. hirsutum	G. hirsutum	G. hirsutum	G. hirsutum	G. hirsutum
Cultivar	TM-1	TM1	ZM24	TM-1	TM1	TM-1	TM1	TM1
Number of contigs	1235	1283	3718	4831	6733	4746	265,279	44,816
N50 of Contigs (kb)	5020	4760	1976	113.02	7839	1891(L50)	34	80
No. of scaffolds	342	599	2238	48	1025	2190	40,407	8591
N50 of scaffolds (Mb)	—	—	—	15.510	108.1	97.73 (L50)	1.6	0.764
Total assembled genome size (Mb)/ Scaffold length (Mb)	2290	2286	2309	2295.26	2305.2	2347.01	2432.7	2173
Number of annotated protein coding genes	74,350	73,624	73,707	72,761	75,376	70,199	70,478	76,943

Reference	[50]	[49]	[51]	[54]	[47]	[55]	[56]	[47]
Species	G. barbadense	G. barbadense	G. barbadense	G. barbadense	G. arboreum	G. arboreum	G. arboreum	G. herbaceum
Cultivar	3–79	Hai7124	3–79	3–79	SXY1	SXY1	SXY1	Mutema
Number of Contigs	4766	6902	4930	—	2432	8223	40,381	1781
N50 of Contigs (kb)	1800	77.66	2151.56 (L50)	—	1832	1100	72	1915
No. of scaffolds	2048	29	3032	29,751	1269	—	7914	732
N50 of scaffolds (Mb)	93.8	23.44	92.88 (L50)	0.260	—	—	0.665	—
Total assembled genome size (Mb) / Scaffold length (Mb)	2195.8	2224.98	2266.65	2573.19	1637	—	1694	1556
Number of annotated protein coding genes	74,561	75,071	71,297	80,876	43,278	40,960	41,330	43,278

Reference	[57]	[58]	[46]	[59]	[50]	[60]	[50]	[50]	[57]
Species	G. raimondii	G. raimondii	G. raimondii	G. australe	G. darwinii	G. longicalyx	G. mustelinum	G. tomentosum	G. turneri
Cultivar	D5-4	D5-3 (CMD10)	—	G2-lz	AD5-32, no. 1808015.09	F1-1	1408120.09, 1408120.10, 1408121.01, 1408121.02, 1408121.03	7179.01, 02, 03	D10-3
Number of contigs	187	41,307	19,735	2598	821	17	2147	750	220
N50 of contigs (kb)	6291.83	44.9	135.6	1825.35	9100	95,880	2300	10,000	7909.23
No. of scaffolds	—	4715	1033	650	334	—	383	319	—
N50 of scaffolds (Mb)	58.81	2.2	6.0	143.60	101.9	—	106.8	102.9	60.46
Total assembled genome size (Mb)/ Scaffold length (Mb)	734.88	775.2	761.4	1752	2183	1190.67	2315	2193.6	755.20
Number of annotated protein coding genes	40,743	40,976	37,505	40,694	78,303	38,378	74,699	78,338	38,489

6. TRANSCRIPTOME STUDIES IN COTTON

The technique of isolating and characterizing the Spatio-temporal pool of mRNA to study the differential gene expression patterns between contrasting genotypes and understanding underlying alternative pathways for specific trait supremacy has been well implemented in cotton. Many characters have been targeted to find the key responsible genes and pathways, especially fiber initiation, and elongation being highly focused up on [61, 62, 63, 64, 65, 66, 67, 68]. Other characters like Green and Brown colored cotton [69, 70, 71], Cadmium tolerance [72], Cold stress [73], Drought stress [74], Nematode resistance [75], Semigamy in Pima cotton [76], Whitefly mediated cotton leaf curl infection transcriptome [77], the transcriptome of Mepiquat chloride-induced compact types using in cotton [78] have also been done in cotton. Many studies have been carried out to identify genes that are differentially expressed, however paucity of the causes of the differential expression, hints us towards epigenetic regulations and transcription factors as the probable cause. Very few methylation studies have been carried out in cotton for fiber quality [79], male sterility [80], cold stress [81], salt tolerance [82], fruiting branch development [83]. Thus, there is huge scope for studies like differential methylation, studies on transcription factor, and correlating them with the differential expression patterns.

7. GENETIC MARKERS IN COTTON

In cotton, various markers like restriction fragment length polymorphism (RFLP) [84], random amplified polymorphic DNA (RAPD) [85], amplified fragment length polymorphism (AFLP) [86], simple sequence repeats microsatellites (SSRs) [87, 88], sequence-related amplified polymorphism (SRAP) [89, 90], target region amplified polymorphism (TRAP) [91], inter simple sequence repeats (ISSRs) [92], expressed sequence tag-Simple Sequence Repeat (EST-SSRs) [93] and single nucleotide polymorphism (SNP) [94, 95, 96, 97] have been used for various genomic studies with each marker system having its advantages and disadvantages. However, SSRs were initially thought that they were sufficiently polymorphic but with the advent of high throughput SNPs, they are being less used. In the era of sequencing, the availability of the cotton reference genome is a boon to cotton researchers as a large number of SNPs are identified using whole-genome re-sequencing and transcriptome sequencing. Further, reduction in cost and advent of reduced representation sequencing methods like Genotype by sequencing (GBS) and Specific-locus amplified fragment sequencing (SLAF) provide scope for high throughput genotyping. To date, in CottonGen, 7,870,031 SSRs and 17,224,361 SNP markers are available for researchers for various studies [98]. There are few SNP arrays developed in cotton like 63k cotton array [99], 80k SNP array [100], and 50k array by Samir Sawanth and I.S. Katageri (unpublished) which are being utilized for linkage and association mapping studies. However, the SNPs associated with various traits identified using different techniques can't be used in a minimalistic laboratory with minimal cost involved, thus it is necessary to exploit trait-associated SNPs through different marker systems like CAPS (Cleaved Amplified Polymorphic Sequences) and dCAPS (derived CAPS) which require minimal laboratory set up. CAPS and dCAPS can be used as dominant marker systems and can be carried out in simple agarose gel electrophoresis. They are highly stable as they are specifically designed for certain genomic targets [101, 102, 103]. However, there are only a few CAPs and dCAPS markers developed in cotton. There are huge prospects for developing simple PCR-based markers in cotton so that the breeders working in a remote research station with minimal laboratory can take advantage of DNA markers in cotton.

8. MOLECULAR MAPPING AND QUANTITATIVE TRAIT MAPPING

The Quantitative Trait Loci identification helps in finding the association between a marker and measurable phenotype at the genomic level or understanding the genetics of traits under study. Various types of populations like F2 [104], Recombinant inbred lines (RILs) [105], Backcross inbred lines (BILs) [106] and Multi-parent Advanced Generation Inter Cross (MAGIC) [107] are commonly used in cotton. Bi-parental RIL Mapping is one of the most common methodologies successfully employed for identifying QTLs in cotton for various traits. Genome-wide association study is also used for developing genetic maps and developing an association between the trait and DNA markers in cotton germplasm. This technique allows detecting association among various markers and traits through assessing Linkage disequilibrium (LD-mapping). In cotton the construction of linkage maps and detection of QTLs for various economic traits has been in progress since 1994 with the first RFLP linkage map [84] being published after which many maps have been constructed [94, 96, 97, 105, 108]. Many genome-wide association studies have also been carried out [95, 107, 109]. Currently, there are around 249 QTL mapping and association studies using various populations and germplasm (Table 4), QTLs identified using Bi-parental mapping/GWAS are presented in Table 5. However, QTLs discovered for various studies indicate that Chromosomes 5, 7, 10 and 25 are harboring many QTLs for fiber length, similarly Chromosomes 7 and 21, for fiber strength. For yield (Seed cotton yield/Lint yield) Chromosomes 1, 13 and 26 seem to be very important. For boll weight, Chromosomes 7 and 13 harbor many QTLs as reported from various studies. For Boll number, Chromosome 25 and for Lint percentage Chromosomes 16 and 13 are over represented (Figure 6). Efforts have been made to develop linkage maps in wild species like *G. hirsutum* X *G. anomalum* [110], *G. trilobum* X *G. raimondii* [111], *G. nelsonii* x *G. austral* [112], *G. hirsutum* X *G.*

darwinii [113], *G. hirsutum* X *G. mustelinum* [114], *G. darwinii* X *G. darwinii* [115], *G. klotzschianum* X *G. davidsonii* [116], *G. hirsutum* x *G. tomentosum* [117, 118, 119] and *G. thurberi* x *G. trilobum* [104]. QTLs after validation can be used directly for marker-assisted selection. Transfer of QTL/pyramiding of QTLs is one way of realizing targeted trait introgression [120] or these QTLS can be utilized for fine mapping and map-based cloning before marker-assisted selection. However, only a few validation studies are done for the Virescent gene in Virescent mutants [121, 122], the fuzzless gene in the fuzzless mutant [123], traits like fiber length [124], Fiber strength [125], leaf shape [126] and QTL affecting root-knot nematode multiplication [127] etc. Though fine-mapping is done it would require still more concentrated efforts to dissect out the traits. There are no successful cotton cultivars deployed in the field that are developed using the identified QTLs unlike in crops like Rice (MAS 946-1, Swarna Sub-1 and Cadet) and Wheat (Patwin, Expresso and AGS2026). Now that the marker development and QTL mapping has been done to a greater depth in cotton, at least for major traits like fiber quality and yield, the focus around the world should now be on utilizing all the major QTLs identified in fine mapping and then in marker assisted selection.

Sl. no.	Genome	Number of maps
1.	*G. hirsutum* x *G. barbadense*	57
2.	*G. barbadense* x *G. hirsutum*	9
3.	*G.hirsutum* x *G.hirsutum*	99
4.	*G. barbadense* x *G. barbadense*	4
5.	*G. hirsutum* x *G. anomalum*	1
6.	*G. trilobum* x *G. raimondii*	1
7.	*G. australe* x *G. nelsonii*	1
8.	*G. hirsutum* x *G. darwinii*	1
9.	*G. hirsutum* x *G. mustelinum*	2
10.	*G. darwinii* x *G. darwinii*	1
11.	*G. davidsonii* x *G. klotzschianum*	1
12.	*G. hirsutum* x *G. tomentosum*	4
13.	*G. thurberi* x *G. trilobum*	1
14.	MAGIC (Multi parent advanced generation intercross)	6
	Total	**188**

Table 4. Number of Documented Quantitative Mapping studies in cotton.

Sl. no.	Trait	Documented QTLs
1.	Fiber strength	1123
2.	Fiber length	1147
3.	Boll weight	676
4.	Boll number	186
5.	Yield	275
6.	Lint Percentage	878
Total		4285

Table 5. Number of QTLs identified for major quality and yield traits in cotton.

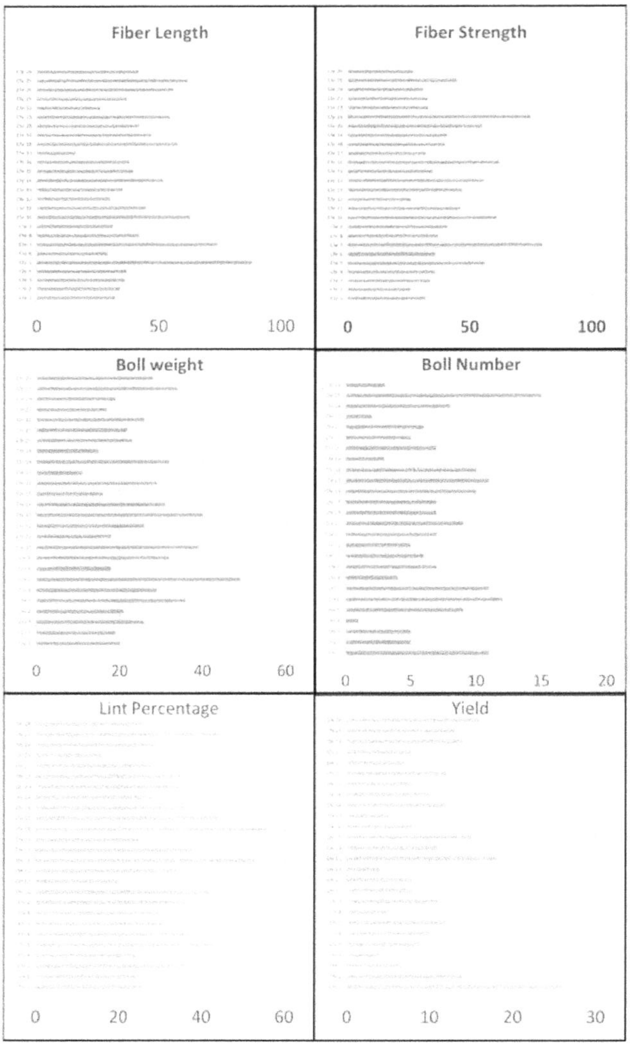

Figure 6. QTLs identified for fiber length, strength, boll weight and yield in cotton.

9. GENOMIC SELECTION IN COTTON

QTL mapping and genome-wide association studies have identified many genomic regions responsible for the important agronomic and fiber quality traits in cotton. Among them only a few traits like disease resistance and pest resistance were qualitatively governed by a few genes/QTLs with a major effect. Marker-assisted selection (MAS) is well-suited for handling these traits. But in the majority of the crops and also cotton, most of the yield, yield contributing and fiber quality traits are quantitatively governed by one or few QTLs with relatively large effects along with several QTLs with small effects, which are not captured through QTL mapping [128, 129]. Hence targeted phenotype has not been achieved successfully through Marker Assisted Selection. Under such a situation, genomic selection (GS) would seem to be a promising and powerful tool of genomics to breed for these traits. GS is a unique form of MAS, here the basis of selection is the genotypic data on marker alleles covering the entire genome, irrespective of whether the effects associated with these marker loci are significant or not [130]. Based on these marker effect estimates, genomic estimated breeding values (GEBVs) of different individuals/ lines will be calculated without actually phenotyping them, which forms the basis of selection (Figure 7). GS empirical studies in maize (*Zea mays*; [132, 133, 134, 135]), rice (*Oryza sativa*; [136, 137, 138, 139]), wheat (*Triticum aestivum*; [140, 141, 142, 143, 144]), and sorghum (*Sorghum bicolor*; [145, 146, 147]) have all recently shown how GS has become an efficient approach in crop breeding with recent developments in the implementation of various high-density array-based DNA marker technologies and their reduced genotyping costs. There are many marker effects estimation models that have been developed for the GS. Their predictability mainly depends on factors like marker density, training population size, and the relationship between training and breeding populations [131, 148]. Hence, the model which is capable of giving the highest GEBV accuracy will be selected. To date only two cotton GS studies have been reported. Islam et al. (2020) compared prediction ability (PA) and prediction accuracy (PACC) of Several GS models in cotton including genomic BLUP (GBLUP), ridge regression BLUP (rrBLUP), BayesB, Bayesian LASSO, and reproducing kernel Hilbert spaces (RKHS). And reported BayesB predicted the highest accuracies among the five GS methods tested and also the same model is suggested by Gapare et al. in 2018 in cotton. In many field crops for different traits, GS prediction accuracies of >0.80 have been reached [149, 150], but now in cotton, the accuracy of 0.71 and 0.59 for fiber length and strength has been achieved, respectively [150]. The prediction ability (PA) and prediction accuracy (PACC) was 0.65, and 0.69, respectively for fiber elongation [148]. In most plant breeding programs, especially in cotton, GS is still in its infancy and one of the biggest barriers to the implementation of GS in practical plant breeding is the high start-up cost required for accurate phenotyping, maintaining a large training population and costs of genotyping entire breeding populations. However, nowadays the genotyping costs are continually decreasing and genotyping of large plant populations is much more manageable than going in for conventional phenotyping. Soon, at points in the breeding program where selection using conventional methods is too costly and time-consuming, GS may have its greatest potential usage.

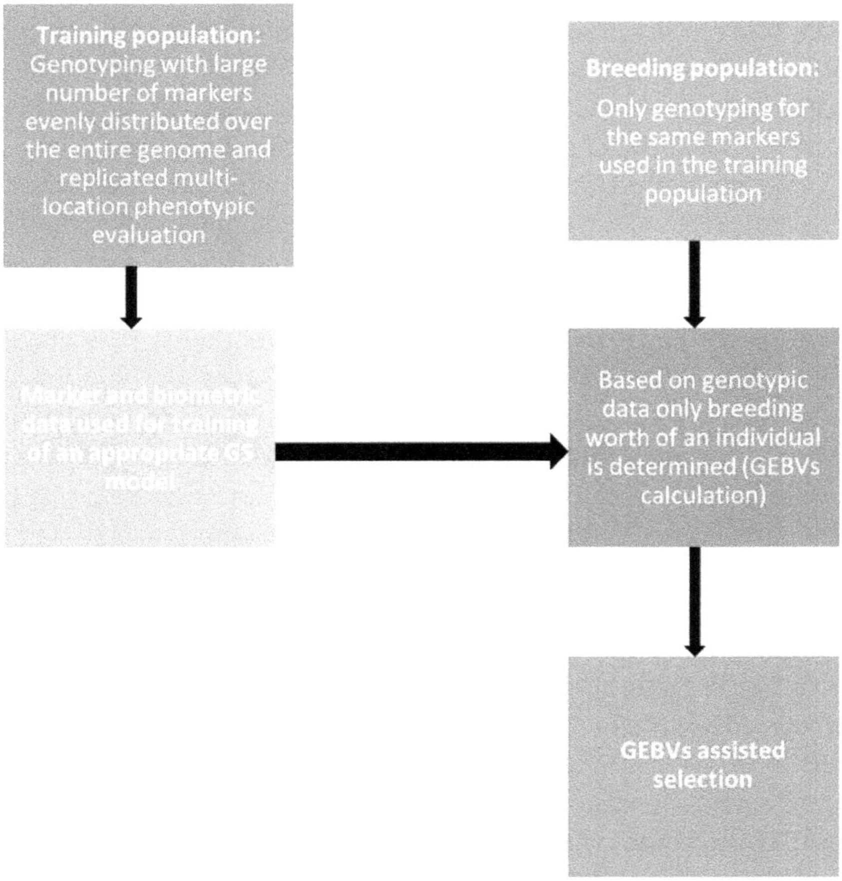

Figure 7.: Schematic representation of genomic selection (GS) scheme (based on [131]).

10. TRANSGENICS

With the advent of recombinant DNA technology in the 1970s, the genetic manipulation of plants entered a new age. Genes and traits previously unavailable through traditional breeding became available through DNA recombination and with greater specificity than ever before. This modern genetic engineering technology allows the transfer of genetic material across a wide range of species and has removed the traditional limits of crossbreeding. It involves the transfer of desired genes into the plant genome, and then regeneration of a whole plant from the transformed tissue/cell. For successful development of transgenic plants, identification of suitable target tissue and efficient gene transfer protocol are essential. Therefore, understanding the genetic variability of different crop plants and genotypes for in vitro regeneration and optimization of routine regeneration protocol is pre-requisite for the utilization of transformation technology in any crop. Currently, the most widely used method for transferring genes into plants is Agrobacterium-mediated transformation [151, 152, 153] and particle bombardment method [154]. Other methods, such as polyethylene glycol (PEG) mediated transformation [155] and electroporation [156] have also been used to transfer genes into plants. Cotton is a recalcitrant crop to generate from in vitro tissue cultures. Compared with many other crops, it is more difficult to obtain somatic embryogenesis, shoot multiplication and plant regeneration in cotton. The nature of tissue explants, the genetic makeup of the crop plant and presence of different growth hormones have a direct effect on regeneration potential. Genotype dependent genetic transformation is well studied and used commercially in cotton. Coker genotypes, which are amenable for regeneration in vitro by somatic embryogenesis, are widely used in genetic transformation experiments [151, 152, 153, 157]. Genotype independent genetic transformation techniques although developed [152, 158] show very low frequency of heritable gene incorporation. In the beginning, the two major goals of genetic engineering in cotton were to confer insect resistance and tolerance to more environmentally acceptable herbicides [159]. To date, 65 plus transgenic cotton events approved in India and all over the world. Continuous exposure of bollworms to BT cotton has led to resistance in them and thereby affecting the efficiency of controlling them. Cotton bollworm P450 monooxygenase gene (CYP6AE14) gene was silenced to impair larval tolerance to gossypol through the plant-mediated RNAi approach [160]. Genetic engineering is a remarkable breakthrough in modern crop improvement. Bt cotton came at the most opportune time when bollworms were causing a lot of destruction to the cotton crop making farmers helpless. Since its release in the USA in 1995, China during 1997 and in India during 2002 the Bt. technology has had a significant impact on bollworm control and reduction in usage of pesticides has been seen.

Acceptance of genetically altered cotton in various regions of the world is offering new opportunities for improvement of cotton yield and quality. Over-expression of GhUGP1 (Cotton uridine diphosphate glucose pyrophosphorylase) in upland cotton improves the fiber quality and reduces fiber sugar content [161]. Overexpression of novel sucrose synthase GhSusA1 gene leads to a considerable increase in biomass and fiber length with a moderate increase in fiber strength [162]. A silkworm fibroin gene was used to improve the fiber structure and quality [163]. The transgenic cotton plants expressing the fiberexpansin gene (GhEXPA8) showed a significant improvement in fiber lengths and micronaire values [63]. The fiber quality QTL-associated phytochrome PHYA1 gene was targeted through RNAi to explore the biological roles of PHYA1 and (indirectly) other phytochrome genes in cotton [164]. The elimination of gossypol from cottonseed has been a long-standing goal of geneticists. A cotton variant was obtained using antisense technology against (+)-delta-cadinene gene to suppress terpenoid aldehydes (gossypol) but with lysigenous glands [165]. RNAi-knockdown of delta-cadinene synthase gene(s) was used to engineer plants that produced ultra-low gossypol cottonseed (ULGCS) [166]. Recently, ultra-low gossypol cottonseed (ULGCS) was obtained by using PTGS and seed-specific promoter (α-globulin) through suppression of CDN genes and these lines are under field evaluations [167]. In the future, increased research investment on biotic and abiotic stresses through a transgenic approach is needed. Much focus is required for exploiting and improving cotton fiber and yield traits with the help of alien gene incorporation. In regards to public acceptance and questions, there is a need to carry on a massive public awareness campaign i.e. benefits, biosafety and risk assessment.

11. CRISPR CAS SYSTEM FOR CROP IMPROVEMENT IN COTTON

In the last decade, there is a revolution in the field of genome modification and continuous advancement in the targeted genome modification technologies. Genome editing tools like zinc-finger nucleases (ZFN), transcription activator-like effectors nucleases (TALENs) were extensively used before the advent of CRISPR Cas9 technology. These ZFN and TALENs technologies didn't become as popular as that of CRISPR Cas9 due to low efficiency, low specificity, low engineering feasibility and low design simplicity. CRISPR/Cas9 (Clustered Regularly Interspaced Short Palindromic Repeats/CRISPR-associated protein) system is the latest gene-editing technology, which has the power to alter the DNA or the code of life. Scientists predict that CRISPR has enormous potential for the next green revolution by 2050. To date, the CRISPR/Cas9 system has been successfully applied to efficient genome editing not only in model plants species but also in crop plant species. Many institutions and several groups all over the world are studying the feasibilities of the CRISPR/Cas9 system in cotton (*Gossypium hirsutum* L.). Successful use of CRISPR/Cas9 system in cotton still relies on Agrobacterium-mediated transformation and tissue culture, a genotype-dependent and low-efficiency process, but it provides a powerful tool for cotton functional genomics as it seems to be more efficient than RNA interference (RNAi) and virus-induced gene silencing (VIGS) in terms of knocking out the function of target genes [168]. The applications of the CRISPR/Cas9 system in cotton is been presented in Table 6. Chen and coworkers [169] demonstrated for the first time that the CRISPR/Cas9 can be used for advanced functional genomic research in cotton through targeted mutagenesis of two endogenous genes [Cloroplastosalterados 1 (GhCLA1) and vacuolar H + -pyrophosphatase (GhVP)]. Many factors influence the efficiency of the CRISPR/Cas9 system to obtain high mutation rates. [174] studied the sgRNA expression and mutagenesis efficiency by taking the endogenous U6 promoter over the existing one (Arabidopsis AtU6-29 promoter). Improved mutagenesis efficiency (4 to 6 times) was obtained by the use of an endogenous U6 promoter to drive the sgRNA expression. This study provided a fast and effective method to validate sgRNA mutagenesis efficiency in cotton using CRISPR/Cas9. Gao and coworkers [168] analyzed the nature of mutations induced by the CRISPR/Cas9 system through transient expression study of two genes Translation elongation factor 1 (GhEF1) and Phytoene desaturase (GhPDS) in cotton. The CRISPR/Cas9 system has been used for multiple sites targeting and simultaneously editing of multiple genes. Wang and his colleagues [171] successfully utilized the CRISPR/Cas9 system in allotetraploid cotton and accomplished multiple sites genome editing by targeting the exogenously transformed gene Discosoma red fluorescent protein2 (DsRed2) and an endogenous gene Cloroplastosalterados 1 (GhCLA1).

Table 6. The applications of the CRISPR/Cas9 system in cotton (*Gossypium hirsutum* L.).

Sl. no.	Gene	Mutation type	Method of Cas9 system delivery	Phenotype	Gene function	Reference
1	GhCLA1 (Chloroplasts alterados 1)	Nucleotide insertion and substitution	Agrobacterium-mediated transformation	Albino phenotype was observed	A novel gene for chloroplast development	[169]
	GhVP (vacuolar H+-pyrophosphatase)			—	Involved in both acidify intracellular compartments and to transport protons across the plasma membrane.	
2	GhMYB25-like A & GhMYB25-like D	Nucleotide insertions and deletions (indels)	Agrobacterium-mediated transformation	—	GhMYB25-like is involved in the development of cotton fiber.	[170]
3	Arginase (ARG)	Nucleotide insertions and deletions (indels)	Agrobacterium-mediated transformation	Improved lateral root system	Plays an important role in the regulation of lateral root formation.	[171]
4	GhPDS, GhCLA1 & GhEF1	Deletions (64%)	Agrobacterium mediated transformation	Albino phenotypes observed	A novel gene for chloroplast development.	[168]
5	An endogenous gene GhCLA1 and DsRed2 (Discosoma red fluorescent protein2)	Nucleotide insertions and deletions (indels)	Agrobacterium tumefaciensmediated transformation	Disappeared red fluorescence and showed albino phenotype	AtCLA1 is involved in the development of chloroplast. DsRed2 protein is utilized as a reporter due to its different benefits over other report proteins.	[171]

Sl. no.	Gene	Mutation type	Method of Cas9 system delivery	Phenotype	Gene function	Reference
6	ALARP	Nucleotide insertions and deletions	Agrobacterium-mediated transformation	—	A gene encoding alanine-rich protein that is preferentially expressed in cotton fibers	[172]
7	Cotton Gland Formation (CGF3)	Nucleotide insertions and deletions	Agrobacterium-mediated transformation	Glandless phenotype	plays a critical role in the formation of glands in the cotton plant	[162]
8	Cotton Gland Pigmentation 1 (CGP1)	Nucleotide insertions and deletions	Agrobacterium-mediated transformation	Decreased accumulation of gossypol and related terpenoids, as well as the color intensity in glands	CGP1 is an MYB Transcription Factor that regulates gossypol accumulation but not gland morphogenesis.	[173]

CRISPR/Cas9 has been used to edit a couple of agronomically important cotton genes, such as the genes involved in fiber development (GhMYB25-like A and GhMYB25-like D) [170] and a gene encoding arginase (ARG) for the increased lateral root formation [160]. Zhu and his colleagues [172] demonstrated the high editing efficiency of the CRISPR/Cas9 system in cotton by targeting-ALARP, a gene encoding an alanine-rich protein that is preferentially expressed in cotton fibers. CRISPR/Cas9 knockout of the Cotton Gland Formation (CGF3) gene resulted in a glandless phenotype in cotton. Gao and coworkers [168] confirmed the important role of Cotton Gland Pigmentation 1 (CGP1) in gland biology through CRISPR knockout of CGP1. Decreased accumulation of gossypol and of related terpenoids was observed in the CRISPR knockout plants. The above successful studies indicate that the CRISPR Cas9 system can further be effectively utilized in the functional genomics of cotton research. However, there are some limitations of the CRISPR/Cas9 system, including off-target effects, difficulties in PAM (protospacer adjacent motif) sequence selection for fewer potential target sites, and difficulties in generating homozygous mutations in the offspring [175, 176, 177]. Therefore, there is a lot of scope for the modification of the CRISPR Cas9 system or finding new alternative CRISPR systems. Zeng and co-workers [178], for the first time, established an efficient CRISPR/LbCpf1 system to expand the scope of genome editing in allotetraploid cotton by targeting the cotton endogenous gene Cloroplastosalterados (GhCLA). In addition to CRISPR/Cas9 & CRISPR/LbCpf1 system, a new effector with a single nuclease domain, a relatively small size, with low-frequency off-target effects, and cleavage capability under high temperature has been recently established and designated CRISPR/Cas12b (C2c1) [179]. CRISPR/Cas12b is a heat-induced system which requires a temperature ranging between 40 and 55°C for effective cleavage, when the temperature is lower than 40°C, cleavage cannot be accomplished [180, 181]. Recently the manipulation of the Cloroplastosalterados (GhCLA) gene in cotton plants using AacCas12b has been successfully established with no off-target effects. This system is ideal for plant species that can tolerate temperatures above 40°C, such as cotton that can grow well at temperatures reaching 45°C [171]. Some researchers are deactivating one or both of the Cas9's cutting domains and fusing new enzymes onto the protein. Cas9 can then be used to transport those enzymes to a specific DNA sequence. In one example, the Cas9 is fused to an enzyme, a deaminase, which mutates specific DNA bases eventually replacing cytidine with thymidine. [182] developed a new base editor system (GhBE3) consisting of a cytidine deaminase domain fused with nCas9 and uracil glycosylase inhibitor (UGI), for use in allotetraploid cotton, and obtained high base-editing efficiency with no detectable off-target effects. From all the above studies, it is indicated that CRISPR/Cas9 and its alternatives are potential gene-editing tools which would be superior to RNAi for cotton functional genomics. In future, this technology will have much scope for targeting tolerance to sucking pests, increased fiber yield and improved fiber quality traits in cotton.

12. CONCISE SUMMARY

Conventional Cotton breeders around the world had made a significant impact on cotton productivity improvement and germplasm conservation by their meticulous research that was developed during the early time is invaluable and highly significant. The advent of newer technology is an added advantage to the new young breeders since they can take advantage of newer technologies, but however prior knowledge of thorough conventional breeding, its limitations, and advantages of advanced molecular breeding, its limitation has to be kept in mind. A small mistake made initially during early molecular breeding may make us pay a heavy price in the end. The wealth of data like QTLs/Transgenic events/Gene-edited lines already developed can be cautiously used in crop improvement programs and further research in advanced technologies like genomic selection, fine mapping, and gene editing has to be the priority area of research for the sustainable cotton production.

AT A GLANCE

Conventional breeding interventions in cotton have been successful and these techniques have doubled the productivity of cotton, but it took around 40 years. One of the techniques of molecular biology i.e., genetic engineering has brought significant improvement in productivity within the year of introduction. With cotton genomics maturing, many reference genomes and related genomic resources have been developed. Newer wild species have been discovered and many countries are conserving genetic resources within and between species. This valuable germplasm can be exchanged among countries for increasing cotton productivity. As many as 249 Mapping and Association studies have been carried out and many QTLs have been discovered and it is high time for researchers to get into fine-mapping studies. Techniques of genomic selection hold valuable trust for deciphering quantitative traits like fiber quality and productivity since they take in to account all minor QTLs. There are just two studies involving genomic selection in cotton, underlining its huge prospects in cotton research. Genome editing and transformation techniques have been widely used in cotton with as many as 65 events being developed across various characters, and eight studies carried out using crisper technology. These promising technologies have huge prospects for cotton production sustainability.

REFERENCES

1. Shikha A, Bhuyan S. Cotton Crop: Various Aspects and Transition from Past, Present and Future. International Journal of Agriculture & Environmental Science. 2017;4
2. Moulherat C, Tengberg M, Haquet J-F, Mille Bt. First evidence of cotton at Neolithic Mehrgarh, Pakistan: analysis of mineralized fibres from a copper bead. Journal of Archaeological Science. 2002;29(12):1393-401
3. Krifa M, Stevens SS. Cotton utilization in conventional and non-conventional textiles—a statistical review. Agricultural Sciences. 2016;7(10):747-58
4. Mahir FI, Keya KN, Sarker B, Nahiun KM, Khan RA. A brief review on natural fiber used as a replacement of synthetic fiber in polymer composites. Materials Engineering Research. 2019;1(2):86-97
5. FAO I. Measuring Sustainability in Cotton Farming Systems: Towards a Guidance Framework. Rome (Accessed 05 July 18). 2015
6. Faostat. Food and Agriculture Organisation of the United Nations 2020. Available from: http://www.fao.org/faostat/en/#data
7. FicciaWA. Global Shifts in Textile Industry & India's Position 2016. 1-29. Available from: http://ficci.in/spdocument/20817/3-FICCI-TAG-2016-Whitepaper.pdf
8. Lutz W, KC S. Dimensions of global population projections: what do we know about future population trends and structures? Philosophical Transactions of the Royal Society B: Biological Sciences. 2010;365(1554):2779-91
9. CICR. Vision 2050: Indian Council of Agricultural Research, New Delhi; 2015. Available from: https://www.cicr.org.in/
10. Chen ZJ, Scheffler BE, Dennis E, Triplett BA, Zhang T, Guo W, et al. Toward sequencing cotton (Gossypium) genomes. Plant physiology. 2007;145(4):1303-10
11. Wendel JF, Brubaker C, Alvarez I, Cronn R, Stewart JM. Evolution and natural history of the cotton genus. Genetics and genomics of cotton: Springer; 2009. p. 3-22
12. Wendel JF, Grover CE. Taxonomy and evolution of the cotton genus, Gossypium. Cotton. 2015;57:25-44
13. Craven L, Stewart JMD, Brown A, Grace J, editors. The Australian wild species of Gossypium. Challenging the Future: Proceedings of the World Cotton Research Conference I; 1994: CSIRO
14. Khadi B, Santhy V, Yadav M. Cotton: an introduction. Cotton: Springer; 2010. p. 1-14

15. Rapp RA, Alvarez I, Wendel JF. Molecular confirmation of the position of Gossypium trifurcatum Vollesen. Genetic Resources and Crop Evolution. 2005;52(6):749-53

16. Brubaker CL, Wendel JF. On the specific status of Gossypium lanceolatum Todaro. Genetic resources and crop evolution. 1993;40(3):165-70

17. Grover C, Zhu X, Grupp K, Jareczek J, Gallagher J, Szadkowski E, et al. Molecular confirmation of species status for the allopolyploid cotton species, Gossypium ekmanianumWittmack. Genetic Resources and Crop Evolution. 2015;62(1):103-14

18. Gallagher JP, Grover CE, Rex K, Moran M, Wendel JF. A new species of cotton from Wake Atoll, Gossypium stephensii (Malvaceae). Systematic Botany. 2017;42(1):115-23

19. Shim J, Mangat P, Angeles-Shim R. Natural variation in wild Gossypium species as a tool to broaden the genetic base of cultivated cotton. J Plant Sci Curr Res. 2018;2(005)

20. Ware JO. Plant Breeding and the Cotton Industry: US Department of agriculture; 1936

21. Islam MS. Muslin! Our story: Drik Picture Library; 2016

22. Wendel JF, Brubaker CL, Percival AE. Genetic diversity in Gossypium hirsutum and the origin of upland cotton. American Journal of Botany. 1992;79(11):1291-310

23. Gumber R, Gill M, Dharminder Pathak GJ, Sarlach R. History and status of cotton. Cotton Research in Punjab. 2008:1-2

24. Manickam S, Gururajan K, Gopalakrishnan N. Development of isogenic restorer line in extra long staple cotton variety Suvin. Electronic Journal of Plant Breeding. 2010;1(4):632-6

25. CICR. Central Institute for Cotton research Nagpur2015. Available from: https: //www.cicr.org.in/CropImprovement.html

26. Abdurakhmonov IY. World cotton germplasm resources: BoD–Books on Demand; 2014

27. Percy R, Frelichowski J, Arnold M, Campbell B, Dever J, Fang D, et al. The US National cotton germplasm collection—its contents, preservation, characterization, and evaluation. World cotton germplasm resources InTech, Rijeka. 2014:167-201

28. Jia Y, Sun J, Du X. Cotton germplasm resources in China. World cotton germplasm resources. 2014:35-53

29. Campbell B, Saha S, Percy R, Frelichowski J, Jenkins JN, Park W, et al. Status of the global cotton germplasm resources. Crop science. 2010;50(4):1161-79
30. Stiller WN, Wilson IW. Australian cotton germplasm resources. World Cotton Germplasm Resources. 2014:1-34
31. Jenkins JN, McCarty JC, Hayes RW, Stelly DM, Saha S. Registration of RMBHMTUP-C4, a Random-Mated Cotton Population Containing Alleles from Four Gossypium Species. Journal of Plant Registrations. 2019;13(3):411-5
32. Jenkins J, McCarty Jr J, Gutierrez O, Hayes R, Jones D. Registration of RMBUP-C4, a Random-Mated Population with Gossypium barbadense L. Alleles Introgressed into Upland Cotton Germplasm. Journal of Plant Registrations. 2013;7(2):224-8
33. Constable G, Reid P, Thomson N. Approaches utilized in breeding and development of cotton cultivars in Australia. 2001
34. Patel C. Evolution of hybrid-4 cotton. Current Science. 1981;50(8):343-6
35. Katarki B. Varalaxmi hybrid cotton--a valuable import substitute. Cotton Dev. 1972
36. Kadapa S, Malgar S, Onkarappa K. Variability for number of glands in the leaves of a few cotton varieties (Gossypium hirsutum L.). Current Research. 1980;9(7):121-2
37. Kadapa S, Thimmaiah G, editors. Breeding for Bollworm Resistance in Cotton (G. hirsutum L.). Proceedings of the IV International Congress of Genetics, New Delhi, Abs; 1983
38. Rao P, Kadappa S, Katageri I. Variability in morphological characters of bollworm tolerant and susceptible cotton genotypes. Journal of Maharashtra Agricultural Universities (India). 1991
39. KatageriISaK, S.N. Heterosis for yield and yield contributing characters in bollworm tolearant interspecific hybrids (G. hirsutum x G. barbadense). Indian J Genet. 1989;49(1):107-11
40. Katageri IS, Kadapa, S.N. and Goud, J.V. A case study of exploitation of heterosis using multiple pest tolerant G. hirsutum cotton strains. Indian J Genet. 1990;50(4):314-9
41. Sundaram V. Handbook of cotton in India. 1999
42. Meredith W. Registration of 'MD51ne'cotton. Crop science. 1993;33(6):1415-
43. Meredith W. Registration of eight sub-okra, semi-smooth, and nectariless near-isolines of DES 119 cotton germplasm. Crop science. 1998;38(6):1725-

44. Ibragimov PS, Avtonomov VA, Amanturdiev AB, Namazov SE, Zaurov DE, Molnar TJ, et al. Uzbek scientific research institute of cotton breeding and seed production: breeding and germplasm resources. J Cotton Sci. 2008;12:62-72

45. Brandies P, Peel E, Hogg CJ, Belov K. The value of reference genomes in the conservation of threatened species. Genes. 2019;10(11):846

46. Paterson AH, Wendel JF, Gundlach H, Guo H, Jenkins J, Jin D, et al. Repeated polyploidization of Gossypium genomes and the evolution of spinnable cotton fibres. Nature. 2012;492(7429):423-7

47. Huang G, Wu Z, Percy RG, Bai M, Li Y, Frelichowski JE, et al. Genome sequence of Gossypium herbaceum and genome updates of Gossypium arboreum and Gossypium hirsutum provide insights into cotton A-genome evolution. Nature genetics. 2020;52(5):516-24

48. Yang Z, Ge X, Yang Z, Qin W, Sun G, Wang Z, et al. Extensive intraspecific gene order and gene structural variations in upland cotton cultivars. Nature communications. 2019;10(1):1-13

49. Hu Y, Chen J, Fang L, Zhang Z, Ma W, Niu Y, et al. Gossypium barbadense and Gossypium hirsutum genomes provide insights into the origin and evolution of allotetraploid cotton. Nature genetics. 2019;51(4):739-48

50. Chen ZJ, Sreedasyam A, Ando A, Song Q, De Santiago LM, Hulse-Kemp AM, et al. Genomic diversifications of five Gossypium allopolyploid species and their impact on cotton improvement. Nature genetics. 2020;52(5):525-33

51. Wang M, Tu L, Yuan D, Zhu D, Shen C, Li J, et al. Reference genome sequences of two cultivated allotetraploid cottons, Gossypium hirsutum and Gossypium barbadense. Nature genetics. 2019;51(2):224-9

52. Zhang T, Hu Y, Jiang W, Fang L, Guan X, Chen J, et al. Sequencing of allotetraploid cotton (Gossypium hirsutum L. acc. TM-1) provides a resource for fiber improvement. Nature biotechnology. 2015;33(5):531-7

53. Li F, Fan G, Lu C, Xiao G, Zou C, Kohel RJ, et al. Genome sequence of cultivated Upland cotton (Gossypium hirsutum TM-1) provides insights into genome evolution. Nature biotechnology. 2015;33(5):524-30

54. Yuan D, Tang Z, Wang M, Gao W, Tu L, Jin X, et al. The genome sequence of Sea-Island cotton (Gossypium barbadense) provides insights into the allopolyploidization and development of superior spinnable fibres. Scientific reports. 2015;5:17662

55. Du X, Huang G, He S, Yang Z, Sun G, Ma X, et al. Resequencing of 243 diploid cotton accessions based on an updated A genome identifies the genetic basis of key agronomic traits. Nature genetics. 2018;50(6):796-802

56. Li F, Fan G, Wang K, Sun F, Yuan Y, Song G, et al. Genome sequence of the cultivated cotton Gossypium arboreum. Nature genetics. 2014;46(6):567-72
57. Udall JA, Long E, Hanson C, Yuan D, Ramaraj T, Conover JL, et al. De novo genome sequence assemblies of Gossypium raimondii and Gossypium turneri. G3: Genes, Genomes, Genetics. 2019;9(10):3079-85
58. Wang K, Wang Z, Li F, Ye W, Wang J, Song G, et al. The draft genome of a diploid cotton Gossypium raimondii. Nature genetics. 2012;44(10):1098-103
59. Cai Y, Cai X, Wang Q, Wang P, Zhang Y, Cai C, et al. Genome sequencing of the Australian wild diploid species Gossypium australe highlights disease resistance and delayed gland morphogenesis. Plant biotechnology journal. 2020;18(3):814-28
60. Grover CE, Pan M, Yuan D, Arick MA, Hu G, Brase L, et al. The Gossypium longicalyx genome as a resource for cotton breeding and evolution. G3: Genes, Genomes, Genetics. 2020;10(5):1457-67
61. Shi Y-H, Zhu S-W, Mao X-Z, Feng J-X, Qin Y-M, Zhang L, et al. Transcriptome profiling, molecular biological, and physiological studies reveal a major role for ethylene in cotton fiber cell elongation. The plant cell. 2006;18(3):651-64
62. Lu Q, Shi Y, Xiao X, Li P, Gong J, Gong W, et al. Transcriptome analysis suggests that chromosome introgression fragments from sea island cotton (Gossypium barbadense) increase fiber strength in upland cotton (Gossypium hirsutum). G3: Genes, Genomes, Genetics. 2017;7(10):3469-79
63. Bajwa KS, Shahid AA, Rao AQ, Bashir A, Aftab A, Husnain T. Stable transformation and expression of GhEXPA8 fiberexpansin gene to improve fiber length and micronaire value in cotton. Frontiers in plant science. 2015;6:838
64. Padmalatha KV, Dhandapani G, Kanakachari M, Kumar S, Dass A, Patil DP, et al. Genome-wide transcriptomic analysis of cotton under drought stress reveal significant down-regulation of genes and pathways involved in fibre elongation and up-regulation of defense responsive genes. Plant molecular biology. 2012;78(3):223-46
65. Padmalatha KV, Patil DP, Kumar K, Dhandapani G, Kanakachari M, Phanindra ML, et al. Functional genomics of fuzzless-lintless mutant of Gossypium hirsutum L. cv. MCU5 reveal key genes and pathways involved in cotton fibre initiation and elongation. BMC genomics. 2012;13(1):1-15
66. Hande AS, Katageri IS, Jadhav MP, Adiger S, Gamanagatti S, Padmalatha KV, et al. Transcript profiling of genes expressed during fibre development in diploid cotton (Gossypium arboreum L.). BMC genomics. 2017;18(1):675

67. Li P-t, Wang M, Lu Q-w, Ge Q, Liu A-y, Gong J-w, et al. Comparative transcriptome analysis of cotton fiber development of Upland cotton (Gossypium hirsutum) and Chromosome Segment Substitution Lines from G. hirsutum× G. barbadense. BMC genomics. 2017;18(1):705

68. Yu S, Cheng G, Zhang L, Wei H, Wang H, Lu J. Transcriptome analysis reveals gene expression associated with fuzz fiber initiation regulated by high-temperature in Gossypium barbadense. 2020

69. Malik W, Khan AA, Cheema HMN, Aslam U, Iqbal MZ, Qayyum A, et al. Transcriptome analysis of pigment related genes in colored cotton. International Journal of Agriculture and Biology. 2015;17(1):205-10

70. Tang Z, Fan Y, Zhang L, Zheng C, Chen A, Sun Y, et al. Quantitative metabolome and transcriptome analysis reveals complex regulatory pathway underlying photoinduced fibercolor formation in cotton. Gene. 2020:145180

71. Gong W, He S, Tian J, Sun J, Pan Z, Jia Y, et al. Comparison of the transcriptome between two cotton lines of different fibercolor and quality. PloS one. 2014;9(11):e112966

72. Han M, Lu X, Yu J, Chen X, Wang X, Malik WA, et al. Transcriptome Analysis Reveals Cotton (Gossypium hirsutum) Genes That Are Differentially Expressed in Cadmium Stress Tolerance. International journal of molecular sciences. 2019;20(6):1479

73. Li Z-B, Zeng X-Y, Xu J-W, Zhao R-H, Wei Y-N. Transcriptomic profiling of cotton Gossypium hirsutum challenged with low-temperature gradients stress. Scientific data. 2019;6(1):1-7

74. Bowman MJ, Park W, Bauer PJ, Udall JA, Page JT, Raney J, et al. RNA-Seq transcriptome profiling of upland cotton (Gossypium hirsutum L.) root tissue under water-deficit stress. PLoS One. 2013;8(12):e82634

75. Li R, Rashotte AM, Singh NK, Lawrence KS, Weaver DB, Locy RD. Transcriptome analysis of cotton (Gossypium hirsutum L.) genotypes that are susceptible, resistant, and hypersensitive to reniform nematode (Rotylenchulusreniformis). PLoS One. 2015;10(11):e0143261

76. Curtiss J, Rodriguez-Uribe L, Stewart JM, Zhang J. Identification of differentially expressed genes associated with semigamy in Pima cotton (Gossypium barbadensel.) through comparative microarray analysis. BMC plant biology. 2011;11(1):49

77. Naqvi RZ, Zaidi SS-e-A, Mukhtar MS, Amin I, Mishra B, Strickler S, et al. Transcriptomic analysis of cultivated cotton Gossypium hirsutum provides insights into host responses upon whitefly-mediated transmission of cotton leaf curl disease. PloS one. 2019;14(2):e0210011

78. Wang L, Yin Y, Wang L-F, Wang M, Zhao M, Tian Y, et al. Transcriptome profiling of the elongating internode of cotton (Gossypium hirsutum L.) seedlings in response to mepiquat chloride. Frontiers in plant science. 2019;10:1751

79. Shreeraksha R. J. ISK, S. A. Gowda and N. V. M. Kumar. DNA methylation in differential gene expression during fiber initiation of cotton (Gossypium spp.). Journal of Farm science. 2020;33(3):319-25

80. Zhang M, Guo L, Qi T, Zhang X, Tang H, Wang H, et al. Integrated Methylome and Transcriptome Analysis between the CMS-D2 Line ZBA and Its Maintainer Line ZB in Upland Cotton. International journal of molecular sciences. 2019;20(23):6070

81. Fan HH, Wei J, Li TC, Li ZP, Guo N, Cai YP, et al. DNA methylation alterations of upland cotton (Gossypium hirsutum) in response to cold stress. Acta physiologiae plantarum. 2013;35(8):2445-53

82. Wang B, Zhang M, Fu R, Qian X, Rong P, Zhang Y, et al. Epigenetic mechanisms of salt tolerance and heterosis in Upland cotton (Gossypium hirsutum L.) revealed by methylation-sensitive amplified polymorphism analysis. Euphytica. 2016;208(3):477-91

83. Sun Q, Qiao J, Zhang S, He S, Shi Y, Yuan Y, et al. Changes in DNA methylation assessed by genomic bisulfite sequencing suggest a role for DNA methylation in cotton fruiting branch development. PeerJ. 2018;6:e4945

84. Reinisch AJ, Dong J-M, Brubaker CL, Stelly DM, Wendel JF, Paterson AH. A detailed RFLP map of cotton, Gossypium hirsutum x Gossypium barbadense: chromosome organization and evolution in a disomic polyploid genome. Genetics. 1994;138(3):829-47

85. Kohel RJ, Yu J, Park Y-H, Lazo GR. Molecular mapping and characterization of traits controlling fiber quality in cotton. Euphytica. 2001;121(2):163-72

86. Lacape J-M, Nguyen T-B, Thibivilliers S, Bojinov B, Courtois B, Cantrell RG, et al. A combined RFLP SSR AFLP map of tetraploid cotton based on a Gossypium hirsutum× Gossypium barbadense backcross population. Genome. 2003;46(4):612-26

87. Lacape J-M, Jacobs J, Arioli T, Derijcker R, Forestier-Chiron N, Llewellyn D, et al. A new interspecific, Gossypium hirsutum× G. barbadense, RIL population: towards a unified consensus linkage map of tetraploid cotton. Theoretical and applied genetics. 2009;119(2):281-92

88. Yu Y, Yuan D, Liang S, Li X, Wang X, Lin Z, et al. Genome structure of cotton revealed by a genome-wide SSR genetic map constructed from a BC 1 population between Gossypium hirsutum and G. barbadense. BMC genomics. 2011;12(1):1-14

89. Lin Z-x, He D, Zhang X-l, Nie Y, Guo X, Feng C, et al. Linkage map construction and mapping QTL for cotton fibre quality using SRAP, SSR and RAPD. Plant breeding. 2005;124(2):180-7

90. Zhang Z-S, Xiao Y-H, Luo M, Li X-B, Luo X-Y, Hou L, et al. Construction of a genetic linkage map and QTL analysis of fiber-related traits in upland cotton (Gossypium hirsutum L.). Euphytica. 2005;144(1-2):91-9

91. Hu J, Vick BA. Target region amplification polymorphism: a novel marker technique for plant genotyping. Plant Molecular Biology Reporter. 2003;21(3):289-94

92. Liu B, Wendel JF. Intersimple sequence repeat (ISSR) polymorphisms as a genetic marker system in cotton. Molecular Ecology Notes. 2001 Sep;1(3):205-8

93. Sekhar L, Khadi B, Patil RS, Katageri I, Mukri G. Biochemical and molecular dissection of thermo-sensitive genetic male sterility in diploid cotton (Gossypium arboreum L.). Journal of Environmental Biology. 2016;37(4):579

94. Kumar NM, Katageri IS, Gowda SA, Adiger S, Yadava SK, Lachagari VR. 63K SNP chip based linkage mapping and QTL analysis for fibre quality and yield component traits in Gossypium barbadense L. cotton. Euphytica. 2019;215(1):6

95. Handi SS, Katageri IS, Adiger S, Jadhav MP, Lekkala SP, Reddy Lachagari VB. Association mapping for seed cotton yield, yield components and fibre quality traits in upland cotton (Gossypium hirsutum L.) genotypes. Plant Breeding. 2017;136(6):958-68

96. Ramesh U, Methre R, Kumar N, Katageri IS, Gowda SA, Adiger S, et al. Genome mapping and molecular markers identification for yield, yield component and fibre quality traits in tetraploid cotton. Plant Breeding. 2019;138(6):880-96

97. Sankeshwar M, Jadhav M, Adiger S, Patil RS, Katageri I. Mapping of QTLs for traits related to leaf pubescence, jassid resistance and yield in cotton (Gossypium spp.). Indian J Genet. 2018;78(2):252-60

98. Yu J, Jung S, Cheng C-H, Ficklin SP, Lee T, Zheng P, et al. CottonGen: a genomics, genetics and breeding database for cotton research. Nucleic acids research. 2014;42(D1):D1229-D36

99. Hulse-Kemp AM, Lemm J, Plieske J, Ashrafi H, Buyyarapu R, Fang DD, et al. Development of a 63K SNP array for cotton and high-density mapping of intraspecific and interspecific populations of Gossypium spp. G3: Genes, Genomes, Genetics. 2015;5(6):1187-209

100. Cai C, Zhu G, Zhang T, Guo W. High-density 80 K SNP array is a powerful tool for genotyping G. hirsutum accessions and genome analysis. BMC genomics. 2017;18(1):654

101. Lee G-A, Koh H-J, Chung H-K, Dixit A, Chung J-W, Ma K-H, et al. Development of SNP-based CAPS and dCAPS markers in eight different genes involved in starch biosynthesis in rice. Molecular Breeding. 2009;24(1):93-101

102. Lestari P, Koh HJ. Development of new CAPS/dCAPS and SNAP markers for rice eating quality. HAYATI Journal of Biosciences. 2013;20(1):15-23

103. Obala J, Saxena RK, Singh VK, Kumar CS, Saxena K, Tongoona P, et al. Development of sequence-based markers for seed protein content in pigeonpea. Molecular Genetics and Genomics. 2019;294(1):57-68

104. Li P, Kirungu JN, Lu H, Magwanga RO, Lu P, Cai X, et al. SSR-Linkage map of interspecific populations derived from Gossypium trilobum and Gossypium thurberi and determination of genes harbored within the segregating distortion regions. PloS one. 2018;13(11):e0207271

105. Zhang K, Kuraparthy V, Fang H, Zhu L, Sood S, Jones DC. High-density linkage map construction and QTL analyses for fiber quality, yield and morphological traits using CottonSNP63K array in upland cotton (Gossypium hirsutum L.). BMC genomics. 2019;20(1):889

106. Ma J, Pei W, Ma Q, Geng Y, Liu G, Liu J, et al. QTL analysis and candidate gene identification for plant height in cotton based on an interspecific backcross inbred line population of Gossypium hirsutum× Gossypium barbadense. Theoretical and Applied Genetics. 2019;132(9):2663-76

107. Naoumkina M, Thyssen GN, Fang DD, Jenkins JN, McCarty JC, Florane CB. Genetic and transcriptomic dissection of the fiber length trait from a cotton (Gossypium hirsutum L.) MAGIC population. BMC genomics. 2019;20(1):112

108. Wang W, Sun Y, Yang P, Cai X, Yang L, Ma J, et al. A high density SLAF-seq SNP genetic map and QTL for seed size, oil and protein content in upland cotton. BMC genomics. 2019;20(1):599

109. Zhang C, Li L, Liu Q, Gu L, Huang J, Wei H, et al. Identification of loci and candidate genes responsible for fiber length in upland cotton (Gossypium hirsutum L.) via association mapping and linkage analyses. Frontiers in plant science. 2019;10:53

110. Zhang T, Yuan Y, Yu J, Guo W, Kohel RJ. Molecular tagging of a major QTL for fiber strength in Upland cotton and its marker-assisted selection. Theoretical and Applied Genetics. 2003;106(2):262-8

111. Rong J, Abbey C, Bowers JE, Brubaker CL, Chang C, Chee PW, et al. A 3347-locus genetic recombination map of sequence-tagged sites reveals features of genome organization, transmission and evolution of cotton (Gossypium). Genetics. 2004;166(1):389-417

112. Brubaker CL, Brown AH. The use of multiple alien chromosome addition aneuploids facilitates genetic linkage mapping of the Gossypium G genome. Genome. 2003;46(5):774-91

113. Chen H, Khan MKR, Zhou Z, Wang X, Cai X, Ilyas MK, et al. A high-density SSR genetic map constructed from a F2 population of Gossypium hirsutum and Gossypium darwinii. Gene. 2015;574(2):273-86

114. Wang B, Zhuang Z, Zhang Z, Draye X, Shuang L-S, Shehzad T, et al. Advanced Backcross QTL Analysis of Fiber Strength and Fineness in a Cross between Gossypium hirsutum and G. mustelinum. Frontiers in plant science. 2017;8:1848

115. Kushanov FN, Buriev ZT, Shermatov SE, Turaev OS, Norov TM, Pepper AE, et al. QTL mapping for flowering-time and photoperiod insensitivity of cotton Gossypium darwinii Watt. PloS one. 2017;12(10):e0186240

116. Kirungu JN, Deng Y, Cai X, Magwanga RO, Zhou Z, Wang X, et al. Simple sequence repeat (SSR) genetic linkage map of D genome diploid cotton derived from an interspecific cross between Gossypium davidsonii and Gossypium klotzschianum. International journal of molecular sciences. 2018;19(1):204

117. Keerio AA, Shen C, Nie Y, Ahmed MM, Zhang X, Lin Z. QTL mapping for fiber quality and yield traits based on introgression lines derived from Gossypium hirsutum× G. tomentosum. International journal of molecular sciences. 2018;19(1):243

118. Magwanga RO, Lu P, Kirungu JN, Diouf L, Dong Q, Hu Y, et al. GBS mapping and analysis of genes conserved between Gossypium tomentosum and Gossypium hirsutum cotton cultivars that respond to drought stress at the seedling stage of the BC2F2 generation. International journal of molecular sciences. 2018;19(6):1614

119. Zheng J, Oluoch G, Riaz MK, Wang X, Cai X, Zhou Z, et al. Mapping QTLs for drought tolerance in an F2: 3 population from an inter-specific cross between Gossypium tomentosum and Gossypium hirsutum. Genetics and molecular research: GMR. 2016;15(3)

120. Mekonnen T, Haileselassie T, Tesfaye K. Identification, mapping and pyramiding of genes/quantitative trait loci (qtls) for durable resistance of crops to biotic stresses. Journal of Plant Pathology and Microbiology. 2017;8(6)

121. Mao G, Wei H, Hu W, Ma Q, Zhang M, Wang H, et al. Fine mapping and molecular characterization of the virescent gene vsp in Upland cotton (Gossypium hirsutum). Theoretical and Applied Genetics. 2019;132(7):2069-86

122. Mao G, Ma Q, Wei H, Su J, Wang H, Ma Q, et al. Fine mapping and candidate gene analysis of the virescent gene v 1 in Upland cotton (Gossypium hirsutum). Molecular genetics and genomics. 2018;293(1):249-64

123. Feng X, Cheng H, Zuo D, Zhang Y, Wang Q, Liu K, et al. Fine mapping and identification of the fuzzless gene GaFzl in DPL972 (Gossypium arboreum). Theoretical and Applied Genetics. 2019;132(8):2169-79

124. Xu P, Gao J, Cao Z, Chee PW, Guo Q, Xu Z, et al. Fine mapping and candidate gene analysis of qFL-chr1, a fiber length QTL in cotton. Theoretical and Applied Genetics. 2017;130(6):1309-19

125. Fang X, Liu X, Wang X, Wang W, Liu D, Zhang J, et al. Fine-mapping qFS07. 1 controlling fiber strength in upland cotton (Gossypium hirsutum L.). Theoretical and applied genetics. 2017;130(4):795-806

126. Andres RJ, Bowman DT, Kaur B, Kuraparthy V. Mapping and genomic targeting of the major leaf shape gene (L) in Upland cotton (Gossypium hirsutum L.). Theoretical and applied genetics. 2014;127(1):167-77

127. Kumar P, He Y, Singh R, Davis RF, Guo H, Paterson AH, et al. Fine mapping and identification of candidate genes for a QTL affecting Meloidogyne incognita reproduction in Upland cotton. BMC genomics. 2016;17(1):567

128. Bernardo R, Yu J. Prospects for genomewide selection for quantitative traits in maize. Crop Science. 2007;47(3):1082-90

129. Nakaya A, Isobe SN. Will genomic selection be a practical method for plant breeding? Annals of botany. 2012;110(6):1303-16

130. Hayes B, Goddard M. Prediction of total genetic value using genome-wide dense marker maps. Genetics. 2001;157(4):1819-29

131. Heffner EL, Sorrells ME, Jannink JL. Genomic selection for crop improvement. Crop Science. 2009;49(1):1-12

132. Shikha M, Kanika A, Rao AR, Mallikarjuna MG, Gupta HS, Nepolean T. Genomic selection for drought tolerance using genome-wide SNPs in maize. Frontiers in plant science. 2017;8:550

133. Vélez-Torres M, García-Zavala JJ, Hernández-Rodríguez M, Lobato-Ortiz R, López-Reynoso JJ, Benítez-Riquelme I, et al. Genomic prediction of the general combining ability of maize lines (Zea mays L.) and the performance of their single crosses. Plant Breeding. 2018;137(3):379-87

134. Liu X, Hu X, Li K, Liu Z, Wu Y, Wang H, et al. Genetic mapping and genomic selection for maize stalk strength. BMC plant biology. 2020;20:1-16

135. Guo R, Dhliwayo T, Mageto EK, Palacios-Rojas N, Lee M, Yu D, et al. Genomic Prediction of Kernel Zinc Concentration in Multiple Maize Populations Using Genotyping-by-Sequencing and Repeat Amplification Sequencing Markers. Frontiers in Plant Science. 2020;11

136. Spindel J, Begum H, Akdemir D, Virk P, Collard B, Redona E, et al. Genomic selection and association mapping in rice (Oryza sativa): effect of trait genetic architecture, training population composition, marker number and statistical model on accuracy of rice genomic selection in elite, tropical rice breeding lines. PLoS Genet. 2015;11(2):e1004982

137. Xu Y, Wang X, Ding X, Zheng X, Yang Z, Xu C, et al. Genomic selection of agronomic traits in hybrid rice using an NCII population. Rice. 2018;11(1):1-10

138. Cui Y, Li R, Li G, Zhang F, Zhu T, Zhang Q, et al. Hybrid breeding of rice via genomic selection. Plant biotechnology journal. 2020;18(1):57-67

139. Yu S, Ali J, Zhang C, Li Z, Zhang Q. Genomic breeding of green super rice varieties and their deployment in Asia and Africa. Theoretical and Applied Genetics. 2020:1-16

140. Crossa J, de Los Campos G, Pérez P, Gianola D, Burgueño J, Araus JL, et al. Prediction of genetic values of quantitative traits in plant breeding using pedigree and molecular markers. Genetics. 2010;186(2):713-24

141. Dong H, Wang R, Yuan Y, Anderson J, Pumphrey M, Zhang Z, et al. Evaluation of the potential for genomic selection to improve spring wheat resistance to Fusarium head blight in the Pacific Northwest. Frontiers in plant science. 2018;9:911

142. Norman A, Taylor J, Edwards J, Kuchel H. Optimising genomic selection in wheat: Effect of marker density, population size and population structure on prediction accuracy. G3: Genes, Genomes, Genetics. 2018;8(9):2889-99

143. Veenstra LD, Poland J, Jannink JL, Sorrells ME. Recurrent genomic selection for wheat grain fructans. Crop Science. 2020

144. Lozada DN, Ward BP, Carter AH. Gains through selection for grain yield in a winter wheat breeding program. PLoS One. 2020;15(4):e0221603

145. de Oliveira AA, Pastina MM, da Costa Parrella RA, Noda RW, Simeone MLF, Schaffert RE, et al. Genomic prediction applied to high-biomass sorghum for bioenergy production. Molecular Breeding. 2018;38(4):49

146. Hunt CH, van Eeuwijk FA, Mace ES, Hayes BJ, Jordan DR. Development of genomic prediction in sorghum. Crop Science. 2018;58(2):690-700

147. Sapkota S, Boatwright JL, Jordan KE, Boyles R, Kresovich S. Multi-trait regressor stacking increased genomic prediction accuracy of sorghum grain composition. bioRxiv. 2020

148. Islam MS, Fang DD, Jenkins JN, Guo J, McCarty JC, Jones DC. Evaluation of genomic selection methods for predicting fiber quality traits in Upland cotton. Molecular Genetics and Genomics. 2020;295(1):67-79

149. Lenz PR, Beaulieu J, Mansfield SD, Clément S, Desponts M, Bousquet J. Factors affecting the accuracy of genomic selection for growth and wood quality traits in an advanced-breeding population of black spruce (Picea mariana). BMC genomics. 2017;18(1):335

150. Gapare W, Liu S, Conaty W, Zhu Q-H, Gillespie V, Llewellyn D, et al. Historical datasets support genomic selection models for the prediction of cotton fiber quality phenotypes across multiple environments. G3: Genes, Genomes, Genetics. 2018;8(5):1721-32

151. Sunilkumar G, Rathore KS. Transgenic cotton: factors influencing Agrobacterium-mediated transformation and regeneration. Molecular Breeding. 2001;8(1):37-52

152. Prashanth S, Katageri I, Vamadevaih H, Khadi B. Effect of external damage on regeneration of cotton explants (Gossypium arboreum and G. barbadense). Karnataka Journal of Agricultural Sciences. 2011;24(5):629-32

153. Jadhav M, Katageri I. Agrobacterium tumefaciens Mediated Genetic Transformation in Coker-312 (Gossypium hirsutum L.) Using Hypocotyls Explants. Int J Curr Microbiol App Sci. 2017;6(12):2771-9

154. Klein TM, Wolf ED, Wu R, Sanford J. High-velocity microprojectiles for delivering nucleic acids into living cells. Nature. 1987;327(6117):70-3

155. Datta SK, Peterhans A, Datta K, Potrykus I. Genetically engineered fertile indica-rice recovered from protoplasts. Bio/technology. 1990;8(8):736-40

156. Fromm M, Taylor LP, Walbot V. Expression of genes transferred into monocot and dicot plant cells by electroporation. Proceedings of the National Academy of Sciences. 1985;82(17):5824-8

157. Leelavathi S, Sunnichan V, Kumria R, Vijaykanth G, Bhatnagar R, Reddy V. A simple and rapid Agrobacterium-mediated transformation protocol for cotton (Gossypium hirsutum L.): embryogenic calli as a source to generate large numbers of transgenic plants. Plant cell reports. 2004;22(7):465-70

158. Sumithra S, Katageri I, Vamadevaiah H. Factors influencing regeneration and Agrobacterium mediated transformation of Gossypium herbaceaum and Gossypium hirsutum cotton genotypes. Karnataka Journal of Agricultural Sciences. 2010;23(2):222-6

159. John ME, Stewart JM. Genes for jeans: biotechnological advances in cotton. Trends in biotechnology. 1992;10:165-70

160. Mao Y-B, Cai W-J, Wang J-W, Hong G-J, Tao X-Y, Wang L-J, et al. Silencing a cotton bollworm P450 monooxygenase gene by plant-mediated RNAi impairs larval tolerance of gossypol. Nature biotechnology. 2007;25(11):1307-13

161. Li B, Yang Y, Hu WR, Li XD, Cao JQ, Fan L. Over-expression of Gh UGP 1 in upland cotton improves fibre quality and reduces fibre sugar content. Plant Breeding. 2015;134(2):197-202

162. Jiang Y, Guo W, Zhu H, Ruan YL, Zhang T. Overexpression of GhSusA1 increases plant biomass and improves cotton fiber yield and quality. Plant biotechnology journal. 2012;10(3):301-12

163. Li F, Wu S, Lü F, Chen T, Ju M, Wang H, et al. Modified fiber qualities of the transgenic cotton expressing a silkworm fibroin gene. Chinese science bulletin. 2009;54(7):1210-6

164. Abdurakhmonov IY, Buriev ZT, Saha S, Jenkins JN, Abdukarimov A, Pepper AE. Phytochrome RNAi enhances major fibre quality and agronomic traits of the cotton Gossypium hirsutum L. Nature Communications. 2014;5(1):1-10

165. Benedict CR, Martin GS, Liu J, Puckhaber L, Magill CW. Terpenoid aldehyde formation and lysigenous gland storage sites in cotton: variant with mature glands but suppressed levels of terpenoid aldehydes. Phytochemistry. 2004 May 1;65(10):1351-9

166. Rathore KS, Sundaram S, Sunilkumar G, Campbell LM, Puckhaber L, Marcel S, et al. Ultra-low gossypol cottonseed: generational stability of the seed-specific, RNAi-mediated phenotype and resumption of terpenoid profile following seed germination. Plant biotechnology journal. 2012;10(2):174-83

167. Palle SR, Campbell LM, Pandeya D, Puckhaber L, Tollack LK, Marcel S, et al. RNA i-mediated Ultra-low gossypol cottonseed trait: performance of transgenic lines under field conditions. Plant biotechnology journal. 2013;11(3):296-304

168. Gao W, Long L, Tian X, Xu F, Liu J, Singh PK, et al. Genome editing in cotton with the CRISPR/Cas9 system. Frontiers in plant science. 2017;8:1364

169. Chen X, Lu X, Shu N, Wang S, Wang J, Wang D, et al. Targeted mutagenesis in cotton (Gossypium hirsutum L.) using the CRISPR/Cas9 system. Scientific reports. 2017;7:44304

170. Li C, Unver T, Zhang B. A high-efficiency CRISPR/Cas9 system for targeted mutagenesis in cotton (Gossypium hirsutum L.). Scientific Reports. 2017;7:43902

171. Wang P, Zhang J, Sun L, Ma Y, Xu J, Liang S, et al. High efficient multisites genome editing in allotetraploid cotton (Gossypium hirsutum) using CRISPR/ Cas9 system. Plant biotechnology journal. 2018;16(1):137-50

172. Zhu S, Yu X, Li Y, Sun Y, Zhu Q, Sun J. Highly efficient targeted gene editing in upland cotton using the CRISPR/Cas9 system. International journal of molecular sciences. 2018;19(10):3000

173. Gao W, Xu FC, Long L, Li Y, Zhang JL, Chong L, et al. The gland localized CGP1 controls gland pigmentation and gossypol accumulation in cotton. Plant biotechnology journal. 2020;18(7):1573

174. Long L, Guo D-D, Gao W, Yang W-W, Hou L-P, Ma X-N, et al. Optimization of CRISPR/Cas9 genome editing in cotton by improved sgRNA expression. Plant Methods. 2018;14(1):85

175. Pattanayak V, Lin S, Guilinger JP, Ma E, Doudna JA, Liu DR. High-throughput profiling of off-target DNA cleavage reveals RNA-programmed Cas9 nuclease specificity. Nature biotechnology. 2013;31(9):839-43

176. Ran FA, Hsu PD, Wright J, Agarwala V, Scott DA, Zhang F. Genome engineering using the CRISPR-Cas9 system. Nature protocols. 2013;8(11):2281-308

177. Zeng D, Li X, Huang J, Li Y, Cai S, Yu W, et al. Engineered Cas9 variant tools expand targeting scope of genome and base editing in rice. Plant Biotechnology Journal. 2020;18(6):1348-50

178. Li B, Rui H, Li Y, Wang Q, Alariqi M, Qin L, et al. Robust CRISPR/Cpf1 (Cas12a)-mediated genome editing in allotetraploid cotton (Gossypium hirsutum). Plant biotechnology journal. 2019;17(10):1862

179. Liu L, Chen P, Wang M, Li X, Wang J, Yin M, et al. C2c1-sgRNA complex structure reveals RNA-guided DNA cleavage mechanism. Molecular cell. 2017;65(2):310-22

180. Shmakov S, Abudayyeh OO, Makarova KS, Wolf YI, Gootenberg JS, Semenova E, et al. Discovery and functional characterization of diverse class 2 CRISPR-Cas systems. Molecular cell. 2015;60(3):385-97

181. Yang H, Gao P, Rajashankar KR, Patel DJ. PAM-dependent target DNA recognition and cleavage by C2c1 CRISPR-Cas endonuclease. Cell. 2016;167(7):1814-28. e12

182. Qin L, Li J, Wang Q, Xu Z, Sun L, Alariqi M, et al. High-efficient and precise base editing of C• G to T• A in the allotetraploid cotton (Gossypium hirsutum) genome using a modified CRISPR/Cas9 system. Plant biotechnology journal. 2020;18(1):45-56

SECTION-4

CROP BIOTECHNOLOGY

CHAPTER-15

THE CONTRIBUTION OF POST-HARVEST PHYSIOLOGY TO THE DEVELOPMENT OF TRANSGENIC CROPS

1.INTRODUCTION: AN OVERVIEW

The evolution of transgenic crops has been a pivotal aspect of modern agricultural advancements, revolutionizing crop production and addressing various challenges in global food security. Amidst the multifaceted factors contributing to this evolution, the role of post-harvest physiology emerges as a critical determinant in shaping the success and impact of transgenic crops. Post-harvest physiology, encompassing the intricate processes that occur after crop harvest, plays a substantial role in determining the quality, shelf life, and nutritional attributes of agricultural produce. Understanding and harnessing the interactions between transgenic modifications and post-harvest physiological responses are key to unlocking the full potential of genetically modified crops.

WHAT IS POST HARVEST?

Post harvest physiology is the science which deals with quantitative and qualitative study of physiology of agricultural products (especially living plant tissues) after harvesting. The technologies used in postharvest physiology mainly consists of the techniques applied in agricultural produce after picking for the purpose of preservation, conservation, quality control/increment, processing, packaging, storage and many more. Postharvest technologies are concerned with enhancing nutritional value of food products as in order to meet consumer's needs. India is one of the largest producer of agricultural products and commodities in the world. According to FAO, the total foodgrain production in 2017-18 was estimated around 275 million tonnes and is the largest producer (25% global production), consumer

(27% world's consumption) and importer (14%) of pulses in the world. Statistics presented by National Horticultural Board also reveals that India stands second after China in the production of fruits and vegetables accounting average global production 13% and 21% respectively [1]. About one-third of agricultural produce available for human consumption in the world per year gets wasted. Food loss can be defined as condition where food is available for human consumption but not being consumed. The qualitative and quantitative losses of food commodities often occurs during postharvest operations commonly refers as postharvest losses. With the global increment in human population, the demand for food supply has been enhanced over few decades. Thus in order to increase food availability the losses of food commodities due to post harvest operations are needed to be minimize in order to provide substantial solution to food crisis, reduce pressure on natural resources, eradicate hunger and increase farmer's income [2]. The postharvest losses can be due to loss of weight, loss of quality and nutritional value, loss of viability and commercial loss. Postharvest losses of agricultural commodities may often contribute to the deterioration of quality and quantity as well. The deterioration of quality refers to various attributes including weight loss, change in color and visible quality, change in nutrient content and flavor whereas quantity refers to loss of amount of product respectively [3]. Therefore it can be concluded that postharvest losses are one of the major factor that not only affects agricultural production but also influences food supply chain and economic growth globally. Thus, postharvest loss can be summarized in Figure 1.

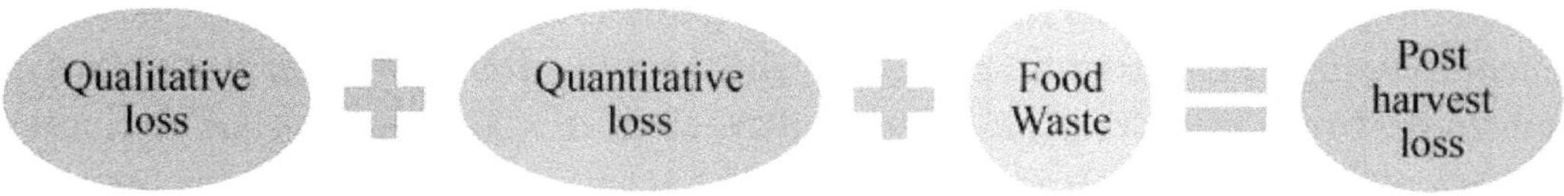

Figure 1. Summary of postharvest losses.

2. CAUSES OF POSTHARVEST LOSSES

The postharvest losses of agricultural commodities may be due to various factors that affects the quality of produce.

2.1 Primary causes

Agricultural commodities like food grains are when grown and transported to consumers are subjected to various agricultural operations like harvesting, threshing, milling, packing, storage etc. The large number of losses may occur when harvesting is not done at adequate moisture content and time of harvesting. Delayed harvesting or too early harvesting may subject the crop severe losses by various factors like attack of birds, rodents, microbes, natural calamities etc [4, 5]. Threshing and cleaning are usually done to separate grains from panicles. However, threshing losses can occur in case of seed splitting, incomplete separation of seed from source, seed cracking due to excessive force [4, 6] . Delayed threshing result in loss of quality and quantity of commodities. In order to prolong the storage of food grains, it is necessary to store the commodities at safer moisture content. Improper drying result in microbe growth in grains and is not desirable for storage and grinding operations. Thus, drying is an important postharvest technology in order to improve quality, prevention from insects and rodents and for transportation [7, 8]. Lack of proper transportation facilities may result in loss of commodities to greater extent. However, the problem of loss due to transportation are relatively less in developed countries due to proper roads, infrastructures and processing equipment.

2.2 Biological cause

- **Respiration**- Respiration is a physiochemical process in which stored organic materials like carbohydrate, lipids, fats are catabolized into simple compounds in order to liberate energy essential for metabolic processes. Respiration is an important phenomenon which influences the physiological and biochemical activities of horticultural produces like fruits and vegetables. In other words, the deterioration of horticultural products is directly related to respiration rate. The respiration rate of horticultural produce can be estimated in terms of O_2 consumed or CO_2 evolved in various processes like development, maturation, ripening etc.
- **Transpiration**- Transpiration is the physiological process which involves loss of water in the form of vapor from living tissues of the plant. The extreme loss of water from harvested produce is the major cause of deterioration which compromises its quality, nutrition, palatability and demand among consumers. Further, the transpiration loss can be ameliorated in storage conditions by:
    - Reducing air movement
    - Lowering air temperature
    - Raising relative humidity
    - Use of protective cover like waxing and protective methods like modified atmospheric packaging, polyethylene films.
- **Microbes**- Stored agricultural produce are often subjected to postharvest diseases caused by bacteria, mould, fungi and incidence of insect pest and rodents. The common pathogen that infects the postharvest produce includes *Penicillum sp., Botrytis sp., Fusarium sp., Phytophthora infestans etc.* Mechanical damages and bruising during

harvesting and other agricultural operations are common point of entry of pathogenic microorganisms which affect the quality and quantity of produce adversely and reduces marketability as well.

- **Ethylene**- Ethylene is a gaseous hormone which plays an active and important role in postharvest technology of agricultural produce. It is a Ripening hormone which controls ripening process in fruits and vegetables. However, it has also some undesirable effects on fruits like premature ripening, skin damage etc.

2.3 Environmental causes

- **Temperature**- Temperature is one of the most crucial environment factor that affect the postharvest life of stored product. Generally, for every 10°C increase in temperature, the rate of deterioration of produce increases to 2-3 folds [9]. High temperature increases transpiration rate, thus increasing water loss whereas low temperature favours microbe development. Undesirable temperature in storage conditions may cause chilling and freezing injury, heat injury which drastically affect the quality of postharvest produce.
- **Relative humidity**- Freshly harvested fruits and vegetable possess 80-95% water by weight. The loss of humidity from horticultural produce is purely dependent on vapor pressure deficit between the surrounding air. The relative humidity is highly influenced by transpiration and respiration processes. In the meantime, where high relative humidity reduces the chances of water loss from produce it also harbors the pathogenic postharvest microorganisms as well.
- **Atmospheric condition**- The composition of gaseous mixture mainly oxygen and carbondioxide plays an important role in controlling the quality of post harvested produce as it controls respiration, temperature, ethylene concentration etc. Therefore, it is necessary to regulate the gaseous composition around produce in order to reduce respiration and enhance shelf-life of produce [10]. Meanwhile the reduction of oxygen and increase of carbon dioxide in storage condition may reduce deterioration of postharvest produce. However, change in gaseous composition in storage chamber can also cause physiological disorder in produce. Eg-Hollow heart in potato can occur due to faulty oxygen balance and during transportation. The unbalanced gaseous composition may also cause other calamities likeirregular fruit ripening, soft texture, poor skin color development etc.

- **Light**- Light exposure may also cause some physiological change in produce along with alteration in biological process. Eg: When potatoes are exposed to light, it forms green tubers due to formation of solanin and chlorophyll, which is toxic for human consumption (Figure 2).

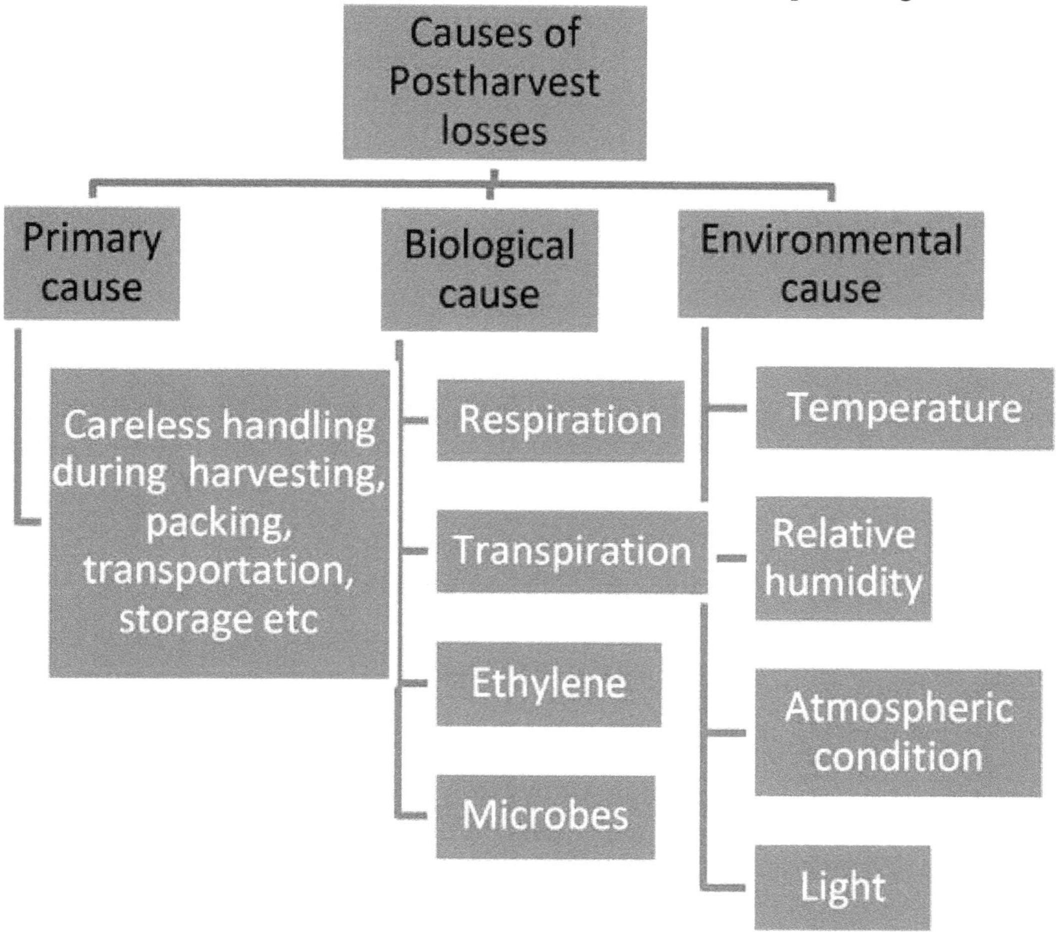

Figure 2. Factors affecting postharvest losses.

3. WHY POST HARVEST TECHNOLOGY?

The food processing industry in India encompasses wide variety of products including fruits and vegetables, milk and poultry, meat and its products, alcoholic beverages, fisheries, grain processing and confectionaries like chocolates, cocoa products, soya based products, high protein products and many more. The agricultural and food products when gets detached from plant or being harvested are prone to various physiological, biochemical and biological events that results in its degradation of loss of commodity. Thus, it is necessary to adopt and practice some measures in order to reduce the loss of food commodities. Several measures can be taken in order to minimize the loss which includes reduction in moisture content, control of microorganisms, denaturation of endogenous enzymes and ensuring proper packaging of food materials. Postharvest technology plays a vital role in flourishing the food industry by minimizing losses and spoilage and value addition of commodities. It ensures the use of optimum harvest factors, utilizes modern machinery to ensure reduced losses due to handling, packaging, storage and transportation as well. It mainly aims to meet the food requirement for growing population by minimizing losses and enhancing nutritive value of food products. Postharvest technology is well established in large as well as small scale industries. The storage of food products can be prolonged by adapting several technologies like use of thermal processing, drying, low temperature, chemical and biological reactions couples with several preservation technologies etc. Postharvest technology opens new opportunities for marketing of food products as processed food commodities are gaining popularity among consumers and it also generate new employment opportunities among individuals. The fundamental objective underlying in postharvest technology sector is to maintain quality of produce after harvest in terms of texture, appearance, flavor etc, to ensure proper safety of food and to reduce loss of food and agricultural commodities from harvest to consumption.

4. ROLE OF ETHYLENE IN POSTHARVEST PHYSIOLOGY

Ethylene is a gaseous hormone which can be produced by almost all parts of higher plants. It is a colourless, odourless gas with solubility in water 20 mg/lit at 200C and 250 mh/lit at 00C [11]. The meristematic and nodal regions of plants play active part in ethylene biosynthesis. However, the production of ethylene gets accelerated during leaf abscission, senescence and ripening processes. Wounding in plant and physiological stresses like chilling, drought, and diseases also enhances ethylene production in plants [11]. The biosynthesis of ethylene initially begins with amino acid methionine (Met) which gets converted to S-Adenosyl methionine (SAM) in presence of enzyme Adomet synthase. The enzyme ACC synthase converts SAM to 1-Aminocyclopropane-1-carboxylic acid (ACC) which further gets converted ethylene by enzyme ACC oxidase which can be described as below (Figure 3).

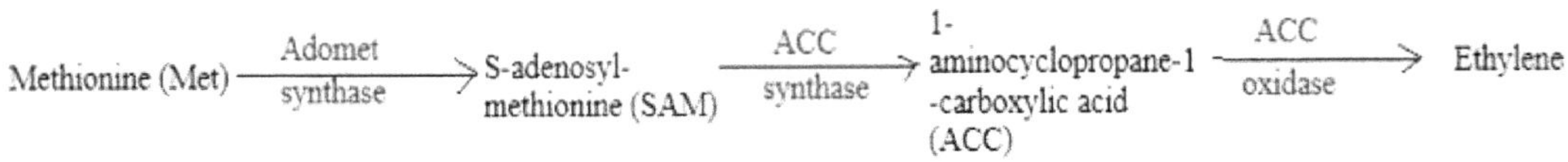

Figure 3. Biosynthesis of ethylene.

Ethylene is considered as multifunctional phytohormone that controls both growth and senescence in plants [12]. It regulates the development of leaves, flowers and fruits and also promotes senescence depending upon the level of ethylene applied to plants [13, 14, 15]. It is known to regulate wide range of responses in plants namely viz., seed germination, cell expansion, cell differentiation, flowering, abscission, senescence etc. Some important physiological role of ethylene is described below (Figure 4).

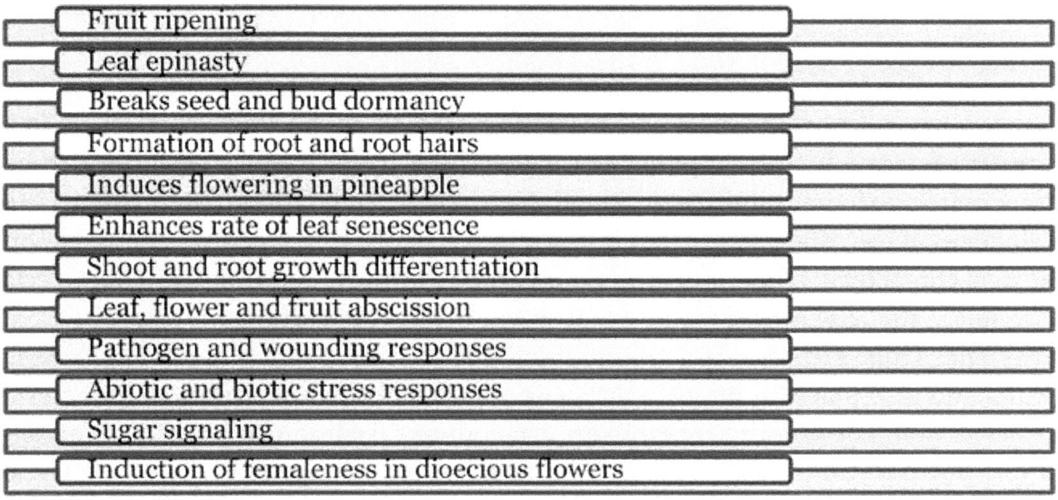

Figure 4. Physiological role of ethylene in plants.

4.1 Ethylene as a fruit ripening hormone

The term fruit ripening involves changes in texture of fruit including softening due to enzymatic breakdown of cell wall, starch hydrolysis, accumulation of sugar and absence of phenolic compounds in fruits that makes the same ready to eat. Ethylene has been identified as ripening hormone since past long years and increase in concentration of ethylene in such fruits accelerates ripening phenomenon. The fruits can be categorized into two major classes on the basis of ethylene production i.e., climacteric and non-climacteric. Climacteric fruits are those which shows sudden characteristics respiratory rise when ripen in response to ethylene. As much treatment with ethylene causes fruit to produce additional amount of ethylene, referred to as autocatalytic. On the other hand, fruits which do not shows the rise in respiration rate upon treatment with ethylene called as non-climacteric fruits.

The respiratory climacteric and ethylene production are two most important and decisive parameters in ripening process. Change in ripening patterns determines the fact that whether fruits has ripen naturally or artificially upon ethylene exposure. Although artificial ripening of climacteric fruits enhances ripening process but also results in spoilage of products as well deteriorating market quality and demand [16]. The elevation in rates of respiration enhances fruit ripening and diminishes the postharvest life of both climacteric as well as non-climacteric fruits. In climacteric fruits, ethylene accelerates the time without modifying magnitude in order to achieve maximum respiration rates while in non-climacteric, since it lacks autocatalytic activities once ethylene is removed the respiration process slows down and respiration rates progresses in concentration dependent manner [17]. Furthermore, it is matter of fact that climacteric respiration does not always associated with increased ethylene responses instead depends on fruit species. Although the biochemical responses on the same are not fully understood yet but the several physiological and molecular studies enlighten that ethylene is a primary factor which is responsible for increased respiration rates and it can be thus concluded that climacteric respiration is an ethylene regulated event [18]. Ethylene on the other hand, does not primarily involved in ripening phenomenon in non-climacteric fruits. But it is noteworthy that despite of lack of climacteric ethylene, the ripening responses of non-climacteric fruits are responsive to ethylene. Thus, ethylene is a crucial regulator of ripening process in climacteric fruits and also plays an important role in regulating the same in non-climacteric fruits as well. The presence of ethylene is not always desirable in entertaining shelf life of postharvest products. However, the extent of damage depends on ethylene concentration, length of exposure time and temperature of the product. Therefore, in order to achieve maximum postharvest benefits of produce the controlled use of ethylene is necessary in order to prevent damage and excessive ripening. Exposure of postharvest products to ethylene accelerates the rate of ageing and senescence (Table 1).

Table 1. Effect of ethylene on postharvest quality.

Produce	Effect of ethylene on postharvest quality
Brinjal	Increase in decay process, abscission of calyx
Cucumber	Yellowing, Softening
Carrot	Development of bitter flavor
Cabbage	Yellowing, leaf abscission
Leafy vegetables	Chlorosis
Lettuce	Russett spotting
Potato	Sprouting
Watermelon	Off flavor, reduced firmness
Brocolli	Yellowing, abscission of florets

Source: Dhall, 2013 [11].

Ethylene biosynthesis in fruits occurs through Yang cycle. The biosynthesis of ethylene is governed by several multi genic families of ACS and ACO enzymes. The expression of ACS and ACO genes are controlled by several environmental & hormonal factors and it consists of positive and negative feedback regulation. As mentioned, fruits are divided into climacteric and non-climacteric on the basis of ripening phenomenon. In climacteric fruits like tomato normal fruit ripening involves ethylene burst. In climacteric plants, two systems of ethylene regulation are identified namely system I and system II. System I is ethylene auto-inhibitory, operates during vegetative growth and stimulates the synthesis of basal ethylene levels, detected in all tissues including non-climacteric fruits. System II remains functional during ripening of climacteric fruit and senescence phenomenon when ethylene synthesis is autocatalytic [19]. Tomato is a model plant to study ethylene biosynthesis and its perception in plants. Several ACS genes are also identified in other plants including apple, melon, pear, banana, citrus, papaya etc. In tomato, ripening results in change in fruit color from green to red, degradation of chlorophyll and accumulation of carotenoids. In tomato, 8 genes have been identified which includes LeACS1A, LeACS1B, LeACS2-7. LeACS2 and LeACS4 are greatly expressed during ripening process whereas LeACS1A and LeACS6 are expressed before onset of ripening. Further studies on mutants revealed that only LeACS6 is ethylene regulated while rest is unaffected. The genes LeACS1A and LeACS6 plays major part in ethylene production in SystemI whereas transition phase includes enhanced expression of LeACS1A and LeACS6 is also induced. The positive feedback regulation of gene LeACS2 maintains System II phase which gets started during transition phase. Similarly, in banana MaACS1 gene is related to ripening phenomenon as its transcript and ACC content enhances during ripening. In *Actinidia chinensis*, the levels of ACS mRNA is upregulated during climacteric ethylene production but ACS itself does not gets affected by exogenous ethylene [20]. Although ACS genes are transcriptionally regulated but post-translation regulation has also been reported. ACC oxidase (ACO) is another crucial enzyme which plays important role in ethylene biosynthesis. The level of ACO is enhanced in pre climacteric fruit before rise in ACS enzyme activity. ACO gene transcripts have been studied in various fruits like tomato, kiwi, pear, apple, banana etc. which regulated temporal and spatial expression. In tomato, so far 3 ACO genes have been identified namely LeACO1 (expressed in ripening fruit), LeACO2 (anther), LeACO3 mRNA (floral organs) with weak expression in fruit at breaker stage. In banana, MaACO1 mRNA is up regulated with onset of ripening but decreases during late ripening stage. In melon, Cm-ACO1 gene is highly expressed during ethylene production but Cm-ACO3 is induced in flowers only. In *A. chinensis*, exposure to ethylene induces up regulation of genes for ACO and Adometsynthetase enzyme and also before respiratory climacteric rise in ethylene biosynthesis.

4.2 Ethylene signaling in fruit ripening

It is well established fact that apart from several crucial role in plants ethylene also induces triple response in several species like Arabidopsis, pea and other plants. Triple response includes inhibition of hypocotyl elongation, initiation of swelling (radial elongation) & inhibition of root elongation and appearance of prominent apical hook. The pathway of ethylene signal transduction has been well established and extensively studied since long time and a number of genes have been identified. In Arabidopsis, the ethylene-resistant1 (ETR1) was first isolated ethylene receptor. Further, 5ethylene receptorsETR1, ETR2, ERS1, ERS 2 and EIN4 has been discovered. These ethylene receptors consist of 2 subfamilies with N-terminal transmembrane ethylene binding domain and C-terminal histidine kinase domain according to amino acid sequence. ETR2, ERS2 and EIN4 are subfamilies of ethylene receptors with extra N- terminal transmembrane domain. ERS1 has only histidine kinase activity whereas ERS1 has both histidine kinase and serine threonine kinase activity. Subfamily1 has major role in ethylene signal transduction as CTR1 (Constitutive triple response1) strongly interact with subfamily1 along with ETR1 & ERS1 as compared to subfamily2 [21, 22]. In tomato researchers have identified 6 ethylene receptors genes [LeETR1, LeETR2,LeETR3 (NR, never ripe), LeETR4-6] in which LeETR1-3 consists of type1 receptors while LeETR4-6 type2 receptor. Gene LeETR1 and LeETR2 are expressed constitutively during developmental phases, LeETR5 in fruits and flowers during biotic stress conditions, NR& LeETR4 in reproductive phase, fruit ripening and senescence. However, ethylene sensitivity can be affected by repression all above genes except LeETR4. Arabidopsis ETR1 forms dimer in endoplasmic reticulum where it requires copper as cofactor for ethylene binding. Responsive-to- antagonist1 (RAN1) acts as copper transporter and its mutation forms inactive receptors lacking copper. CTR1 is a negative regulator of ethylene signaling and shows homology with Raf family of mitogen activated protein kinase kinasekinase (MAPKKK). LeCTR1-4 are homologs of CTR1 identified in tomato where LeCTR1 functionally complement CTR1 in Arabidopsis. Mutation in EIN2 which lies downstream ethylene responses in Arabidopsis plant when ethylene binds to ligand, it trigger activities of several genes via EIN3 &ERF1 family of transcription factors. Primary ethylene response elements (PEREs) stimulate ethylene promoting genes and modulates ripening and senescence related genes as well. However, homodimer of EIN3 binds to PERE which in turn bind to GCC-box present in promoters of stress responsive gene, enabling downstream process. Four EIN3 (LeEIL1-4) genes are identified in tomato in which LeEIL3 functionally complement EIN3 mutation in Arabidopsis. In absence of ethylene, activated form of CTR1 binds to ethylene receptor &retards the downstream ethylene signal transduction process thus, suppressing ethylene stimulated gene expression. When ethylene binds to receptor, conformation change in receptor begins which enables dissociation of CTR1 and CTR deactivation, releasing downstream pathway from suppression. EIN2 protein gets activated to stimulate downstream signal transduction and perception [23].

5. MANAGEMENT OF POSTHARVEST LOSSES

Proper handling of food commodities and minimizing postharvest losses in fruits and vegetables are crucial means for maintain quality of produce, ensuring food security and combating poverty and hunger. The main cause of postharvest loss in developing countries occurs due to lack of infrastructure and agricultural operations while in developed countries it occurs at consumer stage. Postharvest management ensures quality management, increases market share and market value of commodities, ensures proper agricultural production and minimizes postharvest losses of fruits, vegetables, cereals, pulses, oilseeds and many other agricultural commodities. It also safeguards the nutritional quality and health benefits to consumers. Postharvest management of horticultural produces involves them to keep fresh and alive even after harvest to meet consumer demands by keeping a proper care on O_2 and CO_2 concentration. Postharvest handling of harvested produce plays a vital role in maintain quality of produce and managing losses (Figure 5).

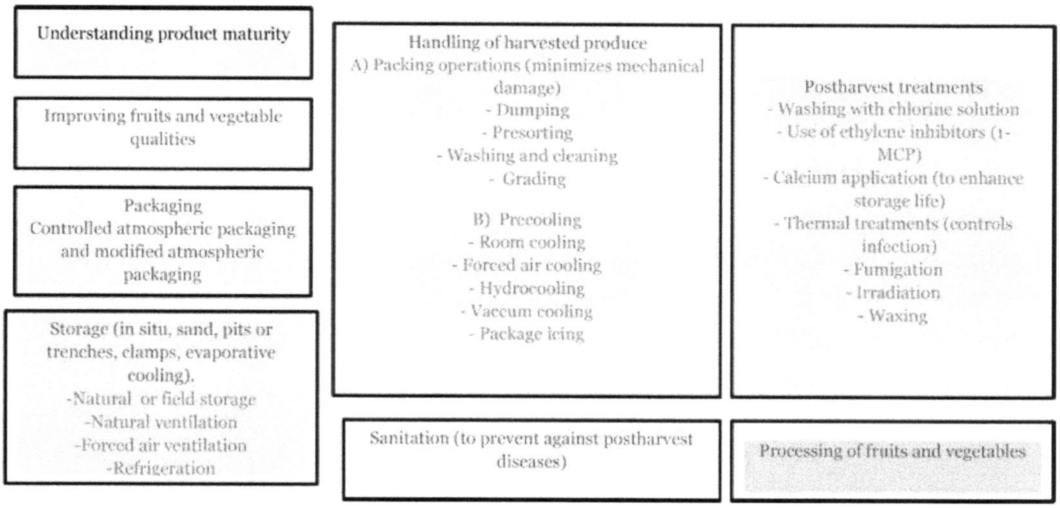

Figure 5. Postharvest management methods.

5.1 Transgenic approaches in postharvest technology

Transgenic or genetically modified crops (GMO) are defined as those species whose DNA has been modified using genetic engineering techniques. Gene of interest is usually identified and taken from another and gets artificially inserted into desired crop species in order to develop genetically modified crop or transgenic crops. The aim is to develop a plant with traits that does not occur naturally in another plant species. The gene of interest which gets inserted in referred as transgene and it may be part of either unrelated plant or completely different plant species. The purpose of developing genetically modified crops is to create desirable and productive product [24]. Transgenic approaches have been commercially used in many plant species like tomato, corn, tobacco, potato, soybean, canola, banana, alfalfa, rice, squash, melon, papaya. Statistics reveals that about 18 million farmers are cultivating GMO crops with total area of 181.5 million hectares in 28 countries in 2014 [25] which has increased to 185.1 million hectares in 2016 and to 191.7 million hectares in 2018 globally. Transgenic crops possess several traits like higher yield, improving shelf life of commodities, quality improvement, resistance to insect-pest, tolerance to abiotic stresses like cold, drought, heat etc. They also possess industrial and pharmaceutical importance (Figure 6).

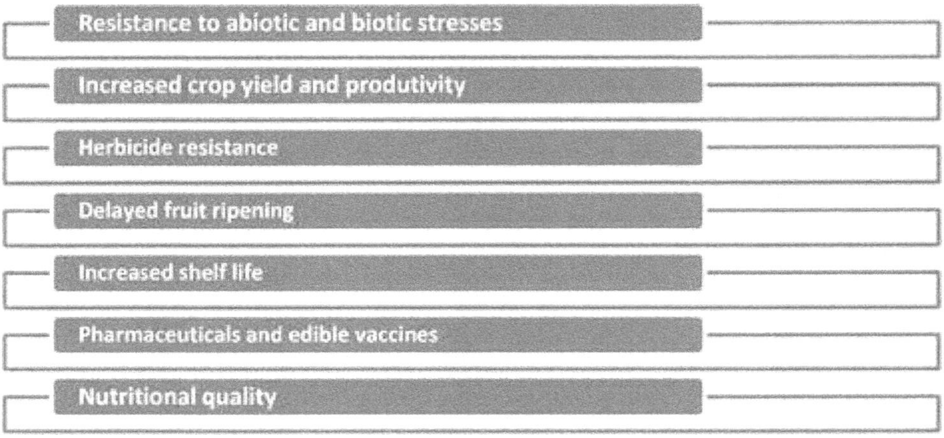

Figure 6. Applications of transgenic in agriculture.

Transgenic plants are usually developed through genetic engineering by altering genetic makeup, adding one or more beneficial genes or removing detrimental genes in plant genome. The detail method for developing transgenic is described as (Figure 7).

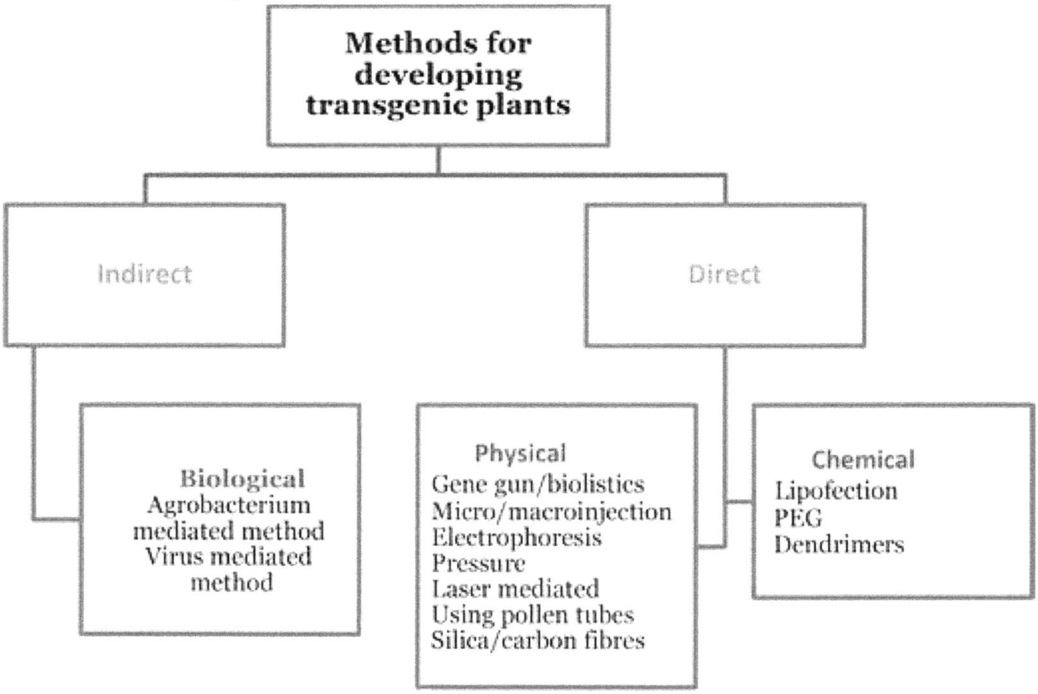

Figure 7. Methods of developing transgenic crops.

Steps involved in developing transgenic plants.

- **Identification, isolation and cloning of genes of desired traits**- It is the initial step of development of transgenic plants. Usually the gene of interest is identified, isolated and cloned in order to transfer to desired plant genome.
- **Design the gene construct for insertion**- The gene construct or gene set must be designed having all possible DNA segment in order to integrate and express into plant genome. The gene construct must have following segments-
  - **The promoter sequence**- The promoter sequence is required for accurate expression of gene as it is on/off switch that controls the when and where the gene will be expressed. Promoter sequence is recognized by transcription factors in DNA transcription process. It is generally located at 5' upstream of gene. Promoters are usually classified as constitutive, tissue specific and inducible in nature. Constitutive promoters express themselves at almost all developmental stages and participate in moderate and constant gene expression. Tissue- specific promoters remains active facilitated gene expression during development- specific stage while inducible promoters are highly influence by environmental stimuli and facilitated gene expression in response to certain external factors Eg: CaMV35s promoter, NtHSP3A etc.
  - **The transgene**- It is generally modified to achieve greater expression in plant.
  - **Termination sequence**- It marks the end of gene sequence
  - **The selectable marker gene**- It is essential in order to identify the particular plant cells that have integrated transgene successfully. Common examples of selectable marker gene are nptII, hpt, acc3, bar, pat etc.
- **Transforming target plant with gene construct**- Genetic transformation is the process of identifying the desired specific gene for a particular trait and isolation of the traits from various plant species. The target plant can be transformed by vector mediated gene transfer (indirect) or vector less gene transfer (direct).
- **Selection of transgenic plants**- Plant tissues or cells are transferred to selective medium containing antibiotic or other components.

- **Regenerating transgenic plants**- Plants are transferred into controlled environmental conditions in order to obtain whole plants of transgenic traits (Figure 8).

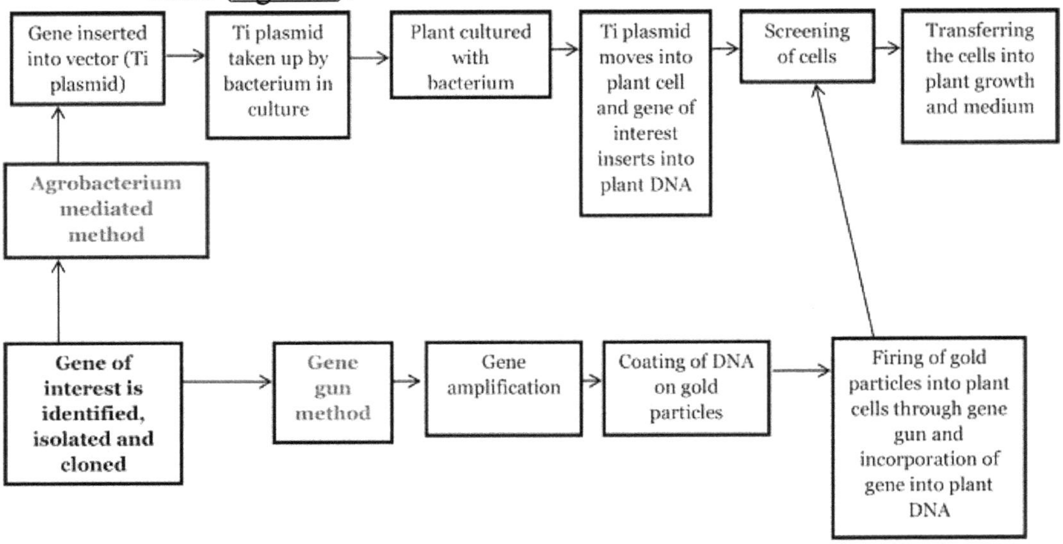

Figure 8. Steps for developing transgenic plant.

Biotechnology and genetic transformation technology is an emerging technology and possess great impact on agricultural production and industry to meet the increasing world demand for food grains and agricultural commodities. The commonly used methods for reducing postharvest losses and increasing shelf life like early harvesting, modulating storage atmosphere, selection of late ripening genotypes are sometimes uneven and not satisfactory for most of the commodities and sometimes can cause loss of flavor, aroma and metabolites. Therefore, transgenic technologies are essential for production of improved crop varieties, enhancing plant tolerance to several biotic and abiotic stresses, securing nutritional status of crop species and many more. Transgenic plants are most promising means to improve crop productivity and production of metabolites and plant products as compared to conventional approaches. Postharvest deterioration of agricultural commodities on the other hand is the matter of great concern globally. As discussed, the main reasons behind this reduced shelf life of food products, rapid spoilage and softening of fruits and vegetables, low temperature, pathogen infection etc. They not only decrease quality of products but also affect their availability adversely. Biotechnological methods for enhancing postharvest factors of food products mainly aims at following perspective i.e., resistance to biotic and abiotic stresses, genetic transformation, secondary metabolites, simultaneous ripening for proper harvesting, improvement in shelf life and organoleptic taste and enhancing nutritional quality [26]. Also, biotechnological approaches are used to create new crop varieties which can acclimate in existing climate change, tolerant to postharvest diseases and acquires extended shelf life. It is well acquainted that the physiological, biochemical and molecular mechanism of plants are directly related to postharvest attributes of fruits and vegetables i.e. storage and shelf life of products. Hence, these traits can be genetically examined and manipulated through molecular and transgenic approaches [27]. Ripening is considered as destructive stage in development of fruits and vegetables and most essential aspect in postharvest study. Ripening process is extremely modulated by various environmental condition and changes in hormone levels but ethylene is a primary hormone that regulate ripening and senescence phenomenon. However, non-climacteric fruits are independent of ethylene production unlike climacteric fruits. Researchers therefore, attempt to manipulate the influence of ethylene on ripening of produce by suppressing it at different growth stages in order to increase storage and shelf life. Ripening in tomato can be delayed by silencing ethylene biosynthetic genes, manipulating ethylene signaling and transduction, discouraging ABA biosynthetic genes etc. TILLING approach is widely used to enhance shelf life of produce.

6. GENOME EDITING TECHNOLOGY

Genetic engineering plays a magnificent role in creation of crops with desired qualities, crop improvement and identification of gene functions. Genome editing technology has been emerged as a new and powerful approach over past few decades which involves the precise editing of gene of interest and utilizes nucleases that target the specific sequence to generate double strand break. The application certain genome editing technologies viz., sequence-specific nucleases (SSN) like ZNFs (zinc-finger nucleases), TALENs (transcription activator-like effector nucleases), CRISPR (clustered regularly interspaced short palindromic repeats)/Cas9 system is a promising and advance technologies that increases crop yield and also confers tolerance to various stress conditions. TALENs has been successfully employed for genetic modification in tomato [28] and potato [29, 30] crops. CRISPR/Cas9 technology has been successfully used as genome editing tool in various vegetables and fruits. In tomato, it has been used with the purpose to increase shelf life and manipulates ripening using RIN, SLALC, lncRNA1459 genes [31, 32, 33] increased resistance to powdery mildew by involving SlMlo1 gene [34] (Table 2) (Figure 9).

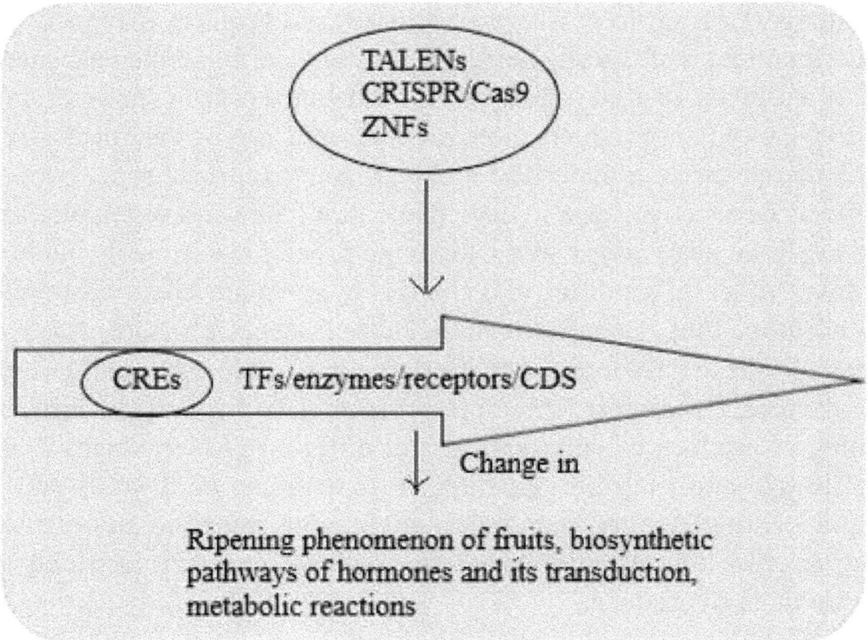

Figure 9. Genome editing technology for postharvest factors (Source: Pizzaro et al., 2018.) [41].

Table 2: Genome editing techniques and their role in various plant species.

Genome editing technique	Species	Gene edited	Physiological responses	References
TALEN	Tomato	PROCERA (PRO)	GA metabolism	Lor et al., 2014 [28]
TALEN	Tomato	LEAFY COTYLEDON1-LIKE4	Pleiotropic response	Hilioti et al., 2016 [35]
Crispr/Cas9	Tomato	RIPENING INHIBITOR(RIN)	Ripening response	Ito et al., 2015, 2017 [33, 36]
Crispr/Cas9	Kiwifruit	AcPDS	Carotenoid synthesis	Wang et al., 2018 [37]
Crispr/Cas9	Apple	PDS	Carotenoid synthesis	Nishitani et al., 2016 [38]
Crispr/Cas9	Apple	DIPM1,2 &4	Pathogen resistance	Malnoy et al., 2016 [39]
Crispr/Cas9	Cucumber	eIFGE	Virus resistance	Chandrasekaran et al., 2016 [40]

6.1 RNA interference

RNA interference is a promising gene regulatory approach in which the molecules of RNA inhibit gene expression or translation process by neutralizing target mRNA molecules. It is also known as post- translational gene silencing (PTGS), quelling.

It permits down regulation of gene expression more precisely without affecting the function of other genes. RNAi has been widely utilized nowadays in enhancing nutritional value of quality, increasing shelf life or perishables, imparting insect-pest, virus and pathogen resistance and abiotic stress tolerance. The fruits and vegetables are highly susceptible to postharvest losses which deteriorate their quality and economic value. The storage life of such produces can be enhanced by RNA interference approach. Transgenic tomatoes containing ACC oxidase dsRNA showed delayed ripening and increased shelf life due to suppression of genes for ethylene synthesis. RNAialso is involved in the enhancing of nutritional quality of food products. In tomato, SINCED1 gene which is important for ABA biosynthesis is suppressed by RNAi and those fruit showed enhanced accumulation of lycopene and beta-carotene [42]. RNAi technology is widely used in Brassica napus to increase the carotenoid content by downregulating the lycopene epsilon cyclase gene (Ɛ-CYC) and possess increased content of β-carotene, violaxanthin and lutein [43]. Jiang et al., 2013 [44] reported that RNAi can be successfully utilized in soybean to increase isoflavone content by silencing flavanone 3-hydroxylase (F3H) gene and the flavone synthase II (GmFNSII) gene, thus regulating flavone and isoflavone production in hairy roots. RNA interference is also applied to produce virus resistant plants. Transgenic tomato produces dsRNA which showed resistance to potato spindle tuber viroid [45]. Transgenic tobacco which expresses the coat protein (CP) gene from Tobacco mosaic virus showed resistance against the same and this technology is used for other crops also like potato against Potato virus Y [46], Prunus domestica resistant to Plum pox virus. The RNAi has been successfully implemented in common bean to induce resistance against Gemini virus Bean golden mosaic virus [47]. Ethylene is a major fruit ripening hormone in postharvest physiology. One of the major factor which is used to extend shelf life of harvested produce is delay in ripening process by downregulation of ethylene action. This can be achieved through antisense RNA technology. In this process, the protein coding region of a gene is inverted in reference to its promoter and thus antisense gene is constructed. The mRNA formed has similar sequence as of antisense gene of natural DNA. Endogenous and antisense RNA present on the same nucleus, their transcription results into antisense and sense RNA transcript. They are complementary to each other and forms RNA molecules having double stranded nature. The duplex thus formed inactivated mRNA and inhibits protein synthesis. Antisense RNA technology suppresses ACC oxidase and ACC synthase genes and plants possessing such techniques are resistant to over ripening and postharvest losses (Figure 10).

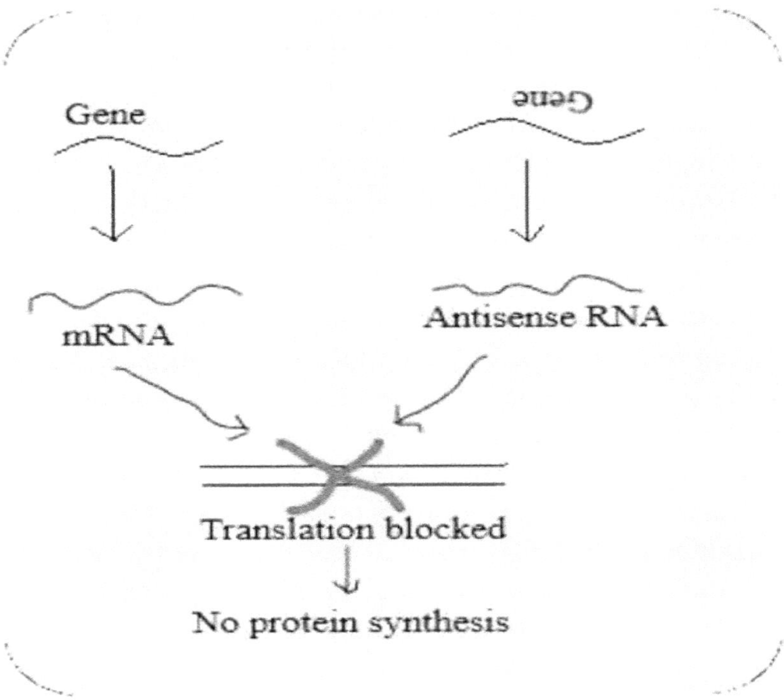

Figure 10. Antisense RNA technology in post harvest physiology (Source: Gupta et al., 2011) [48].

6.2 Breeding methods

Fruits and vegetables along with other agricultural commodities are perishable in nature and they have limited shelf life after harvest. Maintaining freshness and shelf life of produce is great challenge for researchers globally in order to achieve maximum economic value. Modern methodologies like breeding approaches play a vital role in minimizing postharvest losses and management. They include domestication, polyploidy breeding, mutation breeding, selection, hybridization etc. Domestication can be defined as bringing the wild plant species under human cultivation. In includes a series of genetic practices that involves storage and shelf life of species as compared to wild counterparts. Introduction of plant species includes widening of genetic base and many species have been bought through introduction includes Kinnow mandarin, Solo papaya, Jonathan apple etc. Mutations either induced or spontaneous are effective way to extend shelf life and reduce postharvest losses. Mutation has been induced in pear in order to increase shelf life of produce. Polyploids on the other hand also possess greater shelf life and polyploidy is an efficient approach to induce desired traits in plants. Sunny Rogue, a variety of grape possess tolerance to pre and postharvest disease resistance and better shelf life in storage conditions. Hybridization is also an efficient technique to minimize postharvest losses. It not only extend shelf life of fruits but also enhances nutritional quality. Several fruits and vegetables have been developed using this technique which includes mango, banana, papaya, onion, tomato etc.

6.3 Proteomics

Proteins plays an important role in various plant processes including development, growth, ripening and senescence, metabolic functions, resistance to abiotic and biotic stresses. Proteomics and omics approaches has been studied over past few decades and opens a new avenue in understanding ripening and senescence, development, postharvest responses of perishable commodities in many crops like apple, banana, citrus, grapes, strawberry, tomato, peach, papaya, mango etc. They also are crucial in understanding plant mechanisms during pathogen infection as well. Study on proteomics reveals that proteins involved in defense mechanism, energy metabolism and antioxidant system are crucial for food produce in stored condition and elicitor responses [49]. The study of proteomics in postharvest science involves 2-D electrophoresis (classical and/or differential electrophoresis, DIGE) for separating and quantifying proteins based on LC-MS/MS technique either by cross species identification or by species-specific database search. Several postharvest studies and researches also include gel based approach for quantification and identification of proteins. Furthermore, the proteomic research in fruits and vegetables are needed to be extended to other quantitative proteomic approaches like iTRAQ, TMT, peptides demethylation labeling, ICAT, MRM, label-free strategies or the integration of above mentioned approaches with LC–MS/MS. Abdi et al., 2002 [50] studied the postharvest proteomics based on 2D- PAGE related to optimal harvest of peaches. Postharvest proteomics are essential for minimizing postharvest losses by selection of desired quality of cultivars with desired quality of traits. Selection of traits like low susceptibility to bruising and chilling injury to post harvest food products are decisive factors for modulating shelf life of processed products. The combination of transcriptomics, metabolomics and proteomics has been used to study the quality of citrus in storage. Chilling injury is detrimental to produce in storage as it not only affects degradation of proteins, polysaccharides but also affect organic acids and causes injury to postharvest produce (Figure 11).

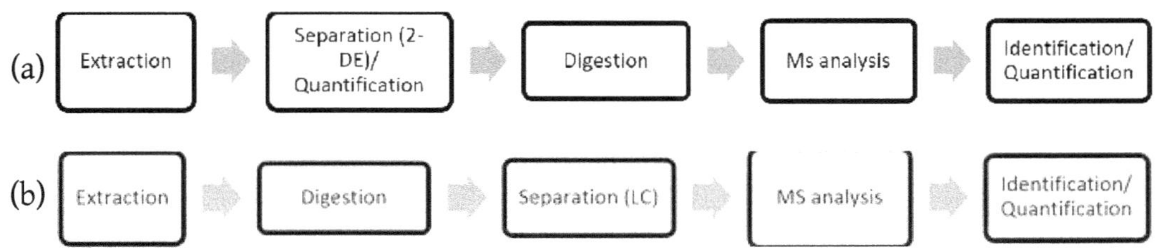

Figure 11. (a) Gel based approach and (b) gel free approach. Source: Pedreschi et al., 2013 [51].

7. CONCISE SUMMARY

India stands second after China, in global production of fruits and vegetables but one-third of agricultural produce available for human consumption in the world per year gets wasted. Agricultural commodities after harvest are liable to decay thus, causing loss of quality and quantity Different factors are responsiple for causing post harvest losses. Ethylene as a gaseous hormone, plays an important physiological role in fruit ripening as well as accelerates the rate of aging and senescence. It also acts as a signaling molecule and controls ripening process in plants. Different transgenic approaches and breeding methods are developed in order to prevent Post harvest losses. Therefore, the significance of post harvest physiology is utmost important in the field of agriculture in developing Genetically modified crops.

AT A GLANCE

The increasing world population over few decades has led to increase in demand of food grains or agricultural commodities, thus possessing great impact on food security. Conventional farming approaches has been traditionally practiced but a lot of effort is required to make to enhance agricultural production. With changing climatic conditions plants are vulnerable to several stress factors. In order to combat such conditions, the agricultural systems are needed to be contemporary with advance and recent technologies. Crops after harvest are liable to a lot of changes which ultimately affect quality and quantity of produce, thus reducing economic value. Postharvest losses are decisive factors in reducing loss of produce and agricultural commodities. Thus in order to achieve maximum production, it is therefore essential to reduce postharvest losses and ensures proper management of postharvest products. Postharvest physiology is the science which deals with quantitative and qualitative study of physiology of agricultural products after harvesting. Biotechnological and transgenic approaches are the recent and emerging technologies that possess great impact on agricultural production. Transgenic technology like genome editing, CRISPR/Cas9, TILLING is successfully used in various species to enhance production, possess resistance to abiotic and biotic stresses, enhance shelf life and improve nutritional quality. Transgenic crops or genetically modified crops (GMO) like tomato, brinjal, soybean, cassava etc are cultivated globally. These techniques therefore are promising means in establishing food security, increasing crop production, reducing postharvest losses, production of secondary metabolites, hormones and plantibodies.

REFERENCES

1. Jha, G.K; Suresh, A; Punera, B; Supriya, P; Growth of horticulture sector in India: Trends and prospects. Indian J. Agric. Sci. 2019, 89 (2): 314-21

2. Sawicka, B; Post-harvest losses of Agricultural Produce. Springer Nature. 2019, doi: 10.1007/978-3-319-69626-3_40-1
3. Hegazy, R; Post-harvest Situation and Losses in India. Technical report, 2013, doi:10.6084/m9.figshare.3206851.v1
4. Baloch, U.K In; Lewis, B; Mejia, D (eds) Wheat: Postharvest operations. Pakistan Agricultural Research Council, Islamabad, 2010, pp 1-21
5. Banjaw, T.D Review of post-harvest loss of horticulture crops in Ethiopia, its causes and mitigation strategies. J Plant SciAgric Res, 2017, 2(1): 1-4
6. Shah, D; Assessment of pre and postharvest losses in tur and soyabean crops in Maharashtra. Agro-Economic Research Centre Gokhale Institute of Politics and Economics, 2013, Pune
7. Gliński, J; Horabikk, J; Lipiec, J; Sławiński, C (eds); Agrophysics. Processes, properties, methods. Institute of Agrophysics BohdanDobrzański, Polish Academy of Sciences, Lublin, 2014, p 135. ISBN: 978-83-89969-34-7
8. Bala, B.K; Haque, M.A; Hossain, M.A; Majumdar, S; Post harvest loss and technical efficiency of rice, wheat and maize production system: assessment and measures for strengthening food security. Bangladesh Agricultural University, Mymensingh, 2010
9. Kader, A.A; Postharvest technology of horticultural crops – An overview from farm to fork. EJST.2013, 1(1): 1-8
10. Mirdehghan, S.H; veGhotbi, F; Effects of salicylic acid, jasmonic acid and calcium chloride on reducing chilling injury of pomegranate (PunicagranatumL.) fruit. J.Agric. Sci. Technol. 2014, 16:163-173
11. Dhall, R.K; Ethylene in Post-harvest Quality Management of Horticultural Crops: A Review. RRJoCST, 2013, pp 9-25
12. Taiz, L.; Zeiger, E.; Plant Physiology (Sinauer Associates, Sunderland, 3rdedn, 2002, pp 690)
13. Iqbal, N; Khan, N.A; Ferrante, A; Trivellini, A; Francini, A; Khan, M.I.R; Ethylene Role in Plant Growth, Development and Senescence: Interaction with Other Phytohormones. Front. Plant Sci.2017, 8:475, doi: 10.3389/fpls.2017.00475
14. Konings, H; Jackson, M. B; A relationship between rates of ethylene production by roots and the promoting or inhibiting effects of exogenous ethylene and water on root elongation. Z. Pflanzen physiol. 1979, 92, 385-397, doi: 10.1016/S0044-328X(79)80184-1
15. Khan NA. The influence of exogenous ethylene on growth and photosynthesis of mustard (Brassica juncea) following defoliation. Sci. Hortic. 2005;105:499-505. DOI: 10.1016/ j.scienta.2005.02.004

16. Nath, P; Trivedi, P.K; Sane, V.A; Sane, A.P; Role of Ethylene in Fruit Ripening. Ethylene action in plants.Springer, 2006, pp-151-176

17. Toivonen, P.M.A; Postharvest physiology of fruits and vegetables. In: Pareek, S. (Ed.), Postharvest Ripening Physiology of Crops. CRC Press, Boca Raton, FL, USA, 2016, pp. 49-80

18. Alós, E; Rodrigo, M.J; Zacarias, L; Ripening and Senescence.Postharvest Physiology and Biochemistry of Fruits and Vegetables.Elsievier, 2019, DOI: https://doi.org/10.1016/B978-0-12-813278-4.00007-5

19. Alexander L, Grierson D. Ethylene biosynthesis and action in tomato: a model for climacteric fruit ripening. J. Exp. Bot. 2002;53:2039-2055

20. Whittaker, D.J; Smith, G.S; Gardner, R.C; Expression of ethylene biosynthetic genes in Actinidiachinensisfruit. Plant Mol Biol,1997, 34:45-55

21. Clark, K.L; Larsen, P.B; Wang, X; Chang, C; Association of the Arabidopsis CTR1 Raf-like kinase with the ETR1 and ERS ethylene receptors. Proc. Natl. Acad. Sci. U.S.A. ,1998, 95: 5401-5406

22. Cancel, J.D; Larsen, P.B; Loss-of-function mutations in the ethylene receptor ETR1 cause enhanced sensitivity and exaggerated response to ethylene in Arabidopsis. Plant Physiol. 2002, 129: 1557-1567

23. Padmanabhan, P; Paliyath,G;Ethylene Signal Transduction During Fruit Ripening and Senescence. Postharvest Biology and Nanotechnology, First Edition, 2019, John Wiley & Sons, Inc

24. Rani, S.J; Usha, R; Transgenic plants: Types, benefits, public concerns and future. Elsevier, 2013, doi: http://dx.doi.org/10.1016/j.jopr.2013.08.008

25. Lucht JM. Public acceptance of plant biotechnology and GM crops. Viruses. 2015;7(8):4254-4281

26. El-Ramady, H.R; Domokos-Szabolcsy, E; Abdalla, N.A; Taha, H.S; Fári, M; Postharvest Management of Fruits and Vegetables Storage. Sustainable Agriculture Reviews. Springer, 2015, doi: 10.1007/978-3-319-09132-7_2

27. Rai, A.C; Rai, A; Singh, M; Singh, A.K; Biotechnological applications in post-harvest management of vegetable crops. APPLICATIONS IN POST-HARVEST MANAGEMENT OF VEGETABLE CROPS.CRC press, Taylor& Francis group, 2019, https://www.researchgate.net /publication/320908501

28. Lor, V.S; Starker, C.G; Voytas, D.F; Weiss, D; Olszewski, N.E.; Targeted mutagenesis of the tomato PROCERA gene using transcription activator-like effector nucleases. Plant Physiol., 2014, 166, 1288-1291

29. Clasen, B.M; Stoddard, T.J; Luo, S; Demorest, Z.L; Li, J; Cedrone, F; et al; Improving cold storage and processing traits in potato through targeted gene knockout. Plant Biotechnol. J. 2016, 14, 169-176

30. Sawai S, Ohyama K, Yasumoto S, Seki H, Sakuma T, Yamamoto T. et al; Sterol side chain reductase 2 is a key enzyme in the biosynthesis of cholesterol, the common precursor of toxic steroidal glycoalkaloids in potato. Plant Cell. 2014;26:3763-3774

31. Li R, Fu D, Zhu B, Luo Y, Zhu H. CRISPR/Cas9-mediated mutagenesis of lncRNA1459 alters tomato fruit ripening. Plant J. 2018;94:513-524

32. Yu, Q.H; Wang, B; Li, N; Tang, Y; Yang, S; Yang, T; et al; CRISPR/Cas9- induced targeted mutagenesis and gene replacement to generate long-shelf life tomato lines. Sci. Rep. 2017, 7, 1874

33. Ito Y, Nishizawa-Yokoi A, Endo M, Mikami M, Toki S. CRISPR/Cas9- mediated mutagenesis of the RIN locus that regulates tomato fruit ripening. Biochem. Biophys. Res. Commun. 2015;467:76-82

34. Nekrasov, V; Wang, C; Win, J; Lanz, C; Weige, D; Kamoun, S; Rapid generation of a transgene-free powdery mildew resistant tomato by genome deletion. Sci. Rep. 2017, 7, 482

35. Hilioti, Z.; Ganopoulos, I; Ajith, S; Bossis, I; Tsaftaris, A; A novel arrangement of zinc finger nuclease system for in vivo targeted genome engineering: the tomato LEC1-LIKE4 gene case. Plant Cell Rep 2016, 35, 2241-2255. doi: 10.1007/s00299-016-2031-x

36. Ito, Y; Nishizawa-Yokoi, A; Endo, M; Mikami, M; Shima, Y; Nakamura, N; et al;. Re-evaluation of the rin mutation and the role of RIN in the induction of tomato ripening. Nat. Plants, 2017, 3, 866-874. doi: 10.1038/s41477-017-0041-5

37. Wang, Z; Wang, S; Li, D; Zhang, Q; Li, L; Zhong, C; et al; Optimized paired-sgRNA/Cas9 cloning and expression cassette triggers high-efficiency multiplex genome editing in kiwifruit. Plant Biotechnol. J. 2018, 16, 1424-1433. doi: 10.1111/pbi.12884

38. Nishitani, C; Hirai, N; Komori, S; Wada, M; Okada, K; Osakabe, K; et al; Efficient genome editing in apple using a CRISPR/Cas9 system. Sci. Rep. 2016, 6:31481. doi: 10.1038/srep31481

39. Malnoy, M; Viola, R; Jung, M.-H; Koo, O.-J; Kim, S; Kim, J.-S; et al; DNA-free genetically edited grapevine and apple protoplast using CRISPR/Cas9 ribonucleoproteins. Front. Plant Sci. 2016, 7:1904. doi: 10.3389/fpls.2016.01904

40. Chandrasekaran, J; Brumin, M; Wolf, D; Leibman, D; Klap, C; Pearlsman, M; et al; Development of broad virus resistance in non-transgenic cucumber using CRISPR/Cas9 technology. Mol. Plant Pathol. 2016, 17, 1140-1153. doi: 10.1111/mpp.12375

41. Pizzaro, C.M; Pose, D; Genome Editing as a Tool for Fruit Ripening Manipulation. Front. Plant Sci. 2018, 9:1415. doi: 10.3389/fpls.2018.01415

42. Choudhary, S; Jain, D; Meena, M.R; Verma, A.K; Sharma, R; GENE SILENCING IN HORTICULTURAL TRANSGENIC CROPS. Genetic Engineering of Horticultural Crops.Elsievier, 2018, doi:http://dx.doi.org/10.1016/B978-0-12-810439-2.00003-9
43. Yu, B; Lydiate, D.J; Young, L.W; Schafer, U.A; Hannoufa, A; Enhancing the carotenoid content of Brassica napusseeds by downregulating lycopene epsilon cyclase. Transgenic Res. 2007,17, 573-585
44. Jiang Y, Hu Y, Wang B, Wu T. Bivalent RNA interference to increase isoflavone biosynthesis in soybean (Glycine max). Braz. Arch. Biol. Technol. 2013;57:163-170
45. Nora S, MichÈLe Z, Asuka I, Biao D, Ming-Bo W, Gabi K, et al. RNAi-mediated resistance to potato spindle tuber viroid in transgenic tomato expressing a viroid hairpin RNA construct. Mol. Plant Pathol. 2009;10:459-469
46. Missiou A, Kalantidis K, Boutla A, Tzortzakaki S, Tabler M, Tsagris M. Generation of transgenic potato plants highly resistant to potato virus Y (PVY) through RNA silencing. Mol. Breed. 2004;14:185-197
47. Bonfim, K; Faria, J.C; Nogueira, E.O; Mendes, E.A; Aragao, F.J; RNAi mediated resistance to bean golden mosaic virus in genetically engineered common bean (Phaseolus vulgaris). Mol. Plant Microbe Interact. 2007, 20, 717-726. http://dx.doi.org/10.1094/MPMI-20-6-0717
48. Gupta, S; Singh, R.P;Rabadia, N; Patel, G; Panchal,H; ANTISENSE TECHNOLOGY. IJPSSR.2011, 9 (2), 38-45
49. Pedreschi, R; Postharvest Proteomics of Perishables.Proteomics in Food Science.Elsievier, 2017, doi: http://dx.doi.org/10.1016/B978-0-12-804007-2.00001-1
50. Abdi N, Holford P, McGlasson B. Application of two dimensional gel electrophoresis to detect proteins associated with harvest maturity in stone fruit. Postharvest Biol. Technol. 2002;26:1-13
51. Pedreschi, R; Lurie, S; Hertog, M; Nicolaï B; Mes, J; Woltering,E; Post-harvest proteomics and food security, Proteomics. 2013, 13, 1772-1783, doi: 10.1002/pmic.201200387

CHAPTER-16

IN VITRO-CULTIVATED CENTAURY PLANTS WITH TRANSGENIC ATCKX

1. INTRODUCTION: AN OVERVIEW

The cultivation of transgenic plants has become a focal point in agricultural research, providing innovative solutions to enhance crop performance and address global challenges. Among the noteworthy developments in this field is the introduction of transgenic plants with altered levels of ATCKX (adenine isopentenyltransferase cytokinin oxidase/dehydrogenase), a key enzyme influencing cytokinin levels. In the context of Centaury plants, this study explores the in vitro cultivation of transgenic variants with modified ATCKX expression. By manipulating the cytokinin pathway, researchers aim to unravel the physiological implications and potential benefits of these genetic modifications, shedding light on the intricate interplay between transgenic technology and the in vitro growth conditions of Centaury plants.

A fundamental biotechnological method, genetic engineering, enables the production of plants with desired properties. Further breeding of obtained plants is possible by another biotechnological tool of great importance such as *in vitro* culture techniques. Applying of tissue culture procedures enables regeneration and multiplication of desired plants in controlled conditions in short time periods. Developing of plant tissue cultures also allows preservation and improvement of plant species endangered in nature, especially medicinal plant species such as e.g. common centaury (*Centaurium erythraea* Rafn.).

1.1 Genetic transformations of plants

Genetic engineering of plants represents a unique field of biotechnology where genes and other DNA sequences are manipulated to obtain plants with certain desired properties. The necessary steps in genetic engineering are the isolation and modification of the gene of interest, the vector construction, transformation and the final selection of transgenic plants. Genetic transformation of plants implies stable incorporation of foreign genes into the plant genome [1]. Genetic transformation of plants can be very successfully achieved in two ways:

- Techniques of direct gene transfer (microbombardment - biolistic method, microinjection, electroporation, polyethylene glycol mediated transformation, protoplast fusion)
- Gene transfer techniques through biological vectors (viruses or bacteria - *Agrobacterium rhizogenes* and *Agrobacterium tumefaciens*)

Direct gene transfer techniques are continually developing and improving over the years. However, a considerable variation in the stability, integration and expression of the introduced gene represent their significant deficiencies. On the other hand, techniques of gene transfer through bacteria of the genus *Agrobacterium* have proven to be very efficient and straightforward. These techniques allow a more precise and stable integration of the desired gene into the plant genome [2].

1.1.1 Plasmids of bacteria of the genus Agrobacterium

The Agrobacterium genus is a group of Gram-negative soil bacteria. Unlike the wild type, laboratory Agrobacterium strains are commonly used for genetic transformations of numerous plant species [3]. Different plant species show different susceptibility to infection by Agrobacterium bacteria [4]. In general, dicotyledonous plants are more susceptible to infection by Agrobacterium bacteria than monocotyledons. Bacteria of the genus Agrobacterium can be present in the vascular tissue of plants (xylem and phloem) but without showing disease symptoms. It indicates that the interaction of bacteria and plant cells does not always occur. Infection and pathological changes of plant tissue occuronly after injury of the plant cells when released phenolic compounds causea positive chemotoxic reaction in Agrobacterium. These phenolic compounds further induce genes, located on chromosomal and plasmid DNA, responsible for bacterial virulence. During genetic transformations, the plasmids in bacterial cells serve as vectors. Bacteria of the genus Agrobacterium have large conjugated plasmids (200–250 kb), parts of which have been successfully integrated into the plant genome during transformation processes for almost half a century [5]. In these bacteria, two basic types of plasmids are identified by the kind of disease they cause: Ti-plasmid (tumor-inducing) in A. tumefaciens and Ri-plasmid (root-inducing) in A. rhizogenes.

Both types of plasmids contain three genetic regions that are necessary for the genetic transformation process:

- **T-DNA (transferred DNA)** – a part of plasmid DNA, size 10–30 Kbp. The transport of T-DNA into a plant cell allows expression of bacterium genes in the plant itself [6].

- *Vir* (virulence) region, size 35 Kbp, which consists of several large loci (*virA*, *virB*, *virC*, *virD*, *virE*, *virG*, *virH* and *virJ*). *Vir* genes encode the synthesis of vir proteins, which are responsible for the process of recognizing specific compounds synthesized by a plant cell after injury and transferring T-DNA to the host cell.
- The bacterial chromosome region, which is made up of *chv* (chromosomal virulence) genes that regulate bacterial chemotaxis and bacterial binding to an injured plant cell [7]. According to Simonović [8], the *chvA* and *chvB* genes are responsible for the production and secretion of cyclic 1,2-glycans while *chvE* product forms a transmembrane receptor complex together with the plasmid *virA* product.

The development of plant tumor tissue occurs after a part of the plasmid DNA (T-DNA) is transported and incorporated into the DNA of the plant. T-DNA of Ti- and Ri-plasmids carries oncogenes, the genes responsible for tumor formation and encoding enzymes that catalyze the synthesis of auxins and cytokinins. Transcription of these genes leads to an increased amount of these plant hormones, which stimulates uncontrolled cell division, tumor development and mass proliferation of transformed cells [9].

1.1.2 Mechanism of genetic transformation of a plant cell using Agrobacterium tumefaciens

Successful genetic transformations of plants using *Agrobacterium* require a bacterial strain with appropriate density, explant tissue culture, determination of the substrate and time for cocultivation, as well as temperature and selection regime [10]. In wild strains of *Agrobacterium*, the Ti- and Ri- plasmids have a T-DNA region divided into two sequences (T_L and T_R) between which is a T_C sequence that is not transmitted to the plant genome [11]. Oncogenes are located on the T_L sequence while genes for opine synthesis are located on the T_R sequence. Opines accumulate in transformed plant cells thus representing metabolic markers for genetic transformations. They seem useless for the plant as no appropriate enzymes for their degradation have been found in plant tissues. Pathogenic bacteria use released opines as a source of carbon and nitrogen. The genes on the plasmid itself encode the enzymes for catabolism of opines whereby the opine, whose synthesis is induced in transformed tissue, is degraded [12]. In laboratory strains of *Agrobacterium*, natural T-DNA can be replaced by a gene of interest that should be introduced into the plant genome by genetic transformation.

Genetic transformation of plants by *Agrobacterium* is a process that, in general, consists of three steps:

Recognition and "attack" of a plant cell by a bacterium

Injured plant cells begin to synthesize specific compounds (amino acids, sugars and organic acids). In dicotyledons these are phenolic derivatives syringone and acetosyringone [13] while in monocotyledons it is ethyl ferrulate [14]. These compounds cause a positive chemotoxic reaction that leads to interaction of plant cells and bacteria. Binding of a bacterium to the plant cell represents the first step in the mechanism of genetic transformation.

Transfer of T-DNA into a plant cell

VirD and VirE proteins play a significant role in the formation of single-stranded T-DNA, its cutting and transfer to the plant cell. The products of the VirD region (VirD1 and VirD2 proteins) are part of an endonuclease complex catalyzing specific single-stranded breaks that release a single-stranded T-DNA segment from the Ti plasmid. After excision, the T-chain shifts polarly and covalently binds the VirD2 protein. This protein protects the T-chain from destruction and remains there during transport to the plant cell. The transport of VirD2/T-DNK complex (immature T-complex) into the cytoplasm of the host cell, together with several other vir proteins (VirE2, VirF and VirE3), is performed through a specialized membrane transport system. In cytoplasm of the plant cell, a mature T-complex is formed, which is protected from the action of nucleases during transport to the nucleus [15].

T-DNA gene incorporation and expression

The last, and probably the most critical step of genetic transformation, is the integration of T-DNA into the host genome. Integration into the plant genome occurs by non-homologous recombination in any part of the genome, taking preferentially place in regions with a higher percentage of adenine-thymine (AT) bonds [16]. Transgene expression depends on numerous factors such as e.g. the number of copies of an introduced gene and the site of transgene integration [17, 18, 19]. Stably integrated T-DNA is transferred to the next generation [20].

The mechanism of genetic transformation today involves the use of so-called "disarmed" *A. tumefaciens* strains, in which oncogenes are removed from the plasmids. The system of binary vectors usually used in genetic transformation processes divides the original Ti-plasmid into two plasmid vectors: an artificial (recombinant) Ti-plasmid carrying the gene of interest and an auxiliary plasmid [21]. In each binary plasmid, additionally to a foreign gene of interest the marker genes are also inserted and divided into selective and reporter genes [22]. The selective marker genes are most often prokaryotic genes that encode resistance to some antibiotics (kanamycin, hygromycin etc.). This negative selection implies using a selective medium which allows the distinction of transformed plant cells from non-transformed ones. Reporter genes are common but not necessary parts of the binary vector. These genes encode products which can be involved in the monitoring of transformation process (growth of transformed cells, regeneration efficiency and growth of transgenic plants). Visual reporter genes have been increasingly used because they do not require special artificial substrates for detection, and their expression is easily visually detected in tissue without destroying.

1.2 Cytokinins

Cytokinins (CKs) are plant hormones that individually or in combination with other hormones regulate numerous developmental and physiological processes in plants [23]. The name cytokinins comes from their originally discovered function, which is the stimulation of cell division - cytokinesis [24]. Cytokinins are divided into two major groups: natural and synthetic.

Natural CKs by chemical structure represent adenine derivatives. Depending on the component that binds to the adenine ring, natural CKs are further categorized into two groups:

- **Isoprenoid CKs** – appear when one isoprenoid C5 unit binds to the N6 atom of adenine [25]. Natural isoprenoid CKs include N6-(Δ2-isopentenyl) adenine (iP), trans-zeatin (tZ), cis-zeatin (cZ) and dihydrozeatin (DHZ) and their derivatives (ribosides, N- and O-glucosides, ribotides) differing in their biological functions. The chemical structures of isoprenoid cytokinins are shown in Figure 1. The contents and distribution of individual isoprenoid CKs in plants vary depending on the plant species, particular plant tissue, developmental stage, etc.
- **Aromatic CKs** – appear when an aromatic ring (either unsubstituted or substituted by a hydroxy or methoxy group) is attached to the N6 atom of adenine [26]. Substitution of the aromatic ring with a hydroxy group produces hydroxy derivatives such as ortho-topolin (oT) and meta-topolin (mT), while substitution with a methoxy group creates methoxy derivatives such as ortho-methoxytopolin (meoT) and meta-methoxytopolin (memT). Hydroxylation of the aromatic ring in meta position is a prerequisite for high CK activity [27, 28] and meta-topolin shows the most considerable effects in plants of all derivatives. The chemical structures of natural aromatic cytokinins are shown in Figure 2. Despite the isolation of the first naturally occurring aromatic CK as early as in the early 1970s [29], the distribution and function of aromatic CKs in the plant kingdom remain still unclear [30].

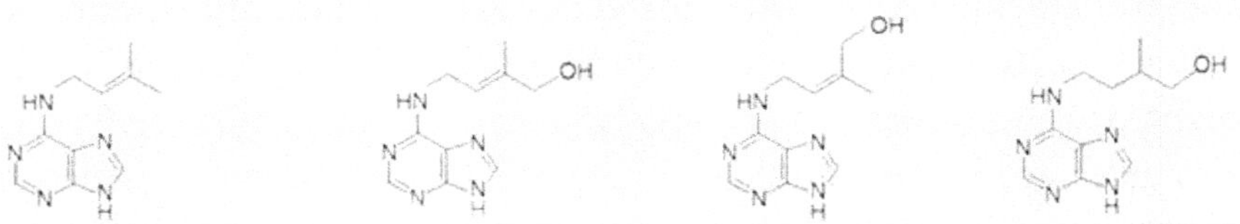

N⁶-(Δ²-Isopentenyl) adenine | trans-Zeatin (*t*Z) | cis-Zeatin (*c*Z) | Dihydrozeatin (DHZ)

Figure 1. Chemical structures of natural isoprenoid cytokinins.

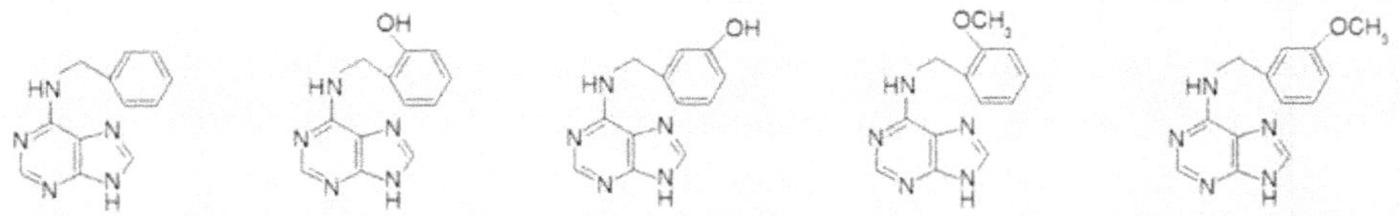

Benzyl-adenine | *ortho*-topolin (oT) | *meta*-topolin (mT) | *ortho*-metoxytopolin (meoT) | *meta*-metoxytopolin (metT)

Figure 2. Chemical structures of natural aromatic cytokinins.

Synthetic CKs by chemical structure represent phenylurea derivatives. These substances show CK effects, however, to date they have not been discovered to be synthesized in plant tissues by normal metabolic pathways. The first identified synthetic CK was diphenylurea (DFU), discovered in 1955 [31]. The unexpected discovery of this compound stimulated the synthesis of numerous analogues such as CPPU (N-phenyl-N'- (2-chloro-4-pyridyl) urea) and TDZ (thidiazuron). These synthetic substances, presented in Figure 3, are highly stable and have stronger CK activity compared to zeatin [32, 33, 34, 35].

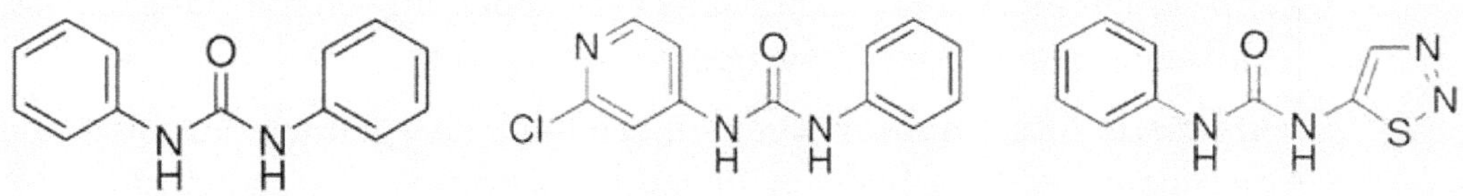

Diphenylurea (DFU) | N-phenyl-N'-(2-chloro-4-pyridyl)urea (CPPU) | Thidiazuron (TDZ)

Figure 3. Chemical structures of synthetic cytokinins.

1.2.1 Cytokinin catabolism

During plant development the levels of endogenous CKs are tightly metabolically regulated in plant tissues. The decreased amount of CKs is directly related to the increased activity of the catabolic enzyme, cytokinin oxidase/dehydrogenase (CKX, EC 1.5.99.12). This enzyme irreversibly degrades CKs by removing the *N*⁶-substituted side-chain to form adenine and the unsaturated aldehyde 3-methyl-2-butenal [36] (Figure 4). To date, CKX is the only known enzyme involved in the catabolism of specific CKs, and thus it represents a crucial factor in the control of CK levels in plant tissues. It is known that especially iP, *t*Z and their ribosides are sensitive to the CKX action. On the other hand, *c*Z is less sensitive [37] even though some CKX isoforms exhibit high affinity for the *c*Z isomer as well [38]. CK *O*-glucosides, DHZ, aromatic CKs and their derivatives are believed to be the non-substrates for CKX [39, 40, 41]. By contrast, *Arabidopsis thaliana*, barley and maize CKX isoforms showed ability to degrade CK *N*9-glucosides [37, 42, 43].

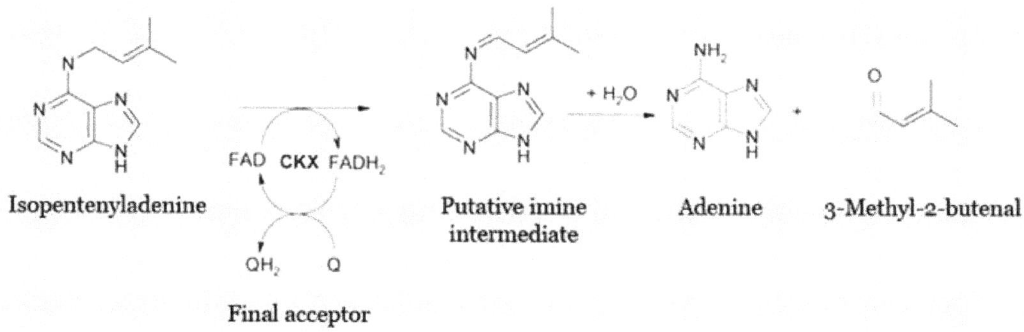

Figure 4. Chemical reaction of irreversible cytokine degradation by CKX enzyme.

According to the chemical structure, CKX enzyme is a flavoprotein with the flavin adenosine dinucleotide (FAD) bound domain [44, 45]. The activity of CKX in plant tissues is induced by exogenously applied CKs, both CKX substrates and non-substrates [46, 47, 48]. The CKX activity was first discovered in tobacco tissue [49], and subsequently this enzyme has been shown widely distributed in numerous plant tissues such as e.g. maize [50], *Vinca rosea* tumor [36], *Phaseolus vulgaris* callus [51, 52], wheat [53], and tobacco and poplar calli [47, 54]. The CKX enzyme is encoded by a small family of genes whose number varies from species to species. In *Arabidopsis thaliana*, seven genes responsible for the synthesis of CKX enzyme (*AtCKX1-7*) have been identified. These genes are expressed variously in different plant tissues [55, 56]. Individual *AtCKX* genes are expressed in the same tissues where CKs are synthesized (axillary buds, trichomes and vascular tissues) or in the nearby of these tissues. Genes encoding the CKX enzyme have been cloned also from numerous plant species (a detailed literature review is presented in Table 1).

Table 1. The list of *CKX* genes cloned and identified in different plant species.

Plant species	*CKX* genes	References
Arabidopsis thaliana	*AtCKX1-7*	[55, 56]
Zea mays	*ZmCKX1-5*	[57]
	ZmCKX10	[41, 58, 59, 60]
Dendrobium sp	*DSCKX1*	[61, 62]
	DhCKX	[63]
Hordeum vulgare	*HvCKX1*	[64]

Plant species	*CKX* genes	References
Oryza sativa	*OsCKX1–11*	[65]
Pisum sativum	*PsCKX*	[66, 67]
Setariaitalica	*SiCKX1–11*	[68]
Triticum aestivum	*TaCKX1–6*	[69, 70, 71, 72, 73]
Brassica napus	*BnCKX1–7*	[74]
Gossypium hirsutum	*GhCKX*	[75]
Glycine max	*GmCKX1–17*	[76]
Jatropha curcas	*JcCKX1–11*	[77]
Medicago sativa	*MsCKX*	[78]

1.3 Genetic transformation of plants using *AtCKX* genes

Genetic transformations of plants with specific genes encoding the CKX enzyme are beneficial for investigations of CK homeostasis. Thus, plants exhibiting increased overexpression of the *CKX* genes and a reduced amount of endogenous CKs can be obtained. Genetically engineered CKX plants can serve as a sound and valuable model system for studying physiological and morphological processes that are under control of CKs. To date, the impact of overexpression of different *CKX* genes on CK metabolism has been investigated in several plant species, as summarized in Table 2. The overexpression of *CKX* genes in tobacco and *A. thaliana* plants was demonstrated to affect significantly the phenotype of transformed plants, causing the *cytokinin deficiency syndrome*. In transgenic plants with this syndrome, apical dominance was decreased, shoots grew slowly, leaf size was reduced, the plants bloomed later and had a smaller number of flowers. Increased root growth size and activity of the root apical meristem (RAM) were observed. At the same time, decreased size and the shoot apical meristem (SAM) activity were also found. These symptoms were more evident in plants overexpressing *AtCKX1* transgene than in those with *AtCKX2* gene [55, 79]. As most of existing knowledge on regulation of CK levels and effects in plants have been obtained from studies of consequences of exogenously applied CKs, the specific CK-deficient transformants are expected to be more informative and to help in elucidation of the processes controlled by this group of phytohormones.

Table 2. The list of plant species genetically transformed using *CKX* genes isolated from *Arabidopsis thaliana*.

Plant species	Initial explants	Genes used for transformation	References
Nicotiana tabacum L cv. Samsun	Leaf explants	AtCKX1-AtCKX4	[79]
Arabidopsis thaliana	In planta transformation	AtCKX1-AtCKX6	[55]
Physcomitrella patens	Protoplast from liquid culture	AtCKX2	[80]
Solanum tuberosum cv. Solara	Leaf explants	AtCKX1	[81]
Solanum tuberosum cv. Désirée	Leaf explants	AtCKX2	[82]
		AtCKX1	[83]
Centaurium erythraea Rafn.	Root explants	AtCKX1, AtCKX2	[84]

In plant shoots, CKs stimulate function of apical meristem and cell divisions. The stimulatory role of CKs on cytokinesis was discovered a long time ago [85]. Recently, it has been confirmed by analyses of transgenic *AtCKX A. thaliana* plants with reduced endogenous CK levels. Ultrastructural analysis of SAM cells showed clear cytological changes indicating detention of cell division and accelerated cell differentiation in transgenic CK-deficient *AtCKX A. thaliana* plants [86]. The development of vascular elements is also regulated by CKs. The optimal concentration of CKs in combination with auxins affects the formation of phloem [87]. In the shoots of transgenic *AtCKX A. thaliana* plants, xylem and phloem were reduced [55]. In addition, CKs control leaf formation and growth. Overexpression of the *CKX* transgenes in *AtCKX A. thaliana* plants significantly reduced leaf surface, which is undoubtedly related to cell divisions as previously described [55]. It has been shown that CKs are not able to completely prevent, but can significantly delay the senescence process [88]. Thus, it can be assumed that a reduced amount of endogenous CKs in CK-deficient plants accelerates senescence. Despite these expectations, the leaves of transgenic *AtCKX A. thaliana* plants did not show accelerated senescence [55]. The role of CKs during the plant reproductive development was approved by transgenic *AtCKX A. thaliana* plants, which formed a smaller number of flowers [55]. Enlarged embryos have also been observed in these plants, indicating a role of CKs in cell division during embryogenesis. The enlargement of embryos could be explained on the basis of *CKX* genes overexpression, occurring only in the later stages of embryogenesis [89].

Unlike the shoots, CKs inhibit the function of the RAM. Using the transgenic *A. thaliana AtCKX* plants, it has been confirmed that reduced CK levels strengthen the root system. The primary root growth is directly related to intense cell divisions in the root meristem of transgenic *A. thaliana AtCKX* plants [55]. The results obtained in these transformants are consistent with those by [90] demonstrating that exogenously applied CKs inhibited root elongation of untransformed plants by reducing the RAM size as well as reducing cell divisions. One would expect that the plants transformed with *CKX* genes have inhibited root branching. However, CK-deficient transgenic *AtCKX A. thaliana* plants surprisingly developed numerous lateral roots that elongated rapidly [55].

1.4 *Centaurium erythraea* Rafn.

Common centaury, *Centaurium erythraea* Rafn. (syn. *C. umbellatum* Gillib and *C. minus* Moench), is the most known species from Gentianaceae family. Thanks to numerous pharmacological activities, centaury is listed as a medicinal plant species. In traditional medicine, centaury is used to treat febrile conditions, anaemia, jaundice, gout and to regulate blood sugar [91]. Bitter centaury juices have also been used for an increment of appetite, digestion stimulation and treatment of gastrointestinal tract diseases [92].

Nowadays, *C. erythraea* is rapidly disappearing from natural habitat and it is marked as an endangered plant species. Fortunately, even almost twenty years ago centaury showed vigorous regenerative potential *in vitro* [93, 94, 95, 96, 97]. The most interesting and essential information is that spontaneous morphogenesis of centaury is possible on nutrition medium without addition of any plant growth regulator [98]. In addition, during the years, centaury has also been used for studies of numerous developmental processes [99, 100, 101]. On the other hand, only few literature data exist describing genetic transformation of centaury. To date, most of Gentianaceae species, including *C. erythraea*, were usually genetically transformed only with *A. rhizogenes* [93, 102]. These investigations have mostly based and described the efficiency of plant regeneration and selection of transformed plant tissues. All of these previous investigations also encouraged us to use *A. tumefaciens* in genetic transformation of centaury for the first time [84]. The successful production of transgenic *AtCKX* centaury plants allowed to use them as a suitable model for studies of numerous physiological and developmental processes under endogenous phytohormonal control.

2. METHODOLOGY

All of the *in vitro* culture experiments were performed with plant material originated from *C. erythraea* Rafn. seeds, obtained from JelittoStaudensamen GmbH, Schwarmstedt, Germany. The plant tissue culture methods were used to establish a solid centaury root culture further utilizable in the genetic transformation process. In all *in vitro* culture experiments, half-strength MS hormone-free medium (½MS) [103] solidified with 0.7% agar and supplemented with 3% sucrose and 100 mg

l^{-1}*myo*-inositol was used. The *A. tumefaciens* strain GV3101 harbouring the pBinHTX plasmid with either the *AtCKX1* or *AtCKX2* gene under the control of the 35S promoter with Triple X enhancer [79] was applied for the transformation of *C. erythraea* roots. Usual molecular biology analyses, including PCR and qPCR reactions, were performed to confirm the insertion of *AtCKX* transgenes into centaury genome. The CKX activity was determined by *in vitro* assays based on the conversion of [2-^{3}H]iP (prepared by the Isotope Laboratory, Institute of Experimental Botany AS CR, Prague, Czech Republic) to [^{3}H] adenine. Detection and quantification of endogenous phytohormones were performed using HPLC/MS system with TSQ Quantum Ultra AM triple-quadrupole high-resolution mass spectrometer (Thermo Electron Corp., San Jose, CA, U.S.A.) operated in the positive-SRM mode. Secondary metabolites including secoiridoids (swertiamarin and gentiopicrin) and xanthones (eustomin and demethyleustomin) were analysed on an Agilent series 1100 HPLC instrument with DAD detector and a reverse phase Zorbax SB-C18 analytical column. The secondary metabolite profile of transgenic centaury plants was determined to as certain whether these plants might serve as potential producers of anti-cancer compounds. The effect of non-transformed and transgenic *AtCKX* centaury methanol extracts on antibacterial and antifungal activity on bacteria and microfungi was also investigated. The salinity tolerance of non-transformed and *AtCKX* transgenic centaury shoots and roots to graded NaCl concentrations were tested separately *in vitro*. All plant material and methods used in investigations are presented in Table 3. The illustrative presentation of transgenic *AtCKX* centaury plants production *in vitro* and all further applied analyses are shown in Figure 5.

Table 3. The list of all plant materials and methods used during the investigation of transgenic *AtCKX* centaury plants grown *in vitro*.

Plant material	Methods	References
C. erythraea Rafn. seeds	Establishment of a solid root culture by plant tissue culture methods	[84]
Centaury root tips (≈10 mm long)	Genetic transformation of cut-edge root explants by immersing into *AtCKX1* and *AtCKX2* bacterial suspensions	
Non-transformed and transgenic *AtCKX* centaury shoots and roots	Molecular biology analyses including PCR and qPCR reactions to confirm insertion of *AtCKX* transgenes into centaury genome	
	Quantification of the CKX activity by *in vitro* assays	
	Detection and quantification of endogenous phytohormones by HPLC/MS	[104]

Plant material	Methods	References
Non-transformed and transgenic *AtCKX1* centaury methanol extracts	Analyses of secondary metabolites by HPLC	[105]
	Determination of the nhibitory capacity of *in vitro* growth of two human colorectal carcinoma cell lines by sulforhodamine B assay	
Non-transformed and transgenic *AtCKX* centaury methanol extracts and pure secoiridod and xanthone compounds	Investigation of antibacterial and antifungal activity on Gram-positive, Gram-negative bacteria and microfungi by microdilution method	[106]
Non-transformed and transgenic *AtCKX* centaury shoots and roots	Evaluation of the shoots and roots salinity tolerance to graded NaCl concentrations by plant tissue culture methods	[107]

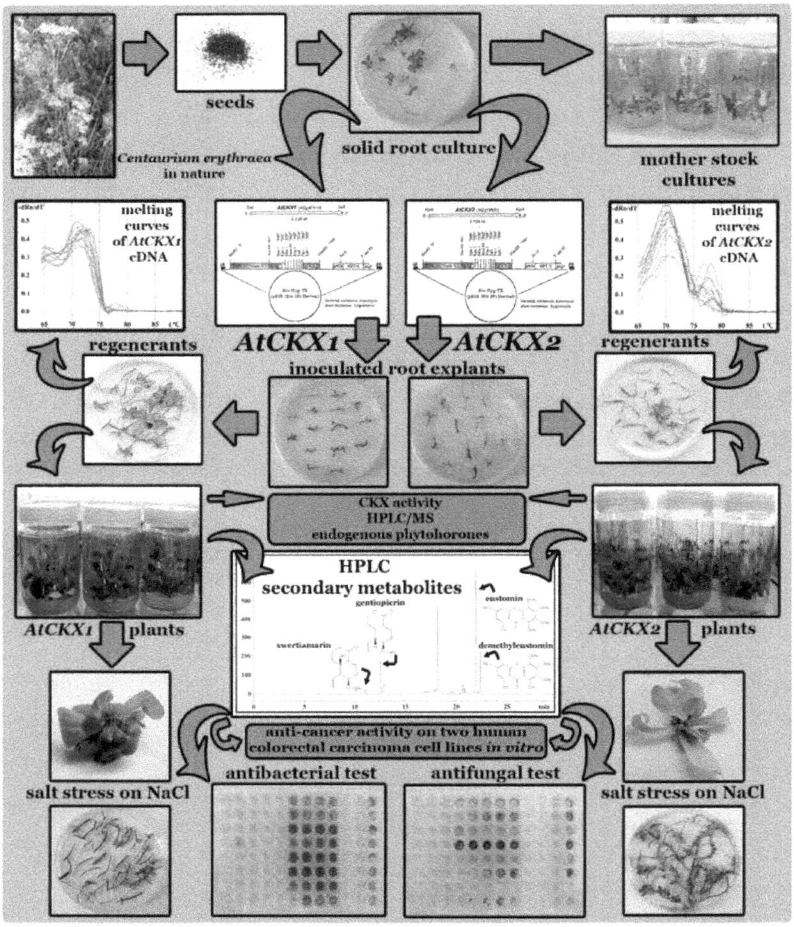

Figure 5: Illustrative presentation of transgenic AtCKX centaury plants production in vitro and all subsequent investigations conducted to date.

3. MOLECULAR AND MORPHOLOGICAL CHARACTERISATION OF TRANSGENIC *ATCKX* CENTAURY PLANTS

The possibility of direct regeneration of centaury shoots and/or somatic embryos makes root cultures a good model for studying *in vitro* morphogenesis of this plant species. Root cultures, in addition to easy manipulation, are characterised by other advantages such as small differences in physiological responses as well as high growth potential and metabolic activity. During the time, it has been shown that root cultures are suitable for production of genetically, biochemically and phenotypically stable plant material without somaclonal variation. The plant cultures, including a callus phase, theoretically promote a higher mutation rate [108]. The first literature data describing centaury root culture appeared almost twenty years ago, when the adventitious buds regeneration was shown to be induced with callus formation on initial root explants [109]. Spontaneous regeneration of adventitious buds in solid root culture was described, for the first time, by Subotić et al. [94]. These results motivated us to use root tips growing on½MS hormone-free medium as initial explants for further spontaneous regeneration and propagation of centaury plants. In our investigations, a solid root culture has been originally proved to be also a suitable and grateful model for genetic transformation of this medicinal plant species using *A. tumefaciens.*

Because of prosperous selection efficiency and adequate regenerative capacity of the transformants, the selection of suitable explants for initial infection with A. tumefaciens represents the most crucial step for a successful process of genetic transformation [110]. The potentially transformed centaury shoots formed directly on the root explants, without callus induction and thus minimized possible somaclonal variation [84]. Genomic PCR with specific primers confirmed the successful integration of the AtCKX1 and AtCKX2 transgenes into centaury genome in 30% and 28.2% of the analysed hygromycin-resistant lines, respectively. The quantification of transgene expression was confirmed by highly sensitive and precise method, qRT-PCR. As transgenic AtCKX1 lines showed low expression and unspecific amplification of cDNA, optimization of PCR and qPCR protocols was necessary. The optimized protocol used for real-time quantification of the transgene expression included combining a gradient of annealing temperatures with the application of seven different PCR enhancers: formamide, DMSO, glycerol, ethylene glycol, trehalose, BSA and Tween-20 [111]. The expression of AtCKX1 and AtCKX2 transgenes was detected and quantified in 50% and 63.6% transgenic lines, respectively. The AtCKX1 roots showed slightly higher expression than shoots whereas the AtCKX2 transgene was better expressed in shoots compared to roots. Evidently, quantitative differences in transgene expression after incorporation into genome of different centaury lines depended on the transgene copy number and positional variations of transgenes.

All of the AtCKX transgenic centaury lines showed decreased regeneration capacity in root culture on the solid hormone-free ½MS medium in comparison to control. However, regenerated shoots spontaneously rooted on the same medium. On the other hand, fresh weight of shoots in most of analysed transgenic centaury lines showed no significant difference in comparison to control. In analogy, the root culture of Dendrobium orchid overexpressing DSCKX1 transgene proved reduced shoot regeneration with decreased biomass compared to the non-transformed controls [112]. Similarly, the increased CKX expression affected on shortened internodes and reduced leaf surface of transgenic Arabidopsis and tobacco shoots [44].

4. CKX ACTIVITY AND ENDOGENOUS PHYTOHORMONE CONTENT IN TRANSGENIC *ATCKX* CENTAURY PLANTS

All centaury transgenic *AtCKX1* and *AtCKX2* shoots and roots showed increased CKX activity. Two to ten times higher CKX activity was detected in roots compared to shoots, which corresponds with previously demonstrated data in other plants such as tobacco [113], maize and barley [41, 64, 114]. In analogy to *AtCKX*-overexpressing *Arabidopsis* plants [55] the higher CKX activity was detected in shoots and roots of *AtCKX2* than *AtCKX1* centaury lines.

As the CKX represents a valuable regulatory enzyme in CK metabolism, CKX activity directly affects the CK amount in plant tissues [115, 116], the contents and profile of endogenous CKs in centaury non-transformed and *AtCKX* transgenic plants have been determined. In general, centaury shoots grown *in vitro* produced higher content of CKs than roots [104]. Considering production and dominant location of CKs in the roots this distribution of endogenous CKs may be a consequence of *in vitro* conditions itself. The roots of centaury plants overexpressing *AtCKX* transgenes showed higher content of total CKs in comparison to non-transformed ones [84]. Previous investigations on *AtCKX* overexpressing *Arabidopsis* and tobacco plants revealed that increased CKX activity affected on the reduction of different endogenous CK metabolites [44, 55, 79]. Theoretically, increased CKX activity should cause a decline in the total CKs amount. Regardless of this, overexpression of *AtCKX* genes did not contribute to lowered total CKs content in transgenic centaury plants but rather to an alteration in the spectrum of particular CK types. In analogy, transgenic potato plants overexpressing *AtCKX* genes were not found to show decreased total endogenous CK contents while bioactive CK forms were predominantly reduced [82, 83]. Accordingly, although enhanced CKX activity results in more substantial CK degradation in plant tissues it might simultaneously serve as an indirect signal for the plant cell to intensify the biosynthesis of CKs. It may lead to higher accumulation of total endogenous CKs and seems to be a mechanism of re-establishment and maintenance of CK homeostasis in plants [117].

Beside the total CKs contents, an altered amounts of individual endogenous CKs groups were specific for*AtCKX* transgenic centaury plants [104]. The reported considerable differences in endogenous CK spectra might reflect the distinct position and/or the number of transgene copies. The most specific characteristic of transgenic centaury plants was a remarkable reduction of bioactive CK forms, including free bases and their corresponding ribosides. Similarly, the *AtCKX* transgenic tobacco and potato plants also showed decreased contents of bioactive CKs [79, 82, 83, 118, 119]. In transgenic centaury plants, the reduction of bioactive CKs was organ-specific. Thus, considerably declined levels of bioactive CK forms were found in the *AtCKX* shoots although their CKX activity was not enhanced. On the other hand, a significant reduction of bioactive CKs levels was shown in *AtCKX* roots together with an increased CKX activity. A possible explanation is that CKs are probably more degraded in roots, the main location of biosynthesis, and accordingly, a smaller content of bioactive CKs is further transported to centaury shoots. Another one attribute of the *AtCKX* transgenic centaury plants was represented by increased amounts of CK storage forms (*O*-glucosides) and/or of irreversibly inactive (or weakly active)CK *N*-glucosides [104]. This is in accordance with recent literature data demonstrating the same CK pattern in *AtCKX1* transgenic potato plants [83]. Simultaneously, the concentrations of CK precursors in transgenic centaury plants were higher than those of bioactive CKs, which also corresponds with the potato plants [82].

To summarize, the AtCKX transgenic centaury plants are characterized by altered CK profiles with reduced levels of bioactive CK forms and, at the same time, increased amounts of storage forms (CK O-glucosides), inactive (or weakly active) forms (CK N-glucosides) andCK nucleotides. Previous investigations showed that overexpression of AtCKX transgene increased the production of storage CK derivatives in transgenic potato plants [81, 82]. It can be assumed that introduction of AtCKX transgenes into centaury genome altered CK metabolism in a way leading to the increased production of endogenous CK conjugates. Simultaneously, the level of bioactive CK forms seems to be controlled, besides the CKX activity, by additional regulatory mechanisms involved in CK metabolism and transport throughout the plant tissues. Considering that bioactive CKs are the preferred substrates of CKX, it is also possible that plant tissues activate new CK biosynthetic pathways and further initiate conversion to storage compounds, deactivation forms and/or CK nucleotides.

Numerous physiological and developmental processes in plants are regulated by the cooperation of CKs and auxins [120, 121, 122, 123]. Overexpression of AtCKX transgenes in centaury plants altered not only the CK homeostasis, but at the same time it affected metabolism of auxins. In transgenic centaury plants, the auxin indole-3-acetic acid (IAA) concentration was significantly decreased compared to the control. In analogy, previous literature data uncovered that an enhancement of the CKX activity led to the decline of endogenous CK levels and simultaneously to the reduction of endogenous IAA content in A. thaliana [79, 122]. In contrast, increased IAA concentration was reported in the AtCKX1-overexpressing potato plants [83]. These findings support the fact that the balance between CKs and auxins represents the most crucial factor in the shoot and root development in plants [55, 79].

In transgenic *AtCKX* centaury roots, the IAA/bioactive CKs ratio was lowered in comparison to control. It might be a consequence of a reduced regeneration capacity of transgenic centaury roots [84]. Considering that endogenous IAA content was lowered in the roots, it is presumable that *AtCKX*-overexpression weakened the regeneration of shoots in solid roots culture. In transgenic centaury shoots, the IAA/bioactive CKs ratios varied considerably. Contrary to the centaury plants, endogenous IAA content was significantly increased in the shoots and roots of transgenic *AtCKX1* potato [83].

An altered CK metabolism resulting from the overexpression of AtCKX transgenes definitely affected IAA content and IAA/ bioactive CKs ratio in transformed centaury plants. Different biosynthetic capacity for CK and IAA production could be a reason for the organ-specific differences among AtCKX transgenic plants.

5. SECONDARY METABOLITES IN TRANSGENIC *ATCKX* CENTAURY PLANTS

It is known that in vitro cultivation, although under controlled temperature and light conditions, stimulates increased secondary metabolites accumulation [124]. The previous investigations also showed that genetic transformation of centaury roots using A. rhizogenes affected the production of secondary metabolites [102, 125]. Considering that centaury plants were transformed using A. tumefaciens carrying AtCKX genes for the first time, it was interesting to reveal the effects of altered levels of endogenous CKs on secondary metabolites production of these transgenic centaury plants. The results of our analyses indicated modifications of the secondary metabolites production in this valuable medicinal plant species as a consequence of the AtCKX transgenes overexpression [105]. In methanol extracts of AtCKX transgenic centaury plants, the presence of bitter secoiridoids (swertiamarin and gentiopicrin) and xanthones (eustomin and demethyleustomin) was detected. The fact that AtCKX transgenic centaury plants produced the same compounds as non-transformed ones confirmed that there were no qualitative differences in specific secondary metabolites resulting from the AtCKX overexpression. On the other hand, quantitative changes in the secondary metabolite contents were found. Considering that gentiopicrin originates from swertiamarin in the metabolic pathway of iridoids, it is possible that increased activity of the swertiamarin → gentiopicrin converting enzyme stimulated gentiopicrin production. In all AtCKX transgenic centaury shoots, secoiridoidswertiamarin represented the predominant component. Still, in almost all centaury AtCKX shoots the content of swertiamarin was lowered in comparison to control shoots from in vitro cultures as well as from natural habitat. On the other hand, the production of the second bitter secoiridoid, gentiopicrin, was increased in the majority of AtCKX transgenic centaury shoots compared to control ones, grown in vitro and collected in natural habitat.

Unlike shoots, the AtCKX transgenic centaury roots produced decreased content of both secoiridoids, swertiamarin and gentiopicrin, in comparison to control roots grown in vitro and from natural habitat. It was shown previously that swertiamarin dominated in centaury shoots and roots from natural habitat while gentiopicrin prevailed in centaury shoots and roots grown in vitro [126, 127, 128, 129].

Beside secoiridoids, the differences in xanthone content were also detected in AtCKX transgenic centaury plants. Most of AtCKX transgenic shoots and roots produced more eustomin and demethyleustomin than shoots from natural habitat but less than control in vitro shoots. The shoots and roots of only one transgenic centaury line, AtCKX1–29, produced significantly increased amount of both xanthones in comparison to both controls, shoots grown in vitro and those collected from natural habitat. It is important to note that shoots and roots of this transgenic line were characterised by a reduced level of bioactive forms of CKs [104].

5.1 Transgenic *AtCKX* centaury plants as potential producers of anti-cancer compounds and antimicrobials

The methanol extracts of in vitro cultured AtCKX1–29 transgenic centaury plants were also tested for their antimicrobial effects [106]. Four Gram-positive, four Gram-negative bacteria and eight species of microfungi were selected for these in vitro investigations. The methanol extracts of non-transformed centaury plants, as well as pure secoiridoids (swertiamarin and gentiopicrin) and xanthones (eustomin and demethyleustomin), were used as a control. In general, the extracts of all tested centaury shoots and roots showed an adequate antibacterial activity on all tested bacteria. Methanol extracts of non-transformed and AtCKX1–29 transgenic centaury roots and pure gentiopicrin had a low antibacterial activity on Micrococcus flavus, Escherichia coli and Enterobacter cloacae. Staphylococcus aureus was found the most sensitive bacterial species in the in vitro assays. It was also shown that the extract of non-transformed centaury shoots was more effective on S. aureus than antibiotics, streptomycin and ampicillin. Interestingly, the pure compounds such as swertiamarin, gentiopicrin, eustomin and demethyleustomin exhibited higher antibacterial activity on all tested bacteria than centaury methanol extracts and commercial antibiotics (streptomycin and ampicillin) used as a positive control. High antimicrobial activity could be ascribed to bitter secoiridoid glycosides, which corresponds with previous literature data [130, 131]. Beside antibacterial impact, all tested methanol extracts of centaury shoots and roots also showed antifungal effects. Most of the applied compounds proved high antifungal activities. All pure secoiridoids and xanthones were more effective than methanol extracts against all fungi. In general, xanthone eustomin showed even 100 times higher antifungal potential than both of the applied mycotics (ketoconazole and bifonazole) used as a positive control. These results are significant as they represent the first report describing centaury xanthones as potential antimicrobials. In general, methanol extracts of non-transformed and transgenic AtCKX1 centaury shoots and roots showed better antibacterial activity, while pure secoiridoids and xanthones were more active against fungi.

Antioxidant properties of centaury plants and pure xanthones were reported almost twenty years ago [132, 133]. Considering that transgenic *AtCKX1*–29 roots produced increased amounts of xanthones, their potential antioxidant activity was evaluated [106]. The transgenic *AtCKX1* centaury shoots and roots were found to exert higher antioxidant activity compared to non-transformed plants. The root extracts were two to five times more effective than those from the shoots. Transgenic *AtCKX1*–29 roots, containing the highest amounts of total phenolics, were the most effective the scavenging of the DPPH radicals. The antioxidant effects of transgenic centaury methanol extracts could be assigned to elevated levels of xanthone compounds. Accordingly, centaury plants with increased content of secondary metabolites could be of practical importance in developing novel drugs with a potential use in agronomy, veterinary, medicine and food industry.

6. TRANSGENIC *ATCKX* CENTAURY PLANTS UNDER SALINITY STRESS *IN VITRO*

Taking into account that centaury plants can inhabit saline soils in natural habitats and because it is known that CKs play an essential role in the salinity stress response, the effect of NaCl-induced stress on regeneration potential of centaury shoots and roots was evaluated. The investigated non-transformed as well as *AtCKX1* and *AtCKX2* transgenic lines showed different salinity tolerance to graded NaCl concentrations. In general, a higher salinity tolerance was found for roots compared to shoots. Furthermore, elevated NaCl concentrations in the culture medium had no inhibitory effect on centaury shoot growth. All centaury shoots, regenerated on media supplemented with different NaCl concentrations, showed similar morphology as shoots regenerated in solid root cultures grown on NaCl-free media. Similarly, Šiler et al. [134] demonstrated that centaury plants keep the rosette forms under salt stress *in vitro*. Variations in the salt tolerance could be explained by differences among centaury genotypes, which is related to the recent report describing this species as a common salt-tolerant medicinal herb from Mediterranean region [135]. Whereas the non-transformed and *AtCKX1* transgenic line showed the same trend of shoot regeneration potential under salt stress conditions, the *AtCKX2* transgenic lines differed exhibiting a gradually decreased frequency of regeneration, the average number of regenerated shoots and fresh shoot weight with increased NaCl in the medium. The *AtCKX* transgenic centaury lines differed in their salinity tolerance from the non-transformed control indicating possible involvement of CKs in this process.

Beside morphological traits, biochemical characterisation of AtCKX transgenic centaury plants was determined. Amount of endogenous proline is considered one of the factors involved in the plant stress tolerance. Accordingly, an increase in the proline content was found in all AtCKX centaury shoots and roots grown on graded NaCl concentrations. Evaluation of malondialdehyde (MDA) level and hydrogen peroxide (H2O2) in centaury plants grown in vitro during salt stress also provided evidence of their oxidative stress tolerance. In all centaury AtCKX shoots and roots, MDA and H2O2 contents were increased at graded NaCl concentrations. These findings correspond well with literature data demonstrating an enhancement of MDA and H2O2 contents under salt stress [136, 137].

On the other hand, AtCKX transgenic centaury plants showed altered reactive oxygen species (ROS) homeostasis. It is rather difficult to summarize and conclude how salt stress affected on antioxidative enzymes activity. An increase of the superoxide dismutase (SOD) activity was detected in all centaury plants grown on graded NaCl concentrations. Interestingly, the SOD activity was always higher in roots than in shoots. Enhanced catalase (CAT) and peroxidase (POX) activities were detected in centaury plants grown on graded NaCl concentrations as well. Similarly to centaury, transgenic AtCKX2 tobacco plants showed increased SOD and CAT activity in comparison to wild type while lower SOD activity was found in roots than in shoots under salt stress ex vitro [118]. Increased SOD, CAT and POX activities in salt-stressed non-transformed and AtCKX transgenic centaury plants indicates a crucial role of these antioxidant enzymes in protecting plant tissues from superoxide radical and hydrogen peroxide and suggests their association with improved tolerance of plants to environmental stress conditions.

Evidently, the salinity stress caused morphological, physiological and biochemical changes in both non-transformed and AtCKX transgenic centaury plants. Elevated proline, MDA and H2O2 contents as well as increased antioxidative enzymes activities confirmed that centaury plants are trying to overcome salt stress conditions to maintain or re-establish their normal growth and development.

7. CONCISE SUMMARY

This chapter presents a survey of obtaining and developing transgenic *AtCKX* centaury plants grown *in vitro* and is intended to provide a synthesis of the existing literature data. In addition, recent achievements in the characterisation of transgenic centaury plants and determination as well as application of their secondary metabolite content is reported in details.

Centaury is a medicinal plant species endangered in natural habitat. A successful protocol for its genetic transformation using A. tumefaciens to introduce foreign, AtCKX, genes is described here. This protocol favours centaury root tips as primary infection explants that can directly regenerate shoots without callus phase. As the in vitro regeneration of centaury is possible without the addition of exogenous plant growth regulator(s), in vitro culture represents an optimal model for studying metabolic pathways of endogenous phytohormones including CKs without any interferences from other exogenous hormones present in nutrition medium. The first step of a successful genetic transformation of centaury was the establishment of stable transformants with AtCKX overexpression and enhanced CKX activity. Molecular, morphometric and developmental characterisation of obtained AtCKX-transgenic plants as well as chemical profiling of CK metabolites represented the following steps. The subsequent secondary metabolite analyses showed antibacterial and antifungal activity of transgenic centaury plants that can also be used as potential producers of anti-cancer compounds. Considering that centaury can inhabit saline soils in natural habitats, salinity tolerance of transgenic AtCKX centaury plants was investigated as well.

All obtained results were summarized and indicated that transgenic AtCKX centaury plants can serve as a suitable material for investigations of numerous physiological and developmental processes that are under endogenous phytohormonal control. Also, further studies of endogenous plant hormone regulations during NaCl-induced stress might represent an useful tool for better understanding the salinity tolerance of non-transformed and CK-deficient transgenic centaury plants grown in vitro. Future research will certainly be focused on the identification of centaury candidate genes specifically expressed during the salt stress to understand complex regulatory pathways in response of centaury plants to salinity in vitro.

AT A GLANCE

The production and breeding of plants with desired properties are possible by a fundamental biotechnological technique, genetic engineering. Applying and developing of genetic engineering procedures also enable preservation and improvement of plant species endangered in nature, including medicinal plant common centaury (Centaurium erythraea Rafn.). Numerous developmental processes in plants are controlled by cytokinins (CKs). The only so far known enzyme involved in CK catabolism is cytokinin oxidase/dehydrogenase (CKX). Genes coding for two Arabidopsis CKX isoforms, AtCKX1 and AtCKX2, were successfully introduced into centaury root explants. Subsequently, the contents of endogenous CKs in AtCKX-overexpressing centaury plants grown in vitro were investigated. Simultaneous secondary metabolite analyses showed antibacterial and antifungal activity of transgenic centaury plants and suggested their use as potential producers of anti-cancer compounds. Considering that centaury can inhabit saline soils in natural habitats, following investigations included evaluation of salinity tolerance in vitro. All obtained and summarized results indicated that transgenic AtCKX centaury plants could serve as a suitable model for studies of numerous physiological and developmental processes under endogenous phytohormonal control.

ABBREVIATIONS

CAT	catalase
CKs	cytokinins
CKX	cytokinin oxidase/dehydrogenase
DPPH	2,2-diphenyl-1-picryl-hydrazyl-hydrate
HPLC	high-performance liquid chromatography
MDA	malondialdehyde
MS	Murashige and Skoog medium
MS spectrometry	mass spectrometry
POX	peroxidase
ROS	reactive oxygen species
SOD	superoxide dismutase

REFERENCES

1. Slater A, Scott NW, Fowler MR. Plant Biotechnology. The Genetic Manipulation of Plants. New York: Oxford University Press; 2004. pp. 35-52

2. Shou H, Frame BR, Whitham SA, Wang K. Assessment of transgenic maize events produced by particle bombardment or Agrobacterium-mediated transformation. Molecular Breeding. 2004;13:201-208. DOI: 10.1023/B:MOLB.0000018767.64586.53

3. Gelvin SB. Agrobacterium-mediated plant transformation: the biology behind the "Gene-Jockeying" tool. Microbiology and Molecular Biology Reviews. 2003;67:16-37. DOI: 10.1128/MMBR.67.1.16-37.2003

4. Cheng M, Lowe BA, Spencer M, Ye X, Armstrong CL. Factors influencing Agrobacterium-mediated transformation of monocotyledonous species. In Vitro Cellular & Developmental Biology – Plant. 2004;40:31-45. DOI: 10.1079/IVP2003501

5. Van Larebeke N, Genetello C, Shell J, Schilperooort RA, Hermans AK, Hernalsteens JP, et al. Acquisition of tumour-inducing ability by non-oncogenic agrobacteria as a result of plasmid transfer. Nature. 1975;255:742-743. DOI: 10.1038/255742a0

6. Baron C, Zambryski PC. Notes from the undergraund: highlights from plant-microbe interactions. Trends in Biotechnology. 1995;13:356-362. DOI: 10.1016/S0167-7799(00)88981-7

7. Watson B, Currier TC, Gordon MD, Chilton MD, Nester EW. Plasmid required for virulence of Agrobacterium tumefaciens. The Journal of Bacteriology. 1975;123:255-264. DOI: 10.1128/JB.123.1.255-264.1975

8. Simonović A. Biotechnology and Genetic Engineering of Plants. 1st ed. NNK internacional: Belgrade; 2011 987-86-6157-003-2

9. Chilton MD, Drummond MH, Merlo DJ, Sciaky D, Montoya AL, Gordon MP, et al. Stable incorporation of plasmid DNA into higher plant cells: the molecular basis of crown gall tumorogenesis. Cell. 1977;11:263-271. DOI: 10.1016/0092-8674(77)90043-5

10. Chandler S, Lu CY. Biotechnology in ornamental horticulture. In Vitro Cellular & Developmental Biology – Plant. 2005;41:591-601. DOI: 10.1079/IVP2005681

11. Veluthambi K, Ream W, Gelvin SB. Virulence genes, borders and overdrive generate single-stranded T-DNA molecules from the A6 Ti plasmid of Agrobacterium tumefaciens. Journal of Bacteriology. 1988;170:1523-1532. DOI: 10.1128/jb.170.4.1523-1532.1988

12. Pettit A, David C, Dahl GA, Ellis JG, Guyon P, Casse-Delbart F, et al. Further extension of opine concept: Plassmids in Agrobacterium rhizogenes cooperate for opine degradation. Molecular Genetics and Genomics. 1983;190:204-214. DOI: 10.1007/BF00330641

13. Stachel SE, Messens E, Van Montagu M, Zambryski P. Identification of the signal molecules produced by wounded plant cells that activate T-DNA transfer in Agrobacterium tumefaciens. Nature. 1985;318:624-629. DOI: 10.1038/318624a0

14. Messens E, Dekeyser R, Stachel SE. A nontransformable Triticum monoccocum monocityledonous culture produces the potent Agrobacteriumvir-inducing

compound ethyleferulate. Proceedings of the National Academy of Sciences of the United States of America. 1990;87:4368-4372. DOI: 10.1073/pnas.87.11.4368

15. Tzfira T, Citovsky V. Agrobacterium-mediated genetic transformation of plants: biology and biotechnology. Current Opinion in Biotechnology. 2006;17:147-154. DOI: 10.1016/j.copbio.2006.01.009

16. Tzfira T, Li J, Lacroix B, Citovsky V. Agrobacterium T-DNA integration: molecules and models. Trends in Genetics. 2004;20:375-383. DOI: 10.1016/j.tig.2004.06.004

17. Matzke AJM, Neuhuber F, Park Y, Ambros PF, Matzke AM. Homology-dependent gene silencing in transgenic plants: epistatic silencing loci contain multiple copies of methylated transgenes. Molecular and General Genetics. 1994;244:219-229. DOI: 10.1007/BF00285449

18. Assad FF, Tucker KL, Singer ER. Epigenetic repeat-induced gene silencing (RIGS) in Arabidopsis. Plant Molecular Biology. 1993;22:1067-1085. DOI: 10.1007/BF00028978

19. Maessen GDF. Genomic stability and stability of expression in genetically modified plants. Acta Botanica Neerlandica. 1997;46:3-24. DOI: 10.1111/plb.1997.46.1.3

20. Hansen G, Wright MS. Recent advances in transformation of plants. Trends in Plant Science. 1999;4:226-231. DOI: 10.1016/s1360-1385(99)01412-0

21. An G. Binary Ti plasmid vectors. In: Gortland KMA, Davey MR, editors. Agrobacterium protocols. Methods in Molecular Biology. Vol. 44. Totowa, New Jersey: Springer; 1995. pp. 47-58. DOI: 10.1385/0-89603-302-3:47

22. Reynaerts A, De Block M, Hernalsteens JP, Van Montagu M. Selectable and screenable markers. In: Gelvin SB, Schilperoort RA, Verma DPS, editors. Plant Molecular Biology Manual. Dordrecht: Springer; 1988. p. A9:1-16. DOI: 10.1007/978-94-009-0951-9_9

23. Wani SH, Kumar V, Shriram V, Sah SK. Phytohormones and their metabolic engeneering for abiotic stress tolerance in crop plants. Crop Journal. 2016;4:162-176. DOI: 10.1016/j.cj.2016.01.010

24. Miller CO, Skoog F, von Saltza MH, Strong M. Kinetin, a cell division factor from deoxyribonucleic acid. Journal of the America. Vol. 77. Chemical Society; 1955. pp. 1329-1334. DOI: 10.1021/ja01610a105

25. Mok DW, Mok MC. Cytokinin metabolism and action. Annual Review of Plant Physiology and Plant Molecular Biology. 2001;52:89-118. DOI: 10.1146/annurev.arplant.52.1.89

26. Strnad M. The aromatic cytokinins. Physiologia Plantarum. 1997;101:674-688. DOI: 10.1111/j.1399-3054.1997.tb01052.x

27. Kamínek M, Vaněk T, Motyka V. Cytokinin activities of N6-benzyladenosine derivatives hydroxylated on the side-chain phenyl ring Journal of Plant Growth Regulation. 1987;6:113-120.DOI: 10.1007/BF02026460

28. Holub J, Hanuš J, Hanke DE, Strnad M. Biological activity of cytokinins derived from Ortho- and Meta-Hydroxybenzyladenine. Plant Growth Regulation. 1998;26:109-115. DOI: 10.1023/A:1006192619432

29. Horgan R, Hewett EW, Purse J, Wareing PF. A new cytokinin from Populus robusta. Tetrahedron Letters. 1973;30:2827-2828. DOI: 10.1016/S0040-4039(01)96062-9

30. Tarkowska D, Dolezal K, Tarkowski P, Astot C, Holub J. Identification of new aromatic cytokinins in Arabidopsis thaliana and Populus canadensis leaves by LC-(+) ESIMS and capillary liquid chromatography/frit-fast atom bombardment mass spectrometry. Physiologia Plantarum. 2003;117:579-590. DOI: 10.1034/j.1399-3054.2003.00071.x

31. Shantz EM, Steward FC. The identification of compound A from coconut milk as 1,3-diphenylurea. Journal of the American Chemical Society. 1955;77:6351-6353. DOI: 10.1021/ja01628a079

32. Takahashi S, Shudo K, Okamoto T, Yamada K, Isogai Y. Cytokinin activity of N-phenyl-N'-(4-pyridyl)urea derivates. Phytochemistry. 1978;17:1201-1207. DOI: 10.1016/S0031-9422(00)94556-4

33. Mok DW, Mok MC, Armstrong DJ, Shudo K, Isogai Y. Cytokinin activity of N-phenyl-N'-1,2,3-thidiazol-5-ylurea (Thidiazuron). Phytochemistry. 1982;21:1509-1511. DOI: 10.1016/S0031-9422(82)85007-3

34. Mok DW, Mok MC, Turner JE, Mujer CV (1987) Biological and biochemical effects of cytokinin-active phenyl-urea derivates in tissue culture systems. Hort Science 22: 1194-1197. DOI:

35. Shudo K. Chemistry of phenylureacytokinins. In: Mok DWS, Mok MC, editors. Cytokinins: Chemistry, Activity and Function. Boca Raton: CRC Press; 1994. pp. 35-42. DOI: 10.1201/9781351071284-3

36. McGaw BA, Horgan R. Cytokinin catabolism and cytokinin oxidase. Phytochemistry. 1983;22:1103-1105. DOI: 10.1016/0031-9422(83)80200-3

37. Galuszka P, Popelková H, Werner T, Frébortová J, Pospíšilová H, Mik V, et al. Biochemical characterization of cytokinin Oxidases/Dehydrogenases from Arabidopsis thaliana expressed in Nicotiana tabacum L. Plant Growth Regulation. 2007;26:255-267. DOI: 10.1007/s00344-007-9008-5

38. Gajdošová S, Spíchal L, Kamínek M, Hoyerová K, Novák O, Dobrev PI, et al. Distribution, biological activities, metabolism, and the conceivable function of cis-zeatin-type cytokinins in plants. Journal of Experimental Botany. 2011;62:2827-2840. DOI: 10.1093/jxb/erq457

39. Armstrong DJ. Cytokinin oxidase and the regulation of cytokinin degradation. In: Mok DWS, Mok MC, editors. Cytokinins: Chemistry, Activity and Function. Boca Raton: CRC Press; 1994. pp. 139-154. DOI: 10.1201/9781351071284-11

40. Van Staden J, Crouch NR. Benzyladenine and derivates - their significance and inter conversion in plants. Plant Growth Regulation. 1996;19:153-175. DOI: 10.1007/BF00024582

41. Bilyeu KD, Cole JL, Laskey JG, Riekhof WR, Esparza TJ. Molecular and biochemical characterization of a cytokinin oxidase from maize. Plant Physiology Plant Physiology. 2001;125:378-386. DOI: 10.1104/pp.125.1.378

42. Mrízová K, Jiskrová E, Vyroubalová Š, Novák O, Ohnoutková L, Pospíšilová H, et al. Overexpression of cytokinin dehydrogenase genes in barley (Hordeum vulgare cv. Golden Promise) fundamentally affects morphology and fertility. PLoS One. 2013;8(11):e79029. DOI: 10.1371/journal.pone.0079029

43. Zalabák D, Galuszka P, Mrízová K, Podlešáková K, Gu R, Frébortová J. Biochemical characterization of the maize cytokinin dehydrogenase family and cytokinin profiling in developing maize plantlets in relation to the expression of cytokinin dehydrogenase genes. Plant Physiology Biochemistry. 2013;74:283-293. DOI: 10.1016/j.plaphy.2013.11.020

44. Schmülling T, Werner T, Riefler M, Krupková E. Bartrina y Manns I. Structure and function of cytokinin oxidase dehydrogenase genes of maize, rice, Arabidopsis and other species. Journal of Plant Research. 2003;116:241-252. DOI: 10.1007/s10265-003-0096-4

45. Popelková H, Galuszka P, Frébortová J, Bilyeu KD, Frébort I (2004) Cytokinin dehydrogenase: characterization and structure homology modeling of the flavoprotein catabolising plant hormones cytokinins. In: Recent Research Developments in Proteins. Pandalai SG (ed), Transworld Research Network, Kerala, India, Vol. II, pp. 63-81

46. Motyka V, Kamínek M. Regulation of cytokinin catabolism in tobacco callus cultures. In: Nijkamp HJJ, Van der Plas LHW, Van Aartrijk J, editors. Progress in Plant Cellular and Molecular Biology. Dordrecht: Springer; 1990. p. 492-497. DOI: 10.1007/978-94-009-2103-0_75

47. Motyka V, Kaminek M. Characterization of cytokinin oxidase from tobacco and poplar callus cultures. In: Kaminek M, Mok DWS, Zazimalova E, editors. Physiology and biochemistry of cytokinins in plants. The Hague: SPB Academic Publishing; 1992. pp. 33-39

48. Conrad K, Motyka V, Schlüter T. Increase in activity, glycosylation and expression of cytokinin oxidase/dehydrogenase during the senescence of barley leaf segments in the dark. Physiologia Plantarum. 2007;130:572-579. DOI: 10.1111/j.1399-3054.2007.00914.x

49. V Werstiuk E, Hall RH. Conversion ofN6-(Δ2-isopentenyl)adenosine to adenosine by enzyme activity in tobacco tissue. Plant Physiology. 1971;48:775-778. DOI: 10.1104/pp.48.6.775

50. Whitty CD, Hall RH. A cytokinin oxidase in Zea mays. Canadian Journal of Biochemistry. 1974;52:789-799. DOI: 10.1139/o74-112

51. Chatfield JM, Armstrong DJ. Cytokinin oxidase from Phaseolus vulgaris callus tissues: enhanced in vitro activity of the enzyme in the presence of copper-imidazole complexes. Plant Physiology. 1987;84:726-731. DOI: 10.1104/pp.84.3.726

52. Chatfield JM, Armstrong DJ. Cytokinin oxidase from Phaseolus vulgaris callus cultures: affinity for concanavalin. Plant Physiology. 1988;88:245-247. DOI: 10.1104/pp.88.2.245

53. Laloue M, Fox JE. Cytokinin oxidase from wheat: partial purufication and general properties. Plant Physiology. 1989;90:899-906. DOI: 10.1104/pp.90.3.899

54. Motyka V, Kamínek M. Cytokinin oxidase from auxin- and cytokinin-dependent callus cultures of tobacco (Nicotiana tabacum L.). Journal of Plant Growth Regulation. 1994;13:1-9. DOI: 10.1007/BF00210700

55. Werner T, Motyka V, Laucou V, Stems R, Van Onckelen H, Schmülling T. Cytokinin deficient transgenic Arabidopsis plant show multiple developmental alterations indicating opposite function of cytokinins in the regulation of shoot and root meristem sctivity. The Plant Cell. 2003;15:2532-2550. : 10.1105/tpc.014928

56. Werner T, Köllmer I, Bartrina K, Holst K, Schmülling T. New insights into the biology of cytokinin degradation. Plant Biology. 2006;8:371-381. DOI: 10.1055/s-2006-923928

57. Houba-Hérin N, Pethe C, d'Alayer J, Laloue M. Cytokinin oxidase from Zea mays: purufication, cDNA cloning and expression in moss protoplasts. The Plant Journal. 1999;17:615-626. DOI: 10.1046/j.1365-313x.1999.00408.x

58. Morris RO, Bilyeu KD, Laskey JG, Cheikh NN. Isolation of a gene encoding a glycosylated cytokinin oxidase from maize. Biochemical and Biophysical Research Communications. 1999;255:328-333. DOI: 10.1006/bbrc.1999.0199

59. Massonneau A, Houba-Hérin N, Pethe C, Madzak C, Falque M, Mercy M, et al. Maize cytokinin oxidase genes: differential expression and cloning of two

new cDNAs. Journal of the Experimental Botany. 2004;55:2549-2557. DOI: 10.1093/jxb/erh274

60. Šmehilová M, Galuszka P, Bilyeu KD, Jaworek P, Kowalska M, Sebela M, et al. Subcellular localization and biochemical comparison of cytosolic and secreted cytokinin dehydrogenase enzymes from maize. Journal of Experimental Botany. 2009;60:2701-2712. DOI: 10.1093/jxb/erp126

61. Yang SH, Yu H. Goh. Isolation and characterization of the orchid cytokinin oxidase DSCKX1 promoter. Journal of Experimental Biology. 2002;53:1899-1907. DOI: 10.1093/jxb/erf055

62. Yang SH, Yu H, Xu Y. Goh. Functional characterization of a cytokinin oxidase gene DSCKX1 in Dendrobium orchid. Plant Molecular Biology. 2003;51:237-248. DOI: 10.1023/A:1021115816540

63. Wang Y, Luo JP, Wei ZJ, Zhang JC. Molecular cloning and expression analysis of a cytokinin oxidase (DhCKX) gene in Dendrobium huoshanense. Molecular Biology Reports. 2009;36:1331-1338. DOI: 10.1007/s11033-008-9316-2

64. Galuszka P, Frébortová J, Werner T, Yamada M, Strnad M, Schmülling T, Frébort I. Cytokinin oxidase/dehydrogenase genes in barley and wheat cloning and heterologous expression. European Journal of Biochemistry. 2004;271:3990-4002. DOI: 10.1111/j.1432-1033.2004.04334.x

65. Ashikari M, Sakakibara H, Lin S, Yamamoto T, Takashi T, Nishimura A, et al. Cytokinin oxidase regulates rice grain production. Science. 2005;309:741-745. DOI: 10.1126/science.1113373

66. Vaseva-Gemisheva I, Lee D, Karanov E. Response of Pisum sativum cytokinin oxidase/dehydrogenase expression and specific activity to drought stress and herbicide treatments. Plant Growth Regulation. 2005;46:199-208. DOI: 10.1007/s10725-005-0143-3

67. Pepper A, Guinel F. Abnormal root and nodule vasculaturein R50 (sym16), a pea nodulation mutant which accumulates cytokinins. Annals of Botany. 2007;99:765-774. DOI: 10.1093/aob/mcm013

68. Wang Y, Liub H, Xin Q. Genome-wide analysis and identification of cytokinin oxidase/dehydrogenase (CKX) gene familyin foxtail millet (Setariaitalica). The Crop Journal. 2014;2:244-254. DOI: 10.1016/j.cj.2014.05.001

69. Zhang L, Zhang BS, Zhou RH, Gao LF, Zhao GY, Song YX, Jia JZ (2007) Cloning and genetic mapping of cytokinin oxidase/dehydrogenase gene (TaCKX2) in wheat. Scientia Agricultura Sinica 33(2007) 1419-1425 (in Chinese with English abstract)

70. Zhang L, Zhang BS, Zhou RH, Kong XY, Gao LF, Jia JZ. Isolation and chromosomal localization of cytokinin oxidase/dehydrogenase gene

(TaCKX5) in wheat. Scientia Agricultura Sinica. 2008;41:636-642 (in Chinese with English abstract). DOI: 10.3864/j.issn.0578-1752.2008.03.002

71. Zhang L, Zhao YL, Gao LF, Zhao GY, Zhou RH, Zhang BS, Jia JZ. TaCKX6-D1, the ortholog of rice OsCKX2, is associated with grain weight in hexaploid wheat, New Phytologist. 2012;195:574-584. DOI: 10.1111/j.1469-8137.2012.04194.x

72. Feng DS, Wang HG, Zhang XS, Kong LR, Tian JC, Li XF. Using an inverse PCR method to clone the wheat cytokininoxidase/dehydrogenase gene TaCKX1. Plant Molecular Biology Reporter. 2008;26:143-155. DOI: 10.1007/s11105-008-0033-8

73. Ma X, Feng DS, Wang HG, Li XF, Kong LR. Cloning and expression analysis of wheat cytokinin oxidase/dehydrogenasegene TaCKX3. Plant Molecular Biology Reporter. 2011;29:98-105. DOI: 10.1007/s11105-010-0209-x

74. Liu P, Zhang C, Ma JQ, Zhang LY, Yang B, Tang XY, et al. Genome-wide identification and expression profiling of cytokinin oxidase/dehydrogenase (CKX) genes reveal likely roles in pod development and stress responses in oilseed rape (Brassica napus L.). Genes (Basel). 2018;9(3):168. DOI: 10.3390/genes9030168. 10.3390/genes9030168

75. Zeng Q-W, Qin S, Song SQ, Zhang M, Xiao YH, Luo M, et al. Molecular cloning and characterization of a cytokinin dehydrogenase gene from upland cotton (Gossypium hirsutum L.). Plant Molecular Biology Reporter. 2012;30:1-9. DOI: 10.1007/s11105-011-0308-3

76. Le DT, Nishiyama R, Watanabe Y, Vankova R, Tanaka M, Seki M, et al. Identification and expression analysis of cytokinin metabolic genes in soybean under normal and drought conditions in relation to cytokinin levels. PLoS One. 2012;7:e42411. DOI: 10.1371/journal.pone.0042411

77. Cai L, Zhang L, Fu Q, Xu ZF. Identification and expression analysis of cytokinin metabolic genes IPTs, CYP735A and CKXs in the biofuel plant Jatropha curcas. Peer J. 2018;6:e4812. DOI: 10.7717/peerj.4812

78. Li S, An Y, Hailati S, Zhang J, Cao Y, Liu Y, et al. Yang. Overexpression of the cytokinin oxidase/dehydrogenase (CKX) from Medicago sativa enhanced salt stress tolerance of Arabidopsis. Journal of Plant Biology. 2019;62:374-386. DOI: 10.1007/s12374-019-0141-z

79. Werner T, Motyka V, Strnad M, Schmülling T. Regulation of plant growth by cytokinin. Proceedings of the National Academy of Sciences of the United States of America. 2001;98:10487-10492. DOI: 10.1073/pnas.171304098

80. Von Schwartzenberg K, Nunez MF, Blaschke H, Dobrev PI, Novák O, Motyka V, et al. Cytokinins in the bryophyte Physcomitrella patens: analyses of activity, distribution and cytokinin oxidase/dehydrogenase overexpression

reveal the role of extracellular cytokinins. Plant Physiology. 2007;145:786-800. DOI: 10.1104/pp.107.103176

81. Hartmann A, Senning M, Hedden P, Sonnewald U, Sonnewald S. Reactivation of meristem activity and sprout growth in potato tuber require both cytokinin and gibberellin. Plant Physiology. 2011;155:776-796. DOI: 10.1104/pp.110.168252

82. Raspor M, Motyka V, Žižková E, Dobrev PI, Trávníčková A, Zdravković-Korać S, et al. Cytokinin profles of AtCKX2-overexpressing potato plants and the impact of altered cytokinin homeostasis on tuberization in vitro. Journal of Plant Growth Regulation. 2012;31:460-470. DOI: 10.1007/s00344-011-9255-3

83. Raspor M, Motyka V, Ninković S, Malbeck J, Dobrev PI, Zdravković-Korać S, et al. Overexpressing AtCKX1 in potato plants grown in vitro: the efects on cytokinin composition and tuberization. Journal of Plant Growth Regulation. DOI: 10.1007/s00344-020-10080-w

84. Trifunović M, Cingel A, Simonović A, Jevremović S, Petrić M, Dragićević IČ, et al. Overexpression of Arabidopsis cytokinin oxidase/dehydrogenase genes AtCKX1 and AtCKX2 in transgenic Centaurium erythraea Rafn. Plant Cell Tissue and Organ Culture. 2013;115:139-150. DOI: 10.1007/s11240-013-0347-6

85. Skoog F, Miller CO. Chemical regulation of growth and organ formation in plant tissue cultured in vitro. Symposium of Society of Experimental Biology. 1957;11:118-131

86. Werner T, Holst K, Pors Y, Guivarc'h A, Mustroph A, Chriqui D, Grimm B, Schmülling T. deficiency causes distinct changes of sink and source parameters in tobacco shoots and roots. Journal of Experimental Botany. 2008;59:2659-2672. DOI: 10.1093/jxb/ern134

87. Aloni R. The induction of vascular tissues by auxin and cytokinin. In: Davies PJ, editor. Plant Hormones: Physiology, Biochemistry and Molecular Biology. Dordrecht, The Netherlands: Kluwer Academic Publishers; 1995. pp. 531-546

88. Gan S, Amasino RM. Inhibition of leaf senescence by autoregulated production of cytokinins. Science. 1995;270(5244):1986-1988. DOI: 10.1126/science.270.5244.1986

89. Custers JBM, Snepvangers SCHJ, Jansen HJ, Zhang L, Van Lookeren Campagne MM. The 35S-CaMV promoter is silent during early embryogenesis but activated during nonembryogenicsporophytic development in microspore culture. Protoplasma. 1999;208(1-4):257-264. DOI: 10.1007/BF01279097

90. Beemster GTS, Baskin TISTUNTEDPLANT. 1 mediates effects of cytokinin, but not of auxin, on cell division and expansion in the root of Arabidopsis. Plant Physiology. 2000;124:1718-1727. DOI: 10.1104/pp.124.4.1718

91. Tucakov J (1990) Healing with plants – phytotherapy. Rad, Belgrade
92. Skrzypczak L, Wesolovska M, Skrzypczak E. Gentiana species: In vitro culture, regeneration and production of secoiridoid glucosides. In: Bajaj YPS, editor. Biotechnology in Agriculture and Forestry. Vol. 21, Medicinal and Aromatic Plants IV. Berlin, Hiedelberg: Springer Verlag; 1993. pp. 172-186
93. Subotić A, Budimir S, Grubišić D, Momčilović I. Direct regeneration of shoots from hairy root cultures of Centaurium erythraea inoculated with Agrobacterium rhizogenes. Biologia Plantarum. 2003/4;47:617-619. DOI: 10.1023/B:BIOP.0000041074.81033.3a
94. Subotić A, Janković T, Jevremović S, Grubišić D (2006) Plant Tissue Culture and Secondary Metabolites Productions of Centaurium erythraea Rafn., a Medical plant. In: Teixeira da Silva JA (ed), Floriculture, Ornamental and Plant Biotechnology: Advances and Topical Issues 1st ed. Global Science Books, London, UK, Vol II, pp. 564-570
95. Subotić A, Jevremović S, Grubišić D. Influence of cytokinins on in vitro morphogenesis in root cultures of Centaurium erythraea - Valuable medicinal plant. Scientia Horticulturae. 2009;120:386-390. DOI: 10.1016/j.scienta.2008.11.034
96. Subotić A, Jevremović S, Grubišić D, Janković T. Spontaneous plant regeneration and production of secondary metabolites from hairy root cultures of Centaurium erythraea Rafn. In: Jain SM, Saxena PK, editors. Protocols for in vitro cultures and secondary metabolite analysis of aromatic and medicinal plants, Methods in Molecular Biology. Vol. 547. Berlin: Springer; 2009. pp. 205-217
97. Subotić A, Jevremović S, Trifunović M, Petrić M, Milošević S, Grubišić D. The influence of gibberellic acid and paclobutrazol on induction of somatic embryogenesis in wild type and hairy root cultures of Centaurium erythraea Gillib. African Journal of Biotechnology. 2009;8:3223-3228. DOI: 10.5897/AJB2009.000-9323
98. Trifunović-Momčilov M, Motyka V, Dragićević IČ, Petrić M, Jevremović S, Malbeck J, et al. Endogenous phytohormones in spontaneously regenerated Centaurium erythraea Rafn. plants grown in vitro. Journal of Plant Growth Regulation. 2016;35:543-552. DOI: 10.1007/s00344-015-9558-x
99. Trifunović M, Tadić V, Petrić M, Jontulović D, Jevremović S, Subotić A. Quantification of arabinogalactan proteins during in vitro morphogenesis induced by β-D-glucosyl Yariv reagent in Centaurium erythraearoot culture. Acta Physiologiae Plantarum. 2014;36:1187-1195. DOI: 10.1007/s11738-014-1495-y

100. Filipović BK, Simonović AD, Trifunović MM, Dmitrović SS, Savić JM, Jevremović SB, et al. Plant regeneration in leaf culture of Centaurium erythraea Rafn. Part 1: The role of antioxidant enzymes. Plant Cell Tissue and Organ Culture. 2015;121:703-719. DOI: 10.1007/s11240-015-0740-4

101. Simonović AD, Filipović BK, Trifunović MM, Malkov SN, Milinković VP, Jevremović SB, et al. Plant regeneration in leaf culture of Centaurium erythraea Rafn. Part 2: The role of arabinogalactan proteins. Plant Cell Tissue and Organ Culture. 2015;21:721-739. DOI: 10.1007/s11240-020-01801-w

102. Piatczak E, Krolicka A, Wysokinska H. Genetic transformation of Centaurium erythraea Rafn by Agrobacterium rhizogenes and the production of secoiridoids. Plant Cell Reports. 2006;25:1308-1315

103. Murashige T, Skoog F. A revised medium for rapid growth and bioassays with tobacco tissue cultures. Physiologia Plantarum. 1962;15:473-479

104. Trifunović M, Motyka V, Cingel A, Subotić A, Jevremović S, Petrić M, et al. Changes in cytokinin content and altered cytokinin homeostasis in AtCKX1 and AtCKX2-overexpressing centaury (Centaurium erythraea Rafn.) plants grown in vitro. Plant Cell Tissue and Organ Culture. 2015;120:767-777

105. Trifunović-Momčilov M, Krstić-Milošević D, Trifunović S, Podolski-Renić A, Pešić M, Subotić A (2016) Secondary metabolite profile of transgenic centaury (Centaurium erythraea Rafn.) plants, potential producers of anticancer compounds. Transgenesis and Secondary Metabolism, Reference Series in Phytochemistry. S Jha (ed), Springer International Publishing, DOI 10.1007/978-3-319-27490-4_5-1. ISBN: 978-3-319-27490-4

106. Trifunović-Momčilov M, Krstić-Milošević D, Trifunović S, Ćirić A, Glamočlija J, Jevremović S, et al. Antimicrobial activity, antioxidant potential and total phenolic content of transgenic AtCKX1 centaury (Centaurium erythraea Rafn.) plants grown in vitro. Environmental Engineering and Management Journal. 2019;18:2063-2072

107. Trifunović-Momčilov M, Paunović D, Milošević S, Marković M, Jevremović S, Dragićević IČ, et al. Salinity stress response of non-transformed and AtCKX transgenic centaury (Centaurium erythraea Rafn.) shoots and roots grown in vitro. Annals of Applied Biology. 2020;177:74-89

108. Krishna H, Alizadeh M, Singh D, Singh U, Chauhan N, Eftekhari M, et al. Somaclonal variations and their applications in horticulturalcrops improvement. 3 Biotech. 2016;6:54

109. Piatczak E, Wysokinska H. In vitro regeneration of Centaurium erythraea Rafn from shoot tips and other seedling explants. Acta SocietatisBotanicorumPoloniae. 2003;72:283-288

110. Kim JB. Development of efficient regeneration and transformation systems in Alstroemeria. PhD thesis. The Netherlands: Wageningen University; 2005. pp. 1-100

111. Simonović A, Trifunović M, Raspor M, Cingel A, Bogdanović M, Dragićević M, et al. Dimethyl sulfoxide improves sensitivity and specificity of RT-PCR and qRT-PCR amplification of low-expressed transgenes. Archives of Biological Sciences. 2012;64:865-876

112. Yang SH, Yu H, Xu Y, Goh YC. Investigation of cytokinin-deficient phenotypes in Arabidopsis by ectopic expression of orchid DSCKX1. FEBS Letters. 2003;555:291-296

113. Motyka V, Faiss M, Strnad M, Kamínek M, Schmülling T. Changes in cytokinin content and cytokinin oxidase activity in response to derepression of ipt gene transcription in transgenic tobacco calli and plants. Plant Physiology. 1996;112:1035-1043

114. Jones RJ, Schreiber BMN. Role and function of cytokinin oxidase in plants. Journal of Plant Growth Regulation. 1997;23:123-134

115. Frébort I, Kowalska M, Hluska T, Frébortová J, Galuszka P. Evolution of cytokinin biosynthesis and degradation. Journal of Experimental Botany. 2011;62:2431-2452

116. Kieber JJ, Schaller GE. Cytokinins. Arabidopsis book. 2014;12:e0168

117. Kamínek M, Motyka V, Vaňková R. Regulation of cytokinin content in plant cells. Physiologia Plantarum. 1997;101:689-700

118. Mytinova Z, Motyka V, Haisel D, Gaudinová A, Lubovská Z, Wilhemová N. Effect of abiotic stresses on the activity of antioxidative enzymes and contents of phytohormones in wild type and AtCKX2 transgenic tobacco plants. Biologia Plantarum. 2010;54:461-470

119. Macková H, Hronková M, Dobrá J, Turečková V, Novák O, Lubovská Z, et al. Enhanced drought and heat stress tolerance of tobacco plants with ectopically enhanced cytokinin oxidase/dehydrogenase gene expression. Journal of Experimental Botany. 2013;64:2805-2815

120. Eklöff S, Astot C, Blackwell J, Mortiz T, Olsson O, Sandberg G. Auxin-cytokinin interactions in wild-type and transgenic tobacco. Plant & Cell Physiology. 1997;38:225-235

121. Nordström A, Tarkowski P, Tarkowska D, Norbaek R, Astot C, Doležal K, et al. Auxin regulation of cytokinin biosynthesis in Arabidopsis thaliana: a factor of potential importance for auxin-cytokinin-regulated development. Proc Natl Acad Sci USA. 2004;101:8039-8044

122. Jones B, Andersson Gunnears S, Petrsson SV, Tarkowski P, Graham N, May S, et al. Cytokinin regulation of auxin synthesis in Arabidopsis involves a homeostatic feedback loop regulated via auxin and cytokinin signal transduction. Plant Cell. 2010;22:2956-2969

123. Liu J, Mehdi S, Topping J, Tarkowski P, Lindsey K. Modelling and experimental analysis of hormonal crosstalk in Arabidopsis. Mol Systems Biol 6. Article number. 2010;373:1-13

124. Hussain MS, Fareed S, Ansari S, Rahman MA, Ahmad IZ, Saeed M. Current approaches toward production of secondary plant metabolites. Journal of Pharmacy &Bioallied Sciences. 2012;4:10-20

125. Janković T, Krstić D, Šavikin-Fodulović K, Menković N, Grubišić D. Xanthones and secoiridoids from hairy root cultures of Centaurium erythraea and C. pulchellum. Planta Med. 2002;68:944-946

126. Van der Sluis WG (1985) Secoiridoids and xanthones in the genus Centaurium Hill (Gentianaceae) - a pharmacognostical study. DrukkerijElinkwijkbv - Utrecht

127. Van der Sluis WG. Chemotaxonomical investigations of the Genera Blackstonia and Centaurium (Gentianaceae). Plant Systematics and Evolution. 1985;149:253-286

128. JankovićT,KrstićD,Šavikin-FodulovićK,MenkovićN,GrubišićD.Comparative investigation of secoiridoid compounds of Centaurium erythraea grown in nature and cultured in vitro. Pharm and Pharmacol Lett. 1997;7:30-32

129. Piatczak E, Wielanek M, Wysokinska H. Liquid culture system for shoot multiplication and secoiridoid production in micropropagated plants of Centaurium erythraea Rafn. Plant Science. 2005;168:431-437

130. Šiler B, Mišić D, Nestorović J, Banjanac T, Glamočlija J, Soković M, et al. Antibacterial and antifungal screening of Centaurium pulchellum crude extracts and main secoiridoid compounds. Natural Products Communications. 2010;5:1525-1530

131. Šiler B, Živković S, Banjanac T, Cvetković J, Nestorović-Živković J, Ćirić A, et al. Centauries as underestimated food additives: Antioxidant and antimicrobial potential. Food Chemistry. 2014;147:367-376

132. Valentão P, Fernandes E, Carvalho F, Andrade PB, Seabra RM, Bastos ML. Antioxidant activity of Centaurium erythraeainfusion evidenced by its superoxide radical scavenging and xanthine oxidase inhibitory activity. Journal of Agricultural and Food Chemistry. 2001;49:3476-3479

133. Pinto MMM, Sousa ME, Nascimento MSJ. Xanthone derivatives: new insights in biological activities. Current Medicinal Chemistry. 2005;12:2517-2538

134. Šiler B, MišićD FB, Popović Z, Cvetić T, Mijović A. Effects of salinity on in vitro growth and photosynthesis of common centaury (Centaurium erythraea Rafn.). Archives of Biological Sciences. 2007;59:129-134

135. Petropoulos SA, Karkanis A, Martins N, Ferreira ICFR. Halophytic herbs of the Mediterranean basin: an alternative approach to health. Food and Chemical Toxicology. 2018;114:155-169

136. Palma F, López-Gómez M, Tejera NA, Lluch C. Involvement of abscisic acid in the response of Medicago sativa plants in symbiosis with Sinorhizobiummeliloti to salinity. Plant Science. 2014;223:16-24

137. Ikbal FE, Hernández JA, Barba-Espín G, Koussa T, Aziz A, Faize M, et al. Enhanced salt-induced antioxidative responses involve a contribution of polyamine biosynthesis in grapevine plants. Journal of Plant Physiology. 2014;171:779-788

CHAPTER-17

UTILIZATION OF GENOME EDITING TECHNIQUES IN THE FIELD OF PLANT BREEDING

1. INTRODUCTION: AN OVERVIEW

The application of genome editing methods has revolutionized the landscape of plant breeding, offering unprecedented precision and efficiency in the manipulation of plant genomes. This innovative approach enables scientists and breeders to target specific genes with a level of accuracy that was once unimaginable. By harnessing technologies such as CRISPR-Cas9, TALENs, and others, researchers can modify plant genomes with remarkable precision, opening new avenues for crop improvement.

Currently the world population is about 8 billion people. According to the UN data, the number of people experience moderate or severe food shortages has reached 2 billion or 26.4% of the world population [1]. Huge efforts are being made to eradicate hunger and malnutrition in the world actually. Many of them are associated with scientific breakthrough in the life science and agriculture area [2]. However, despite the achievements of plant breeding, the issue of short term creation of new high-yielding and stress resistant varieties of crops is still actual. All of this is aimed to challenge the such problems as crop losses due to climatic changes, reducing of cultivated areas and spread of more aggressive and resistant pathogens. No less important reason is world population growth [2, 3]. At the moment, this is impossible without the use of biotechnological and genetic engineering approaches.

In the 20th century a classical crop breeding approaches were based on either natural mutations or artificially induced mutagenesis [4]. However, the traditional breeding methods have sufficient disadvantage such as long-term period to create of new varieties with desired agronomic characteristics of any crops. This depends not only on the duration of growing season and reaching of mature stage of plants (especially the long-period growth plants, e.g. trees), as well as is associated with applying of multiple stages of crossing, selection and testing in breeding process. In addition, the following should be mentioned, both natural mutations and conventional methods of chemical and physical mutagenesis do not permit to target the plant genome [4].

At the turn of XXIth century the development and introduction of molecular DNA markers allow to significantly reduce the time required to create new lines and varieties of agricultural crops. In other words, the marker assisted selection approaches were appeared, thus significantly increase the effectiveness of breeding programs to increment in productivity of crops in a wide range of environmental conditions [5, 6]. However, these approaches also do not enable to target the crop genome.

At the same time, advances of next-generation sequencing, multitude of sequenced genomes of major crops and newly identified genes and their functions motivate researchers to pursue targeted breeding of plants. All of this have significantly promoted the development of targeted genome editing (GE) approaches [7, 8, 9, 10]. One of the first GE technologies was RNA interference (RNAi) [4, 11, 12]. Despite the successful application of this technology in functional genomics and plant breeding [15, 16, 17], this method has a number of disadvantages, such as partial gene function suppression and indefinite insertion place of an RNAi construction into the genome [4].

The solution to these breeding problems was the application of GE methods using sequence-specific nucleases (SSN) to introduce targeted mutations in crops with high efficiency and accuracy [7, 8, 9, 10, 12]. Artificially engineered SSNs such as zinc finger nucleases (ZFNs), transcription activator-like effector nucleases (TALENs), and clustered regulatory interspaced short palindromic repeats (CRISPR) associated with the endonuclease Cas (CRISPR/Cas) have been shown to be highly effective in targeting mutagenesis in a wide range of model plants and crops [7, 8, 9, 10, 12]. In addition, oligonucleotide directed mutagenesis (ODM) allows to edit the genome at the single nucleotide level. Moreover, base editors (BEs) have recently been developed to replace A-T base pairs to G-C base pairs [12, 13].

Nowadays GE technologies are widely applied both in functional genomics and in the development of new varieties of crops with new valuable properties and resistant to various biotic and abiotic stresses [7, 8, 9, 10, 12, 13]. Herewith, despite the fact that modern GE technologies are much more accurate than conventional mutagenesis, the legistation of GE crops remains the main bottleneck [13, 14]. A particular difficulty is associated with the biosafety assessment of such crops, the impossibility of determining the subsequent effect of single base mutations after using ODM and BEs [13, 14].

This review discusses GE mechanisms and their use for crop improvement, as well as the problems associated with these approaches.

2. GENOME EDITING MECHANISMS

2.1 Programmable nucleases

Currently, there are three main GE methods classified according to the mechanism of action. The most commonly one applied for plant genomics is the targeted generation of double-stranded DNA breaks (DSBs) by SSNs [12, 13]. Whereat this DSBs are recovered by the cell's own endogenous repair mechanisms either through non-homologous end joining (NHEJ) or homologous recombination (HDR) [7, 8, 9, 10, 12, 13, 14, 15, 16]. Thereat, the reparation of the target DNA sequence leads to the genesis of single base mutations that changing or shifting of the reading frame and initiating of indels or nucleotide substitutions, as well chromosomal rearrangements [8, 15].

Targeted induction of DSB is possible by programmable nucleases. The prevailing nucleases for GE are zinc finger nucleases (ZFNs), transcription activator-like effector nucleases (TALEN), and clustered regularly interspaced short palindromic repeats (CRISPR) associated with the endonuclease Cas (CRISPR/Cas). These three classes of nucleases are different in structure, activity and action mechanism, that leads to differences in target selection, efficiency, specificity and nature of mutation [7, 8, 9, 10]. Let's take a closer examine each programmable nucleases types.

2.1.1 Zinc finger nucleases (ZFNs)

Zinc finger nucleases (ZFNs) are the first artificial endonucleases designed for GE [17]. Every ZFN is derived by aggregating of DNA-binding domain containing of a few linked zinc finger (ZF) motifs and the nonspecific endonuclease FokI [12, 17]. Association of ZF motifs promotes to develop ZF proteins (ZFP) consisting of approximately 30 amino acids and having ββα structure stabilized by zinc ions chelation [17]. Combination of ZFP with methylase, FokI and transcription activator/repressor gives rise ZFN [12, 13, 17]. FokI is an endonucleases recognizing of 5-mer non-palindromic sequence 5'-GGATG-3' : 5'-CATCC-3' and cleaving DNA 9/13 nucleotides downstream of the recognition site [12, 17, 18].

By intersecting with DNA, each ZF motif is capable to bind one triplet of nucleotides inserting an α-helix into the major groove of the DNA double helix [12, 18]. It should be also noted that one ZF has not sufficient specificity for binding to the target genome. However, artificial ZFN usualy contains three or four ZFs, which permit to bind 18-24-mer site after FokI dimerization which is necessary for efficient DNA restriction [17]. During FokI dimerization, two ZFNs can bind both forward and reverse DNA strands respectively, and two target sequences - forward and reverse should be separated by a spacer sequence of 5 to 7 bp [7]. In this case, ZFN acts like dimer and generates DSBs with short 5'-cohesive overhangs, which are filled by homologous recombination, that gives rise to indels into the genome [7, 8, 9, 10, 12, 17, 18].

It should also be noted that, despite the sufficient binding specificity of ZFNs, they are more likely to make nucleotide mismatch [17]. Heterodimerization of FokI nuclease is used to minimize non-target effects and, accordingly, cellular toxicity of ZFN [19].

According to the first report in 1996, ZFNs have been successfully applied for gene modification in plants [17, 18]. The ZFN technology was used to edit the genome of tobacco and Arabidopsis [18]. In tobacco ZFN technology was used to restore the function of the defective reporter gene GUS:NPTII [18]. In Arabidopsis, the induction of ZFN expression under control of the heat shock protein promoter led to 106 mutations on the DNA, where 83 (78%) were 1-52-mer deletions, 14 (13%) - 1–4-mer insertions and 9 (8%) – deletions accompanied by insertions [18]. Nowadays, there are many studies confirming the possibility of GE in tobacco, Arabidopsis, maize, soya, canola and other plants using ZFNs [8, 15, 18]. At the same time, the use of ZFNs permit to introduce mutations in the endochitinase-50 gene (CHN50) in tobacco to emergence of resistance to pat herbicides [18]. In addition, similar results was got by the target editing of IPK1 (inositol-1, 3, 4, 5, 6-pentakisphosphate kinase 1) gene, responsible for phytic acid biosynthesis in maize. ZFN-based targeting of ABI4 (ABA Insensitive-4) gene in Arabidopsis, Dicer-like genes (DCL4a and DCL4b) in soybean and genes of alcohol dehydrogenase and chalcone synthase in Arabidopsis have been also reported [18].

However, despite the rather successful use of ZFNs, they have not become widespread as a GE tool due to the presence of a number of disadvantages. Main of them is complexity and high cost intensive technology, constructing of protein domains for each specific locus of the genome [18], likelihood of inaccurate cleavage of the target DNA due to single nucleotide substitutions or incorrect interaction between domains [8, 15, 17, 18, 19].

2.1.2 Transcription activator-like effector nucleases (TALENs)

TALENs similar to ZFN are enzymes consisting of specific DNA-binding domains of highly conservative repeats originating from effectors such as transcription activators (TALEs) which associate with FokI [20]. TALEs domains contain 15–30 copies of 33–34 highly conserved amino acid sequences [20]. The exceptions are 12th and 13th amino acid residues, which have high variability (repeat-variable diresidues – RVD) [17, 20]. It permits to establish the recognition code for specific nucleotides using a pair of such amino acids within the repeating peptide chains of a given protein [20]. This code is degenerate, but there is a clearly pronounced preference for some combinations of amino acids [17, 20]. It permit to design recombinant proteins capable of recognizing specific DNA sequences [20]. Activity of TALEN depends on amino acid number between TALE domain and FokI, as well as base number between binding sites [17, 20].

In contrast to ZF each repeat in the TALE domain recognizes one nucleotide [17, 20]. The TALE domains recognize 15-30 nucleotides that is 30-60 nucleotides for each TALEN dimer after FokI dimerization. Moreover, despite the fact that TAL domains have higher binding specificity, they are more likely to allow nucleotide mismatches [12, 13, 17]. As well as ZFN the heterodimerization of FokI is applied to minimize off-target effects by using of TALEN [21].

Analysis of mutations occurred during GE using TALEN shows that deletions are way more than insertions (89% versus 1.6%). The reason is the longer length of the TALEN spacers, which provide more extended protruding ends for the DNA fragments after DSBs [22].

Theoretically, the use of TALEN permit to introduce DSB into any part of the genome. There is one limitation only – the presence of thymidine upstream of the 5' end of the target sequence is needed for the TALEN nuclease recognition sites. However, variation of the spacer length allows to select restriction sites [20].

2.1.3 Clustered regulatory interspaced short palindromic repeats (CRISPR/Cas9)

CRISPR/Cas technology permits to make different changes in the DNA sequence [12, 13, 14, 15, 16]. Moreover, this GE technology is much cheaper, faster, more efficient and simple in practical application in comparing to ZFN or TALEN [17, 23]. This technology is based on the use of mechanisms of adaptive "immunity" discovered in bacteria – a specific antiviral defense of bacterial cells based on the complementary binding of viral DNA and their follow destruction [7, 8, 9, 10, 12, 13, 14].

In this system small guide RNAs (crRNA) are used for sequence-specific interference of foreign nucleic acids. CRISPR/Cas includes a genetic locus so-called CRISPR containing short repeats separated by unique sequences (spacers) [24, 25, 26, 27, 28, 29]. The CRISPR complex is predated by the AT-enriched leader sequence and flanked by cas genes [24, 25, 26, 27].

Depending on the cas genes classification CRISPR/Cas systems are divided into two classes. The class 1 of CRISPR-Cas (types I, III, and IV) uses for interference protein complexes with several cas, while class 2 systems (types II, V, and VI) – single effector protein [26, 28, 29]. Type I system is characterized by the presence of Cas3. Type II systems use Cas1, Cas2, Cas9 and Csn2 or Cas4, and type III systems – Cas10, the role of which has not yet been identified [29].

Currently, CRISPR/Cas type II is most often used for genome editing. This type contains the protein Cas9, which is necessary for interference and bacterial immunity [26]. Let's a closer look at this genome editing system.

To edit target genes, the CRISPR/Cas9-based GE requires the occurrence of CRISPR-associated protein 9 (Cas9), CRISPR RNA (crRNA), transactivating crRNA (tracrRNA) and ribonuclease III (RNase III) [12, 13, 14]. Thereat the crRNA and tracrRNAcoassemble into a single guide RNA (sgRNA) [28].

Cas9 is the endonuclease cleaving a double-stranded DNA (dsDNA) [24, 25, 26, 27]. This nuclease were isolated from various bacteria, such as Brevibacilluslaterosporus, Staphylococcus aureus, Streptococcus pyogenes, Streptococcus thermophilus [24, 25, 26]. It should be noted that Cas9 from Streptococcus pyogenes (SpCas9) is most often used to genome edit [27].

Cas9 contains two domains: His-Asn-His (HNH) and RucV-like domains cleaving dsDNA at 3 bp upstream of the motif adjacent to protospacer (PAM) (5□ NGG or 5□-NAG for SpCas9) [24, 25, 29]. The HNH domain cleaves the complementary crRNA strand, while the RucV-like domain - the opposite strand of dsDNA [47]. Then generated DSBs are repaired by NHEJ or HDR [24, 25, 26, 27, 28, 29].

sgRNA is 100-mer synthetic RNA and consists crRNA and tracrRNA. The 5'-end of sgRNA contains a 20 nucleotide guide sequence to identify the target sequence followed by consensus PAM sequence [27]. 3'-end of sgRNA has loop structure which permit to fix the guide sequence to the target site and interact with Cas9 forming ribonucleoprotein complex (RNP) which generates DSB at the target DNA region [27, 29].

Efficient DNA cleavage are provided by ribonucleoprotein complex (RNP) [29]. crRNA plays an important role in the recognition of target DNA due to the sequence that directs RNP to a specific locus by base pairing with target DNA with formation of R-loop structure [29]. R-loop generation activates the HNH and RuvC-like endonuclease domains of Cas9, which cleave dsDNA, creating blunt-ended DSB at 3 bp upstream of the PAM site [25]. Thus, CRISPR/Cas9 performs gene editing in three stages. At the first stage Cas9 is expressed, at the second stage – generation of sgRNA which containing 20 nucleotides complementary to the target region. The third stage requires an NGG PAM recognition site located closer to the 3' end of the target region. The RNP guided by sgRNA generates a blunt-ended DSB at 3 bp upstream of the PAM site [25].

The one limitation of the CRISPR/Cas9 using is the fact that sgRNA design requires the presence of the targeted PAM sequence at the 3' end [24, 25, 26].

For SpCas9 this sequence is defined as 5'-NGG-3' [25]. Frequency of PAM sites in the plant genomes revealed by in silico is 5-12 sites per 100 bp [30]. This fact underlies a difficulty of identifying target sequence, especially in large genomes with nomerous of repetitive sequences, such as maize, cotton, wheat, etc. [13]. All of mentioned above is one of the factors of CRISPR/Cas9 untargeted effect [30]. To reduce unexpected effects other Cas9 can be used, for example from S. aureus, which recognizing less common NNGRRT-PAM [31] or mutant SpCas9 recognizing non-canonical PAM [32].

2.2 Oligonucleotide-directed mutagenesis (ODM)

ODM is site-directed mutagenesis tool using mutagenic DNA fragments 20-200 nucleotides in length [33]. In this approach, the fragment sequence match with a target sequence in the genome, except of a single base pair, which is a putative mutation introduced into the genome [33, 34].

In eukaryotic cells the oligonucleotide for ODM penetrates into the cytoplasm through the cell membrane, then enters the nucleus and complementarily interacts with the target DNA sequence. Herewith, the mismatched nucleotides contribute to initiate a specific change of the sequence that occur in the target gene due to the errors reparation mechanism in the cell. This is two-stage process initially requires annealing of the specific oligonucleotides with target DNA, and subsequently repair of nucleotide mismatch leading to a directed mutation [33]. This system was first demonstrated on mammalian, after that in plants [35]. However, it should be noted that in plants the oligonucleotide does not integrate into the genome because of modifications of the 5' and 3' ends, that prevent DNA ligation, and due to the activity of endonucleases that destroy oligonucleotides [34].

ODM can be accomplished by single-stranded DNA fragments, but their using is limited by a short intracellular half-life [13]. To overcome this disadvantage, the stabilizing modifications for ssODN are necessary. These include chimeraplasts (DNA/RNA duplexes modified by methylation), modification with phosphorothioate, ssODNs with a 5'-tag Cy3 and modified 3'idC reverse base [33]. Additionally, it should be noted the rather low efficiency of ODM, that positive correlate with the length of oligonucleotide fragments. It was shown that by increasing of ssODN length to 200 nucleotides it allows to increase accuracy of editing up to 0.05% in *Arabidopsis thaliana* [33]. The chimeraplasts using did not lead to an increase of mutation frequency higher than the level of spontaneous mutations in *Nicotiana tabacum* or *Brassica napus* [36]. In this regard, to increase the mutagenesis efficiency and the target mutations frequency ODM is often used in combination with nonspecific reagents inducing DSB, such as antibiotics or TALENs and CRISPR/Cas9 in *Arabidopsis* and *Linum usitatissimum* [37].

2.3 Base editors (BE)

All programmable nucleases generate DSBs that are repaired by either NHEJ or HDR [7, 8, 9, 10]. The common disadvantage of these nucleases is the non-predictable results of DSB repairing due to NHEJ mechanism and the low efficacy of HDR [13]. In this reason it became necessary to develop new methods for the introduction of point mutations excluding of DSBs, which resulted in GE new tools so-called base editors (BE) [38].

BE technologies use nickase Cas9 (nCas9) or functionally inactive Cas9 (dCas9) combined with a cytosine or adenosine deaminase domain, which bring into action transformation bases [39]. For example, cytosine deaminases transform cytosine (C) to uracil (U), which is identified as thymine (T) during following DNA repair and replication, thus providing a C:G to T:A replacement [38, 39]. In the same way, adenine deaminases convert adenine (A) to inosine (I), which polymerases interpreting as guanine (G), creating A:T for G:C replacement [39]. Cytidine deaminase-based BE (CBE) has been used to edit genomes in rice, *Arabidopsis*, wheat, corn, tomato and watermelon [39, 40, 41].

In the lack of known adenine deaminases, the substitution of A:T to G:C is a more difficult task than on the contrary [39]. This problem has been solved by bio-engineering of tRNA adenosine deaminases for their adaptation to DNA as a substrate [39, 40, 41].

Initially developed BE had an editor window of for several base pairs only, that led to the appearance of unexpected random mutations. But further improvement of these methods allows to create high-precision Cas9-based BEs that control the length and flexibility of the linker and thus able to selectively edit the bases in direct position with high accuracy, efficiency and simplified PAM requirements [39, 40, 41].

3. APPLICATION OF GE APPROACHES

3.1 Application of GE in functional genomics

NHEJ- or HDR-mediated repair of DSBs generated by programmed nucleases leads to the appearance of insertions /deletions, which shift the reading frame [7, 8, 9, 10], while the single base changes by ODM [33] or BE [39] promotes targeted replacement of nucleotides. These events lead to changes in the genes activity due to the effects of gene-knockout or gene-knockin, that can be used to reveal their function [17, 29, 34, 42, 43].

Besides gene-knockout and gene-knockin, GE technologies can also be used to regulate gene expression. In this case, genes repression and activation achieved by combining repressors or activators of transcription with the DNA-binding domains of programmed nucleases is most often used [7].

Unlike technologies targeted introducing changes in the DNA nucleotide sequence, gene regulation by GE methods is carried out at the level of transcripts. It allows to reveal the functions of many non-canonical non-coding transcripts without open reading frames [16].

Moreover, the CRISPR/Cas using makes it possible to simultaneously introduce into cells several genetic constructs targeted different regions of the genome [15, 16]. It allows to have an effect on the work of several genes simultaneously and to study intergenic interactions. In this wise, it is possible, for example, to determine the genes involved in the process of crop domestication [29].

In addition, the CRISPR/dCas9 system combined with epigenetic regulatory factors involved in histone acetylation or DNA methylation, can be efficiently used to modulate chromatin activity and gene expression patterns involved in plant development and adaptation to the environment [29].

3.2 Using of GE approaches in plant breeding

Nowadays, GE technologies are effectively used to create new varieties of agricultural crops with improved traits, such as increased yield, product quality and resistance to biotic and abiotic stresses. Such traits improvement is often carried out by introducing target mutations into the corresponding regulatory genes that control the development of undesirable traits leading to the suppression of their activity [7, 8, 9, 10].

In this section, we review key advances in crop trait enhancement using GE techniques and discuss their prospects for improving food security.

3.2.1 Crop yields increase

Productivity is one of the most important economically valuable traits of agricultural crops. At the same time, this trait is also one of the most difficult to improve by conventional breeding methods [44]. It's explained by the fact that yield is often a quantitative multigenic trait, the development of which is controlled by multiple quantitative trait loci (QTL) [45]. Additionally, traditional yield-based selection is complicated by QTL introgression between different varieties, which is especially pronounced in the case of closely linked loci [44, 45].

In this regard, GE technologies represent a promising tool for the rapid and directed mutagenesis of target genes [7, 8, 9, 10, 13]. Herewith, the most effective way to increase yields using genome editing technologies is to knock out ("turn off") genes negatively affecting the yield [44]. For example, CRISPR/Cas9-based "turn off" of the functions of yield negative regulators (*Gnla*, *DEP1* and *GS3*) in rice has led to yield improvement, that manifested itself as increased number of grains in panicles and a larger grain size, respectively. It should also be noted that this gene knockout is inherited and observed at least in the T_2 generation inclusive [10, 44].

Additionally, there is evidence that CRISPR/Cas9-based multiplex knockout of the main negative regulators of rice grain weight (GW2, GW5, and TGW6) allowed to significantly increase the weight of grains. Similar results were obtained by CRISPR/Cas9-mediated knockout of the GASR7 gene (a negative regulator of the wheat grain width and weight). In addition, CRISPR/Cas9-based silencing of *OsGn1a, OsDEP1, OsGW2, OsGW5, OsTGW6, OsGS3, OsIPA1, OsPYLs, OsCCD7, OsLAZY1* and *NtPDR6* genes in wheat allowed to improve the yield-related characteristics [44, 46].

Also, it was shown that the CRISPR/Cas9-xyr5APOBEC1-mediated single base mutations in two rice genes, NRT1.1B and SLR1 improved the efficiency of nitrogen utilization and increased yield [39]. Also, CRISPR/Cas9-mediated knockout of genes contributing to yield improvement allows to amend this economically valuable trait in many other crops [44, 46].

3.2.2 Product quality improving

Products quality is another economically valuable trait, the selection of which by traditional methods is accompanied by significant difficulties. Thereat, selection for this trait is complicated both by the difficulty in obtaining targeted mutations by the methods of chemical and physical mutagenesis, and the presence of negative correlations between the traits of quality and yield [11]. GE technologies allow to cope with the deficiencies of chemical and physical mutagenesis due to the ability to introduce targeted mutations into the genome and improve the nutritional properties of crops [7, 8, 9, 10, 44].

Let us consider some examples of the potential application of GE methods for modifying the chemical composition of plants. For example, silencing one of the key genes of phytate biosynthesis *ZmlPK* by TALEN and CRISPR/Cas9 systems allowed to reduce its content in corn (*Zea mays*) [10, 44]. Herewith, the feed value of such corn grain is much higher due to the fact that phytate is considered an anti-nutritional element, reducing availability for digestion of proteins and minerals. Similar results were obtained in barley with TALEN-mediated knockout of the *HvPAphy* gene, which plays an important role in phytate biosynthesis [10, 44].

Also, TALEN-based "turn off" of *VInv* gene encoding vacuolar invertase allows to obtain potatoes (*Solanum tuberosum*) without the potential carcinogen acrylamide, which is formed during frying as a result of the reaction between reducing sugars and free amino acids [10, 44]. Additionally, TALEN system was used to knock out the *OsBADH2* gene in rice that resulted to an increase in 2-acetyl-1-pyrroline [44], which is responsible for the smell of cooked rice. Along with this, TALEN-mediated mutagenesis of the *FAD2-1A/B* gene in soybeans increased the content of oleic acid [44].

CRISPR/Cas9-based targeting of conserved regions in the □-gliadin genes has allowed to create wheat lines with reduced gluten immunoreactivity [44]. At the same time, CRISPR/Cas9-mediated multiplex mutagenesis of *SGR1*, *LCY-E*, *LCY-B1*, *LCY-B2* and *Blc* genes involved in lycopene biosynthesis contributed to the production of tomato lines with an increased content of lycopene [9, 10].

In addition, it was reported that CRISPR/Cas9-mediated knockout of genes responsible for amylose biosynthesis: *GBSS* gene in potatoes and the *Wx1* gene in maize allowed to obtain potato and maize lines with a reduced amylose content [9, 10, 44]. The opposite result was obtained in rice by silencing of the *SBEI* and *SBEIIb* genes responsible for the biosynthesis of starch [44]. CRISPR/Cas9 system was also used to decrease the level of linolenic acid and to increase the level of oleic acid in *Camelina sativa* by multiplex knockout of *FAD2* homeologues [46].

3.2.3 Herbicide resistance improving

Herbicides are the class of chemical compounds most widely used in agricultural practice. This is due to the fact that weeds cause significant damage to agriculture, reducing yields from competition with crops for resources [47]. However, despite the success of the herbicide using, their main disadvantage is their non-selective effect. To overcome this disadvantage, herbicide-resistant biotechnological varieties were created using genetic engineering methods. Currently all herbicide-resistant varieties approved for use have been obtained by transgenesis [11]. At the same time, GE methods are an effective tool for creating herbicide-resistant crop lines [7, 8, 9, 10].

The main genes targeted by the GE in creating herbicide-resistant lines are the *EPSPS* and *ALS* genes. *ALS* gene encodes acetolactate synthase, participated in the biosynthesis of branched-chain amino acids, and the EPSPS gene encodes 5-enolpyruvylshikimate-3-phosphate synthase involved in the biosynthesis of essential plant aromatic amino acids [47]. Thus, based on the genes functions, it can be assumed that targeting *ALS* gene allows to obtain crop lines resistant to sulfonylurea herbicides, and *EPSPS* gene – to glyphosates [44]. Such lines were obtained by ODM-mediated targeted mutagenesis of the *ALS* gene in tobacco, rice, corn, and wheat [46]. Similar results of obtaining tobacco lines resistant to sulfonylurea were obtained by ZFN [18]. Also, other programmable nucleases (TALEN and CRISPR/Cas9) were used to obtain herbicide-resistant lines of potato, rice, maize, and soybeans [44]. In addition, to obtain herbicide-resistant lines of rice (*Oryza sativa*), wheat (*Triticum aestivum*), potato (*Solanum tuberosum*), and watermelon (*Citrullus lanatus*), single base mutations were introduced into the ALS gene by CBE and ABE [39].

Along with this, CRISPR/Cas9-mediated point replacement of two nucleotides in the *EPSPS* gene allows to obtain glyphosate-resistant flax (*Linum usitatissimum*) and rice lines [44]. The targeting of the protoporphyrinogen oxidase (*PPO*) gene in *Arabidopsis* and exon 2 of inositol-1,3,4,5,6-pentakisphosphate-2-kinase (*IPK1*) gene in maize by ZFN was also used to obtain herbicide-resistant lines [18].

3.2.4 Biotic stress resistance improvement

Biotic stresses are one of the basic factor of crop losses in agriculture [47]. The main biotic stresses affecting crops include phytopathogens (viruses, bacteria, fungi), insect and pests (phytophagous insects, acari or nematodes). The strategy for dealing with biotic stresses is either in increasing of self defense mechanisms in plants or in introducing into the genome of constructs aimed against the pathogens [47, 48].

The producing of crops lines resistant to pathogens and pests using traditional breeding methods is based on increasing the own defense mechanisms in plants, but the introduction of constructs targeted pathogens into the genome is carried out by genetic engineering. The most of biotechnological crop lines resistant to biotic stresses created to date are obtained by transgenesis or RNA interference (RNAi) methods [47].

Nowadays, GE approaches are widely used to create new resistant lines [7, 8, 9, 10, 44]. Herewith, it should be mentioned these methods makes possible to use both strategies to deal with biotic stresses [44, 47, 48]. Let us to consider some examples of the use several GE technologies to develop lines and varieties of crops resistant to biotic stresses.

Targeting on plant susceptibility genes. Plant susceptibility genes are essential to successful infection and development pathogens [49]. Thereat, affect these genes brings about development of plant resistance [7, 44, 49]. As targeting of TALEN- and CRISPR/Cas9-based *MLO* homologues providing resistance to powdery mildew allows obtaining resistant lines of barley, wheat, Arabidopsis, tomato and pea [7, 44, 49]. Additionally, the CRISPR/Cas9-mediated directed affecting of *DMR6* gene allows creating tomato lines resistant to *Pseudomonas syringae*, *Phytophthora capsici*, and *Xanthomonas* spp [49].

Targeting of disease susceptibility factors was also used for creation of virus-resistant crops. As CRISPR/Cas9-based silencing of *eIF4e* factors associated with plant infection by positive sense RNA viruses allows creating virus-resistant plants. Herewith, CRISPR/Cas9-mediated disruption of *eIF4Es* gene function in *Arabidopsis* and cucumber promotes to develope potyvirus resistance [49]. Similarly, CRISPR/Cas9-based mutagenesis of *eIF4G* allows obtaining some rice lines resistant to rice tungro spherical virus (*RTSV*) [49].

To creation of crop lines resistant to biotic stress GE approaches are also used to impact on regulatory elements that can affect the process of pathogen proliferation

[49]. So as, TALEN- and CRISPR/Cas9-mediated mutagenesis in effector binding site of promoter region of *OsSweet14* gene permits to develop *Xanthomonas oryzae* pv. *oryzae* resistant rice lines [7, 44, 49]. Also it was shown that Cas9/sgRNA-based targeting the effector binding element of *CsLOB1* gene in citrus provided the generation cancer-resistant citrus varieties [50].

Targeting on genes of plant hormonal system. It is known that hormonal signalling plays an important role in immune response of plants [51]. Herewith, salicylate-mediated immune response develops in response to biotrophic and hemibiotrophic pathogens infection, and by jasmonate- and ethylene-mediated – against necrotrophic pathogens. Thus, directed effect on the genes of the hormonal response permit to create pathogen-resistant crop lines [49]. Recently this assumption has been confirmed experimentally. Herewith, CRISPR/Cas9-based targeting of the *SlJAZ2* gene in tomato resulted in genesis of resistance to *Pseudomonas syringae* pv. *tomato* (Pto) DC3000 [48].

Targeting of pathogen genomes. Another advanced strategy for pathogens defence is directed effect on pathogen genome. Currently, this strategy is mainly used against viral plant diseases [7, 44]. Thus, CRISPR/Cas9 system has been successfully applied to increase resistance to DNA-containing viruses, including tomato yellow leaf curl virus (*TYLCV*), beet root curl virus (*BCTV*), Merremia mosaic virus (*MeMV*), beans yellow dwarfism virus (*BeYDV*), cotton leaf curl virus and beet severe curly top virus (*BSCTV*) [44, 48]. It should be noted that one sgRNA designed to targeting a conserved region can mediate interference against a numerous DNA-containing viruses [8, 44]. Additionally, CRISPR/LshCas13a system capable to interfere of viral RNA turnip line is successfully applied to creation of resistant to RNA-containing turnip yellow mosaic virus (*TuMV*) [44].

3.2.5 Abiotic stress resistance improvement

Abiotic stresses are the main factors that negatively affect the yield of most crops [52]. In this regard, a creation of resistant to adverse environmental factors crop varieties is the urgent problem. However, the use of traditional breeding methods to develop such varieties is limited by the fact that the traits of resistance to abiotic stress are multigene controlled and have a complex inheritance type [52, 53]. The disadvantages of traditional breeding can be successfully overcome through the use of GE approaches. Herewith, a literature analysis has shown that the GE application in various cultures allowed to increase their resistance to abiotic stresses [7, 44, 53, 54]. Let us consider detailed examples of the GE application to obtain varieties of crops resistant to abiotic stresses.

Targeting of structural genes. Structural genes are one of the most convenient targets to increase plant stress resistance. This class of genes can be divided into tolerance genes (T-genes) and sensitivity genes (S-genes). The T-genes encode enzymes of the antioxidant system, while S genes – negative regulators in plant defense mechanisms [54]. Therefore, "turn off" of S-genes allows to obtain drought-resistant crop varieties. This assumption was confirmed by CRISPR/Cas9-based targeting of *ARGOS8* (a negative regulator of ethylene response) in maize, that permit to obtain drought-resistant lines [53]. Similar results were obtained during directed mutagenesis of the *OST2* gene in *Arabidopsis*, *TaDREB2* gene in wheat (*Triticum aestivum*) and *OsDERFl*, *OsEPSPS* and *OsMSHl* genes in rice (*Oryza sativa* L.). In addition, CRISPR/Cas9-mediated targeting of *OsPDS*, *OsMPK2*, and *OsDEPl*, *OsAOXla*, *OsAOXlb*, *OsAOXlc* and *OsBEL* genes allows to develop rice lines resistant to a wide range of abiotic stresses [10, 44, 53, 54].

Targeting of regulatory genes. Regulatory genes, such as transcription factors, phosphatases and kinases are also involved in the regulation of intracellular signals during abiotic stresses and can be used as target for GE tools to create stress-resistant crop varieties [54]. Presently, CRISPR/Cas9-mediated silencing of *SlMAPK3* regulatory gene in tomato (*Solatium lycopersicum*), *TaERF3* gene in wheat, *OsSAPK2*, *OsPMS3*, *OsRAV2*, and *OsNAC.041* genes in rice permit to create lines of these crops that are resistant to abiotic stresses [44, 53].

Summarize the presented data, it should be noted that GE systems are successfully used to modify a wide range of economically valuable traits in main agricultural crops.

4. SAFETY ASSESSMENT AND REGULATORY FRAMEWORK OF GE SYSTEMS

4.1 Off-target effects

One of the main limitation of GE systems is off-target effects, when nucleases, along with target regions, affect other parts of the genome [7, 8, 9, 10, 39, 44]. It stems from the fact that the efficiency of DNA cleavage by nucleases depends both on nuclease activity and the availability of the target site and also the affinity of the DNA-binding domain. Moreover, the designed nucleases specificity largely depends on the binding affinity of the nuclease to DNA. In addition, FokI domain dimerization and interaction of Cas9 with PAM may also play an important role [17, 55, 56].

The preliminary comprehensive bioinformatics analysis to choose of specific sites for the introducing of DBS can minimize the off-target effects of the GE system. When choosing the desired sites, it should avoid regions with repeated sequences, as well as regions with high homology to other regions of the genome [17, 55, 56]. In addition, to minimize off-target effects and cellular toxicity of ZFN and TALEN heterodimerization of FokI nuclease is used [19, 21]. Another effective way to reduce the frequency of unintentional mutations is the use of the sgRNA/aptazyme system (ligand-dependent ribozyme) [56].

4.2 Regulatory framework

The legal and regulatory framework in regard to GE plants in different countries has a great impact on their competitiveness. In most countries, the current biosafety framework is meant to regulate transgenic GMOs. Currently, only Argentina and Brazil have adopted additional legislation for crops obtained using GE techniques [57].

There are two main approaches to defining the regulatory framework for GMOs in global legislation: one is process-based and the other is product-based. In the European Union, a standard is focused on the first approach, while in Canada – on the second. In the United States, a hybrid system is used: the decision on whether a crop belongs to GMOs is based on the first approach, and the risk assessment – on the second. In this case, GE cultures are considered individually [13]. Thus, the same GE culture can be classified in different ways depending on the regulatory framework. For example, in Argentina, crops that are classified as "zero segregants" are not regulated as GMOs [57]. In the EU, GE crops are subject to Directive 2001/18/EC and must undergo a full biosafety assessment procedure, as well as meet the requirements for GMO products [13, 57].

5. FUTURE PROSPECTS

GE tools are considered one of the most promising tools for practical agricultural biotechnology because of their high efficiency, relatively low cost, ease to use and multiplexing ability [7, 8, 9, 10, 14]. Herewith, directed mutagenesis makes it possible to effectively "turn off" or silencing various genes, that helps to determine the functions of genes [17, 29, 34, 42, 43]. Multiplex targeting allow to reveal the role of individual genes and encoded proteins in intracellular signaling pathways and contributes to the engineering of multigenic agronomic traits in crops [7, 8, 9, 10, 44].

Furthermore, CRISPR/Cas9 system may be used for spatial and temporal control of gene expression, as well as tissue-specific regulation of expression [42]. In the furtherance of this goal, influence on the *cis* regulatory elements of the gene promoter region has large potential and allows to change the expression level, expression of patterns, and tissue-specific expression [58, 59]. At the same time, BEs can be used to replace key nucleotides in *cis* elements to modulate the affinity of transcription factors and, accordingly, the level of expression [58]. Additionally, the introduction of mutations using programmed nucleases or BEs into the binding sites of the promoter regions will disrupt their interaction with the virulence proteins of pathogens and, thus, increase the resistance of crops [48, 58, 60].

Epigenomic studies are another promising uses of GE technologies. Herewith, the insertion of transposons into the promoter region due to the NHEJ or HDR mechanisms can affect the epigenetic status and, accordingly, the level of expression of commercially valuable genes [58]. It can be achieved through the using of such dCas9 systems as targeted DNA methylation and demethylation systems [58].

It should also be noted that an advanced uses of GE methods, in particular, CBE and ABE, is the introduction of single base changes into the microRNA (miRNA) binding site of the target gene without alteration of the amino acid sequence of the encoded protein. It will lead to disruption of miRNA/mRNA base pairing and, consequently, to disruption of mRNA cleavage due to the fact that the position and number of mismatches in the miRNA binding site strongly affect the efficiency of miRNA-mediated mRNA cleavage [58]. In turn, such disturbance of miRNA-mediated mRNA cleavage can be used to proper alignment of the target genes expression and, accordingly, can have effect on the development of many agronomical desired trait [58]. In addition, CRISPR/Cas system can be used for NHEJ- or HDR-mediated introduction of translation enhancers into the initiation codon of open reading frames, which provide fine tuning of gene expression at the translation level [58].

Thus, in summary it should be noted that GE systems, especially CRISPR/Cas and BEs, offer great opportunities for crop improvement.

6. CONCISE SUMMARY

In conclusion, it can be noted that GE technology has enormous potential to create new varieties of crops resistant to biotic and abiotic stresses and improved food value and yield. However, for the effective using of these technologies, it is necessary to resolve issues related to the biosafety assessment, including the revision of regulatory frameworks.

AT A GLANCE

The main task of plant breeding is creating of high-yield, resistant to biotic and abiotic stresses crop varieties with high product quality. The using of traditional breeding methods is limited by the duration of the new crop varieties creation with the required agronomic traits. This depends not only on the duration of growing season and reaching of mature stage of plants (especially the long-period growth plants, e.g. trees), as well as is associated with applying of multiple stages of crossing, selection and testing in breeding process. In addition, conventional methods of chemical and physical mutagenesis do not allow targeting effect to genome. However, the introduction of modern DNA-technology methods, such as genome editing, has opened in a new era in plant breeding. These methods allow to carry out precise and efficient targeted genome modifications, significantly reducing the time required to get plants with desirable features to create new crop varieties in perspective. This review provides the knowledge about application of genome editing methods to increase crop yields and product quality, as well as crop resistance to biotic and abiotic stresses. In addition, future prospects for integrating these technologies into crop breeding strategies are also discussed.

REFERENCES

1. FAO, IFAD, UNICEF, WFP and WHO. The state of food security and nutrition in the World 2020. Transforming food systems for affordable healthy diets. Rome, FAO; 2020. 320 p. DOI: 10.4060/ca9692en
2. van Meijl H, Shutes L, Valin H, Stehfest E, van Dijk M, et al. Modelling alternative futures of global food security: Insights from FOODSECURE. Global Food Security. 2020;25:100358. DOI: 10.1016/j.gfs.2020.100358
3. FAO. Climate change and food security: Risks and responses. Rome, FAO; 2016. 110 p. ISBN: 978-92-5-108998-9
4. Ni Z, Han Q, He Y-Q, Huang Sh. Application of genome-editing technology in crop improvement. Cereal Chemistry. 2018;95:35-48. DOI: 10.1094/CCHEM-05-17-0101-FI
5. Tuberosa R. Marker-assisted breeding in crops. In: Meyers RA, editor. Encyclopedia of Sustainability Science and Technology. New York: Springer; 2012. p. 6425-6448. DOI: 10.1007/978-1-4419-0851-3_393
6. Hickey LT, Hafeez AN, Robinson H, Jackson SA, Leal-Bertioli SCM, et al. Breeding crops to feed 10 billion. Nature Biotechnology. 2019;37(7):744-754. DOI: 10.1038/s41587-019-0152-9
7. Zhang H, Zhang J, Lang Z, Botella JR, Zhu J-K. Genome editing – principles and applications for functional genomics research and crop improvement. Critical Reviews in Plant Sciences. 2017;36:1-19. DOI: 10.1080/07352689.2017.1402989
8. Mishra R, Zhao K. Genome editing technologies and their applications in crop improvement. Plant Biotechnology Reports. 2018;12:57-68. DOI: 10.1007/s11816-018-0472-0
9. Imran M, Butt M, Hannan A, Manzoor A, Qaisar U. Gene editing: A potential tool to enhance field crop production. International Journal of Biotech Trends and Technology. 2020;10(1):72-82. DOI: 10.14445/22490183/IJBTT-V10I1P612
10. Tan Y-Y, Du H, Wu X, Liu Y-H, Jiang M, et al. Gene editing: an instrument for practical application of gene biology to plant breeding. Journal of Zhejiang University-SCIENCE B (Biomedicine & Biotechnology). 2020;21(6):460-473. DOI: 10.1631/jzus.B1900633
11. Abdurakhmonov IY, Buriev ZT, Saha S, Jenkins JN, Abdukarimov A, Pepper AE. RNA interference for functional genomics and improvement of cotton

(Gossypium spp.). Frontiers in Plant Science. 2016;7:202. DOI: 10.3389/fpls.2016.00202

12. Mohanta TK, Bashir T, Hashem A, Abd_Allah EF, Bae H. Genome editing tools in plants. Genes. 2017;8:399. DOI: 10.3390/genes8120399
13. Jansing J, Schiermeyer A, Schillberg S, Fischer R, Bortesi L. Genome editing in agriculture: Technical and practical considerations. Int. J. Mol. Sci. 2019;20:2888. DOI: 10.3390/ijms20122888
14. El-Mounadi K, Morales-Floriano ML, Garcia-Ruiz H. Principles, applications, and biosafety of plant genome editing using CRISPR-Cas9. Front. Plant Sci. 2020;11:56. DOI: 10.3389/fpls.2020.00056
15. Chen K, Gao C. Targeted genome modification technologies and their applications in crop improvements. Plant Cell Rep. 2014;33:575-583. DOI: 10.1007/s00299-013-1539-6
16. Zhang Y, Massel K, Godwin ID, Gao C. Applications and potential of genome editing in crop improvement. Genome Biology. 2018;19:210. DOI: 10.1186/s13059-018-1586-y
17. Gaj T, Gersbach ChA, Barbas CF. ZFN, TALEN and CRISPR/Cas-based methods for genome engineering. Trends Biotechnol. 2013;31(7):397-405. DOI: 10.1016/j.tibtech.2013.04.004
18. Novak S. Plant biotechnology applications of zinc finger technology. In: Kumar S, Barone P, Smith M, editors. Transgenic Plants. Methods in Molecular Biology. USA: Humana Press; 2019. p. 295-310. DOI: 10.1007/978-1-4939-8778-8_20
19. Doyon Y, Vo TD, Mendel MC, Greenberg SG, Wang J, et al. Enhancing zinc-finger-nuclease activity with improved obligate heterodimeric architectures. Nat. Method. 2011;8:74-79. DOI: 10.1038/nmeth.1539
20. Jankele R., Svoboda P. TAL effectors: tools for DNA targeting. Brief Funct Genomics. 2014;13:409-419. DOI: 10.1093/bfgp/elu013
21. Guilinger JP, Pattanayak V, Reyon D, Tsai SQ, Sander JD, Joung JK, Liu DR. Broad specificity profiling of TALENs results in engineered nucleases with improved DNA-cleavage specificity. Nat. Method. 2014;11:429-435. DOI: 10.1038/nmeth.2845
22. Kim Y., Kweon J., Kim J.S. TALENs and ZFNs are associated with different mutation signatures. Nature Methods. 2013;10:185. DOI: 10.1038/nmeth.2364
23. Kurakov FA. The phenomenon of fixation of rights to industrial application of the results of intellectual activity at the stage of fundamental research. Economics Sci. 2017;3(2):116-125. DOI: 10.22394/2410-132X-2017-3-2-116-125
24. Jiang F, Doudna JA. CRISPR-Cas9 structures and mechanisms. Annu Rev Biophys. 2017;46:505-529. DOI: 10.1146/annurev-biophys-062215-010822

25. Hille F, Charpentier E. CRISPR-Cas: biology, mechanisms and relevance. Philosophical Transactions of the Royal Society B: Biological Sciences. 2016;371:20150496. DOI: 10.1098/rstb.2015.0496

26. Hille F, Richter H, Wong SP, Bratovič M, Ressel S, Charpentier E. The biology of CRISPR-Cas: Backward and forward. Cell. 2018;172(6):1239-1259. DOI: 10.1016/j.cell.2017.11.032

27. Liu X, Wu S, Xu J, Sui Ch, Wei J. Application of CRISPR/Cas9 in plant biology. Acta Pharmaceutica Sinica B. 2017;7(3):292-302. DOI: 10.1016/j.apsb.2017.01.002

28. Kumar R, Kaur A, Pandey A, Mamrutha HM, Singh GP. CRISPR-based genome editing in wheat: a comprehensive review and future prospects. Mol. Biol. Rep. 2019;46:3557-3569. DOI: 10.1007/s11033-019-04761-3

29. Manghwar H, Lindsey K, Zhang X, Jin Sh. CRISPR/Cas system: Recent advances and future prospects for genome editing. Trends in Plant Science. 2019;24(12):1102-1125. DOI: 10.1016/j.tplants.2019.09.006

30. Rath J. Safety and security risks of CRISPR/Cas9. In: Schroeder D, Cook J, Hirsch F, Fenet S, Muthuswamy V, editors. Ethics Dumping: Case Studies from North-South Research Collaborations. Cham: Springer; 2018. p. 107-113. DOI: 10.1007/978-3-319-64731-9_13

31. Friedland AE, Baral R, Singhal P, Loveluck K, et al. Characterization of Staphylococcus aureus Cas9: A smaller Cas9 for all-in-one adeno-associated virus delivery and paired nickase applications. Genome Biol. 2015;16:257. DOI: 10.1186/s13059-015-0817-8

32. Hu X, Wang C, Fu Y, Liu Q, Jiao X, Wang K. Expanding the range of CRISPR/Cas9 genome editing in rice. Mol. Plant. 2016;9:943-945. DOI: 10.1016/j.molp.2016.03.003

33. Sauer NJ, Mozoruk J, Miller RB, Warburg ZJ, et al. Oligonucleotide-directed mutagenesis for precision gene editing. Plant Biotechnol. J. 2016;14:496-502. DOI: 10.1111/pbi.12496

34. SongstadDD,PetolinoJF,VoytasDF,ReichertNA.Genomeeditingofplants.Critical Reviews in Plant Sciences. 2017;36(1):1-23. DOI: 10.1080/07352689.2017.1281663

35. Beetham PR, Kipp PB, Sawycky XL, Arntzen CJ, May GD. A tool for functional plant genomics: Chimeric RNA/DNA oligonucleotides cause in vivo gene-specific mutations. PNA. 1999;96(15):8774-8778. DOI: 10.1073/pnas.96.15.8774

36. Ruiter R, van den Brande I, Stals E, Delaure S, Cornelissen M, D'Halluin K. Spontaneous mutation frequency in plants obscures the effect of chimeraplasty. Plant Mol. Biol. 2003;53:675-689. DOI: 10.1023/B:PLAN.0000019111.96107.01

37. Sauer NJ, Narvaez-Vasquez J, Mozoruk J, Miller RB, et al. Oligonucleotide-mediated genome editing provides precision and function to engineered

nucleases and antibiotics in plants. Plant Physiol. 2016;170:1917-1928. DOI: 10.1104/pp.15.01696

38. Komor AC, Kim YB, Packer MS, Zuris JA, Liu DR. Programmable editing of a target base in genomic DNA without double-stranded DNA cleavage. Nature. 2016;533:420-424. DOI: 10.1038/nature17946
39. Mishra R, Joshi RK, Zhao K. Base editing in crops: current advances, limitations and future implications. Plant Biotechnology Journal. 2020;18:20-31. DOI: 10.1111/pbi.13225
40. Hess GT, Tycko J, Yao D, Bassik MC. Methods and applications of CRISPR-mediated base editing in eukaryotic genomes. Mol Cell. 2017;68:26-43. DOI: 10.1016/j.molcel.2017.09.029
41. Zhang Y, Gao C. Recent advances in DNA-free editing and precise base editing in plants. Emerg Top Life Sci. 2017;1:161-168. DOI: 10.1042/ETLS20170021
42. Kumar V, Jain M. The CRISPR-Cas system for plant genome editing: advances and opportunities. J. Exp. Bot. 2015;66(1):47-57. DOI: 10.1093/jxb/eru429
43. Chen K, Wang Y, Zhang R, Zhang H, Gao C. CRISPR/Cas genome editing and precision plant breeding in agriculture. Annu. Rev. Plant Biol. 2019;70:667-697. DOI: 10.1146/annurev-arplant-050718-100049
44. Sedeek KEM, Mahas A, Mahfouz M. Plant Genome engineering for targeted improvement of crop traits. Front. Plant Sci. 2019;10:114. DOI: 10.3389/fpls.2019.00114
45. Abdurakhmonov IY. Genomics era for plants and crop species—advances made and needed tasks ahead. In: Abdurakhmonov I, editor. Plant Genomics. Croatia: InTechOpen; 2016. p. 3-25. DOI: 10.5772/62083
46. Jaganathan D, Ramasamy K, Sellamuthu G, Jayabalan S and Venkataraman G. CRISPR for crop improvement: An update review. Front. Plant Sci. 2018;9:985. DOI: 10.3389/fpls.2018.00985
47. Lombardo L, Coppola G, Zelasco S. New technologies for insect-resistant and herbicide-tolerant plants. Trends Biotechnol. 2016;34:49-57. DOI: 10.1016/j.tibtech.2015.10.006
48. Schenke D, Cai D. Applications of CRISPR/Cas to improve crop disease resistance: beyond inactivation of susceptibility factors. iScience. 2020;23:101478. DOI: 10.1016/j.isci.2020.101478
49. Bisht DS, Bhatia V, Bhattacharya R. Improving plant-resistance to insect-pests and pathogens: The new opportunities through targeted genome editing. Semin. Cell Dev. Biol. 2019;96:65-76. DOI: 10.1016/j.semcdb.2019.04.008

50. Sun L, Nasrullah, Ke F, Nie Zh, Wang P, Xu J. Citrus genetic engineering for disease resistance: Past, present and future. Int. J. Mol. Sci. 2019;20:5256. DOI: 10.3390/ijms20215256

51. Van Wersch S, Tian L, Hoy R, Li X. Plant NLRs: The Whistleblowers of plant immunity. Plant Comm. 2020;1:100016. DOI: 10.1016/j.xplc.2019.100016

52. Singhal P, Jan AT, Azam M, Haq QMR. Plant abiotic stress: a prospective strategy of exploiting promoters as alternative to overcome the escalating burden. Frontiers in Life Science. 2016;9(1):52-63, DOI: 10.1080/21553769.2015.1077478

53. Surabhi GK, Badajena B, Sahoo SK. Genome editing and abiotic stress tolerance in crop plants. In: Wani SH, editor. Recent Approaches in Omics for Plant Resilience to Climate Change. Switzerland: Springer Nature; 2019. p. 35-56. DOI: 10.1007/978-3-030-21687-0_2

54. Zafar SA, Zaidi SSh-e-A, Gaba Y, Singla-Pareek SL. Engineering abiotic stress tolerance via CRISPR-Cas mediated genome editing. Journal of Experimental Botany. 2020;71(2):470-479. DOI: 10.1093/jxb/erz476

55. Lee CM, Cradick TJ, Fine EJ, Bao G. Nuclease target site selection for maximizing on-target activity and minimizing off-target effects in genome editing. Molecular Therapy. 2016;24 (3):475-487. DOI: 10.1038/mt.2016.1

56. Hajiahmadi Z, Movahedi A, Wei H, Li D, et al. Strategies to increase on-target and reduce off-target effects of the CRISPR/Cas9 system in plants. Int. J. Mol. Sci. 2019;20:3719. DOI: 10.3390/ijms20153719

57. Eckerstorfer MF, Engelhard M, Heissenberger A, Simon S, Teichmann H. Plants developed by new genetic modification techniques-comparison of existing regulatory frameworks in the EU and non-EU countries. Front. Bioeng. Biotechnol. 2019;7:26. DOI: 10.3389/fbioe.2019.00026

58. Hua K, Zhang J, Botella JR, Ma Ch, et al. Perspectives on the application of genome-editing technologies in crop breeding. Mol. Plant. 2019;12(8):1047-1059. DOI: 10.1016/j.molp.2019.06.009

59. Birchler JA. Editing the phenotype: a revolution for quantitative genetics. Cell. 2017;171:269-270. DOI: 10.1016/j.cell.2017.09.024

60. Peng A, Chen S, Lei T, Xu L, et al. Engineering canker-resistant plants through CRISPR/Cas9-targeted editing of the susceptibility gene CsLOB1 promoter in citrus. Plant Biotechnol. J. 2017;15:1509-1519. DOI: 10.1111/pbi.12733

INDEX

www.ingramcontent.com/pod-product-compliance
Ingram Content Group UK Ltd.
Pitfield, Milton Keynes, MK11 3LW, UK
UKHW050948290926
15470UKWH00005B/1712

9 781787 152700